Pelican Books
Pelican Geography and Environmental Studies
Editor: Peter Hall

Geology and Scenery in Scotland

After graduating in Geography at Reading University in 1952, John Whittow studied landform evolution in North Wales, for which he was awarded a Ph.D. in 1957. He has held university teaching posts in Ireland, East Africa and Australia, and has also been a research associate at the University of California, Los Angeles. He has contributed to *Collier's Encyclopedia*, *The Grolier Book of Knowledge* and *Encyclopaedia Britannica*. At present he is Senior Lecturer in Geography at the University of Reading. He is the author of *Geology and Scenery in Ireland* (Penguin Books, 1975) and co-author of the revised edition of A. E. Trueman's *Geology and Scenery in England and Wales* (Penguin Books, 1971). A Fellow of the Royal Geographical Society and of the Geological Society of London, he is currently Chairman of the Landscape Research Group, U.K. John Whittow is married, with one daughter.

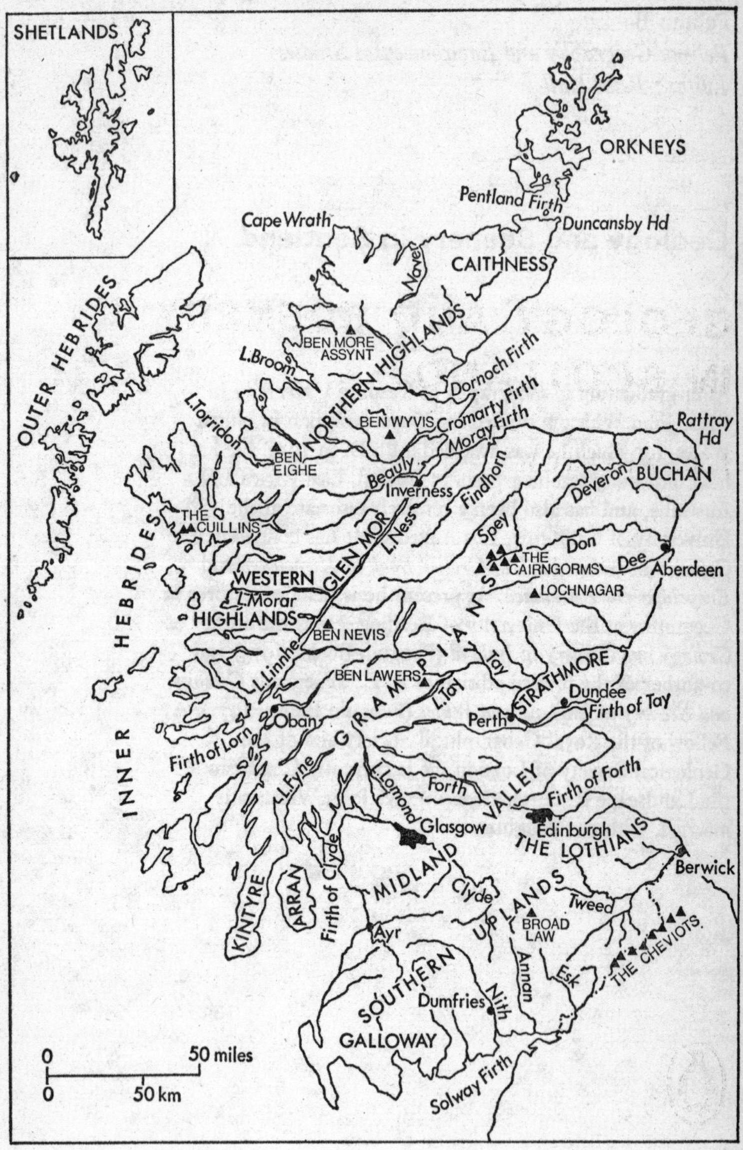
SHETLANDS
ORKNEYS
Pentland Firth
Cape Wrath
Duncansby Hd
CAITHNESS
Naver
OUTER HEBRIDES
BEN MORE
ASSYNT
NORTHERN HIGHLANDS
L.Broom
Dornoch Firth
L.Torridon
BEN WYVIS
Cromarty Firth
Moray Firth
Rattray Hd
BEN
EIGHE
Beauly
Inverness
Findhorn
Deveron
BUCHAN
THE
CUILLINS
GLEN MOR
L.Ness
Spey
Don
THE
CAIRNGORMS
Dee
Aberdeen
INNER HEBRIDES
WESTERN
L.Morar
HIGHLANDS
BEN NEVIS
LOCHNAGAR
GRAMPIANS
Tay
L.Linnhe
BEN LAWERS
STRATHMORE
L.Tay
Dundee
Oban
Perth
Firth of Tay
Firth of Lorn
L.Fyne
L.Lomond
Forth
VALLEY
Firth of Forth
Glasgow
Edinburgh
THE LOTHIANS
Berwick
KINTYRE
ARRAN
Firth of Clyde
MIDLAND
Clyde
UPLANDS
Tweed
BROAD
LAW
Ayr
THE CHEVIOTS
SOUTHERN
Annan
Esk
Nith
Dumfries
GALLOWAY
0
50 miles
0
50 km
Solway Firth

GEOLOGY AND SCENERY IN SCOTLAND

J. B. WHITTOW

PENGUIN BOOKS

Penguin Books Ltd,
Harmondsworth, Middlesex, England
Penguin Books,
625 Madison Avenue, New York, New York 10022, U.S.A.
Penguin Books Australia Ltd,
Ringwood, Victoria, Australia
Penguin Books Canada Ltd,
2801 John Street, Markham, Ontario, Canada L3R 1B4
Penguin Books (N.Z.) Ltd,
182–190 Wairau Road, Auckland 10, New Zealand

First published 1977

Made and printed in Great Britain by
Richard Clay (The Chaucer Press) Ltd
Bungay, Suffolk
Set in Monotype Ehrhardt

To Fiona

Contents

Acknowledgements

The publishers wish to thank the following for permission to use their photographs: Institute of Geological Sciences for Plates 1, 7, 26 and 34, published by permission of the Director: Natural Environment Research Council copyright; John Dewar Studios, Edinburgh, for Plates 2, 11, 21 and 23: copyright reserved; Aerofilms Ltd for Plates 4, 24 and 35: copyright reserved; Professor J. K. S. St Joseph and the Cambridge University Collection for Plates 6, 8, 12, 14, 15, 19, 20, 22, 27, 30, 36, 38 and 39: copyright reserved; Scottish Tourist Board for Plates 31 and 42: copyright reserved; Mr W. A. Poucher for Plate 37: copyright reserved; the *Glasgow Herald* for Plate 9: copyright reserved; W. Ralston Ltd for Plate 10: copyright reserved; Aberdeen Journals Ltd for Plate 17: copyright reserved; Mr Patrick Bailey for Plate 43: copyright reserved; Mr P. A. Macnab for Plate 29: copyright reserved; Dr Alan Stroud for Plates 44 and 45: copyright reserved. The author supplied Plates 3, 5, 13, 16, 18, 25, 28, 32, 33, 40 and 41.

The map on the back cover, together with Figures 7, 9, 10, 13, 16, 18, 26, 29, 34, 35, 36, 38, 39, 40, 41, 42, 44, 53 and 54 are based on material prepared by the Institute of Geological Sciences, by kind permission of the Director: Crown Copyright.

Figures 11, 17, 48, 49 and 50 are based on diagrams in *The Geology of Scotland*, edited by G. Y. Craig, and appear by permission of the publishers, Oliver & Boyd. Figure 28 is based on material published in *The Structure of the British Isles*, by J. G. C. Anderson and T. R. Owen, and is reproduced by permission of Pergamon Press Ltd. Dr J. B. Sissons kindly gave permission to reproduce Figures 20, 21 and 24, which are taken from his publications in various scientific journals and from his book, *The Evolution of Scotland's Scenery* (Oliver & Boyd). Figure 5 is reproduced from *Edinburgh Geology* by kind permission of the Edinburgh Geological Society. Acknowledgement is gratefully given to the late Professor D. L. Linton, to Professor J. A. Steers and to Messrs J. R. Coull, H. A. Moisley and J. L. Roberts, on whose work Figures 17, 24, 25, 30, 31, 32 and 45 are partly based.

The author is pleased to acknowledge with gratitude the assistance of Mrs M. Rolley, who typed the manuscript, and of Mrs K. King, who drew the maps. Thanks are also due to my wife, who drove or navigated along many hundreds of miles of Scottish roads and who tramped scores of Scottish mountains and moorlands during the collection of material for this book.

List of Plates

List of Text Figures

Preface

Almost forty years ago Sir Arthur Trueman wrote a book which set out to describe and explain the *Geology and Scenery in England and Wales*, and this has become something of a classic in its field. When I had completed (with a colleague, Dr John Hardy) a thorough revision of this earlier volume, incorporating fresh material (published in Penguin Books, 1971), it became clear to me that two further volumes were required to complete a survey of the British Isles. In 1975 the second volume of the trilogy was published – *Geology and Scenery in Ireland* (Penguin Books, 1975) – and the present volume is intended to complete the exercise. The two earlier publications were designed to appeal to interested laymen as well as to students in schools and colleges and those beginning at polytechnics and universities, and *Geology and Scenery in Scotland* has been written with the same readership in mind.

Scotland must take a very special place in the annals of geological research, for here many of the major principles of this branch of the earth sciences were worked out. A stream of brilliant Scottish geologists played their parts in the slow unravelling of the complex problems which confronted them in the eighteenth and nineteenth centuries: men such as James Hutton, John Playfair, Charles Lyell, Roderick Impey Murchison and Hugh Miller were some of the more famous pioneers, but above all the name of Archibald Geikie stands pre-eminent among Scottish geologists. Around the turn of the century he became the leading figure in British geology, remaining so for several decades, and he was the only geologist to have been President of the Royal Society. He it was who wrote the first major treatise on Scotland's physical geology, the classic volume entitled *The Scenery and Geology of Scotland* (Macmillan, 1865). Not for another hundred years was any attempt made to emulate Geikie's *tour de force*, at which point J. B. Sissons's *The Evolution of Scotland's*

Scenery (Oliver & Boyd, 1967) incorporated a wealth of new geomorphological ideas in a volume which paid particular attention to landform evolution during the Quaternary. Two years earlier Professor Gordon Craig had edited *The Geology of Scotland* (Oliver & Boyd, 1965), and these two publications remain the standard texts on the more advanced and detailed aspects of the subject.

I have not attempted to write a textbook of geological principles or a geological history of Scotland, for the above-mentioned texts cover these aspects thoroughly. Instead, the present volume has been written in the hope that many of the general queries concerning landscape evolution will be answered and that the better-known aspects of Scottish scenery will appear more interesting to the tourists who visit the beautiful Scottish countryside in ever-increasing numbers. But it is not only the rural landscapes which require an explanation, for the imprint of the Industrial Revolution is written large upon the face of the Midland Valley. Here are found extensive deposits of coal, iron and oil shale which, together with the abundant water power of the Highlands, have helped to influence the making of the Scottish landscape as surely as have the granites and basalts of the Highlands and Islands. Scottish townscapes have also been examined in an attempt to explain their character and layout in terms of their building-stones and their adaptation to the local topography. Latterly, the impact of North Sea oil has begun to affect the landscapes of Scotland in locations as far removed from the industrial heartland as Loch Kishorn and the Shetlands, so that the age-old scenery is now being threatened by the 'erosion' of twentieth-century technology related to the new-found mineral resources.

Although those readers with some geological knowledge may prefer to turn randomly to the regions they know best, it would be advantageous for the layman to read systematically from the beginning, since many of the more important concepts of landform genesis are introduced in the earlier chapters. Despite the complexity of many of the Scottish structures, there is only a short introduction to the basic principles of rock formation, stratigraphy and the geological time-scale. Technical terms, although kept to a minimum, are explained either in the text or in the glossary, whilst many geological concepts are best understood by reference to the text figures.

Eilean Fladda, 1975 J.B.W.

1. The Shaping of Scotland

It is generally true that as one travels north-westwards from London the rocks of the British geological succession become progressively older, and if we were to terminate our journey in the Outer Hebrides we should find there the Lewisian gneiss, one of the oldest rocks known to man. Soon after crossing the border into Scotland it becomes abundantly clear that the Scottish rock formations include only the older series of the geological succession and that these have helped to contribute to the rugged land-forms of a country which lies entirely within the tract known as Highland Britain. It follows, therefore, that Scotland will exhibit close geological and scenic relationships with Ireland and the Lake District of northern England, but that it will have less similarity with the geology and scenery of Lowland England, where the youngest rocks predominate. It is not only the age of the rocks which distinguishes Scotland's geology from that of southern Britain, for we shall soon discover in subsequent chapters that the Scottish Highlands have been carved from some of the most complicated geological structures in the world. In one sense this complexity has stimulated research, so that Scottish geologists have made some of the most fundamental discoveries among their own mountains and moorlands, which have made Scotland a classic region for the study of both geology and scenery.

A glance at a geological map (Figure 1) will demonstrate to the reader that the lines along which the Scottish rocks are orientated exhibit a very clear north-east to south-west trend. Such a trend will be found to dominate the structures of most of the British and Irish uplands, but since it is so marked in Scotland it has been termed the Caledonian trend. Farther south, an east–west trend affects the structures of southern Ireland, South Wales and south-west England, but in Scotland this subsidiary trend (termed the Hercynian or Armorican) can be distinguished only in

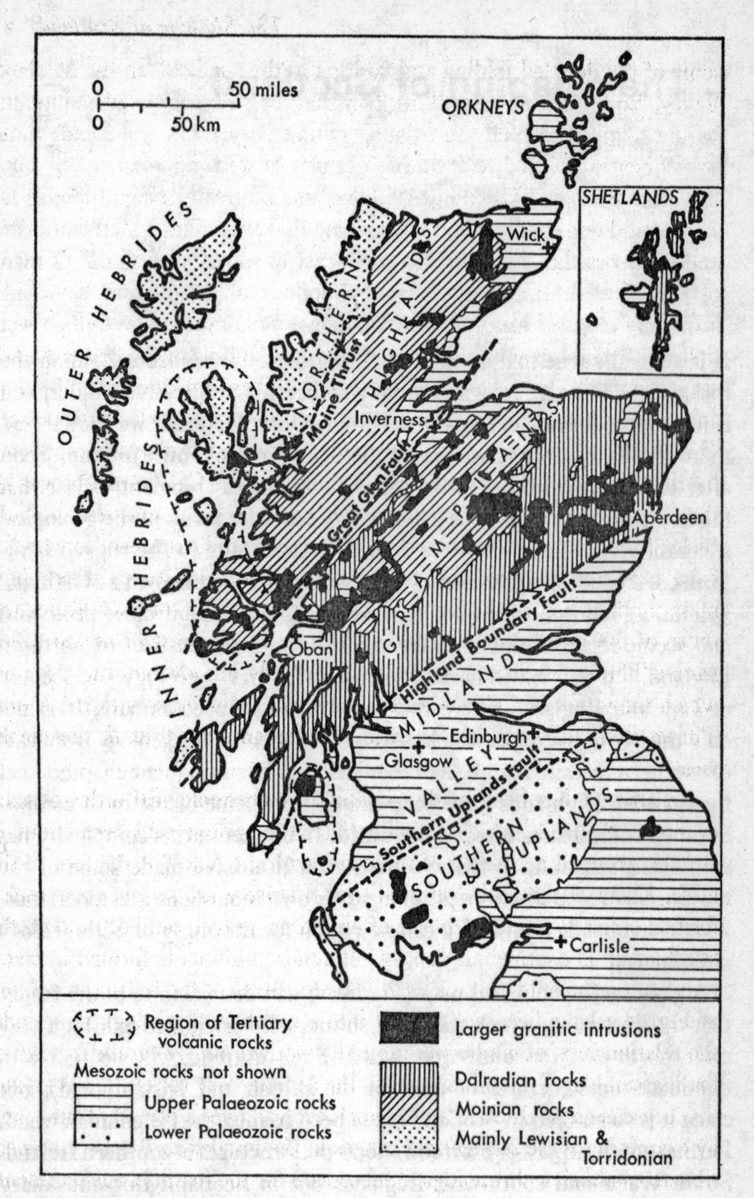

Fig. 1. *The generalized geology of Scotland.*

some of the detailed folding and faulting of the coalfields in the Midland Valley. Both of these trend lines represent bygone periods of mountain-building, during which the existing sedimentary rocks were folded and subsequently uplifted to form high ranges of fold mountains. But since the older period of Caledonian folding was of greater magnitude, as far as Scotland was concerned, it is this structural alignment which dominates and produces the characteristic north-east to south-west 'grain' of many of the Scottish landforms. Lengthy periods of denudation have subsequently reduced many of the Caledonian mountain ranges (which were first uplifted some 400 million years ago) to the hills of the present landscape, so that today we are looking merely at the worn-down stumps of an ancient mountain chain which in its time was more mighty than the present Himalayas. There were, however, parts of Scotland that were not affected by this gigantic period of mountain-building (the Caledonian orogeny): they lay to the north-west and formed an even older massif of ancient rocks. Most geologists refer to this older tract as the Archaean 'foreland' – hard Lewisian gneisses and tough Torridonian sandstones which now make up the whole of the Outer Hebrides, parts of the Inner Hebrides and those of the Northern Highlands which lie to the west of the so-called Moine Thrust (Figure 1). Here was a virtually immovable shield against which the waves of Caledonian folding spent their force, with the Moine Thrust itself marking the limits of Caledonian structures in the north-west.

An attempt has been made to depict the chronological order of such events in Figure 2, which also illustrates the various rock formations in Britain, grouped according to their age, with the oldest at the foot of the table. Many of the rocks referred to in Figure 2 are sedimentary rocks, formed either as marine deposits of limestone, gravel, sand or mud on the sea-floor or as continental deposits of similar sediments formed in lakes, rivers or on the land-surface itself. The four eras of the left-hand column of Figure 2 have been subdivided into a number of geological periods, the relationships of which were initially worked out by careful studies of fossil assemblages contained within the sedimentary rocks. It was known that organisms, both plant and animal, evolved over a period of time, and by examining these sequential changes in the fossil record it became possible to establish a chronological succession for the fossiliferous sedimentary rocks.

Era	Period	Series / Stage	Absolute Time-scale*	Events in Scotland
CAINOZOIC	QUATERNARY	Holocene	12,000 BP	Post-glacial. Growth of peat bogs. Appearance of man in Scotland. Flandrian transgression.
			Millions BP	
		Pleistocene	c.1.5	The Ice Age. Deposition of glacial drifts.
	TERTIARY	Pliocene	c.7	Widespread erosion and drainage development. ? Gravels in Buchan.
		Miocene	26	Widespread Uplift. No sedimentary record.
		Oligocene	38	No sedimentary record in Scotland.
		Eocene	54	Vulcanicity in N.W. Scotland. Intrusion of plutonics in Arran, Inner Hebrides and St Kilda. Interbasaltic lignites and plant remains. (Eocene and Palaeocene)
		Palaeocene	65	
MESOZOIC	CRETACEOUS	Chalk	100	Thin chalk beds and sandstones of Morvern and Mull.
		Upper Greensand and Gault	105	Glauconitic sandstones of Morvern, Mull and Eigg.
		Lower Greensand	112	Not represented in Scotland. (Lower Greensand and Wealden)
		Wealden	136	
	UPPER JURASSIC	Purbeck and Portland		Not represented in Scotland.
		Kimmeridge, Corallian, Oxford and Kellaways Beds	162	Sedimentaries of Skye, Mull, E. Sutherland, Cromarty.
	MIDDLE JURASSIC	Cornbrash, Great Estuarine Series and Oolites	172	Sedimentaries of Skye, Raasay, Mull.
	LOWER JURASSIC	Lias	195	Shales, sandstones and limestones of Skye, Raasay (ironstone), Applecross, Mull, Ardnamurchan, and E. Sutherland.
	TRIASSIC	Rhaetic		Rhaetic bed in Mull.
		Keuper; Bunter (New Red Sandstone)	225	Sandstones and conglomerates of Gruinard Bay, Skye, Applecross, Mull, Elgin, Arran and Solway. ? Stornoway Beds.
	PERMIAN	Magnesian Limestone and Sandstone (New Red Sandstone)	280	Sandstones and breccias of Arran, Ayrshire, Galloway and Southern Uplands.

Era	Period	Subdivision	Age (m. yrs)	Events
PALAEOZOIC	CARBONIFEROUS	Upper {Coal Measures / Millstone Grit	325	Hercynian (Armorican) orogeny. Widespread folding, faulting and uplift. Mineralization. Barren Red Measures overlying valuable coal seams. Restricted gritstones.
		Lower {Carboniferous Limestone Series / Calciferous Sandstone Series	345	Thick limestones with important coal seams and ironstones. Widespread vulcanicity in southern Scotland. Oil shales, cement stones and calciferous sandstones.
	OLD RED SANDSTONE (Devonian)	Upper	359	Widespread deposition of 'continental' sandstones from Southern Uplands to Shetlands.
		Middle	370	Restricted to northern Scotland. Caithness Flagstones. Extensive vulcanicity.
		Lower	395	Sandstones and thick conglomerates formed. Downtonian restricted to southern Scotland.
	SILURIAN	Generally referred to as Lower Palaeozoic	440	Caledonian orogeny and formation of major Scottish granites. Much mineralization, faulting and thrusting. Major metamorphism.
	ORDOVICIAN		500	Formation of thick arenaceous and argillaceous rocks in Southern Uplands and Midland Valley.
	CAMBRIAN		570	Quartzites, grits and limestones of N.W. Scotland.
PROTEROZOIC & ARCHAEOZOIC	PRE-CAMBRIAN AND ARCHAEAN	including Dalradian (Upper Dalradian is of Lower Palaeozoic age)	Pre-600	Dalradian metamorphic rocks of Grampians and S.W. Scotland. Moinian metamorphic rocks of Grampians and Northern Highlands. Torridonian sandstone of N.W. Scotland. Lewisian gneiss of N.W. Scotland and Hebrides. Laxfordian orogeny (1,200–1,600 BP). Scourian orogeny (2.200–2,900 BP).

Fig. 2 *Table of geological succession.*

* BP = Before Present

The simple law of superposition – oldest rocks at the bottom and newest at the top – may have seemed at first sight to provide sufficient evidence on which to base a straightforward geological succession. In Scotland, however, very few of the lithological relationships are straightforward. For one thing, many of the Scottish rocks are unfossiliferous, for another, the normal order of superposition was found to be inapplicable, especially in the Highlands, where severe folding and thrust-faulting had turned the succession upside down. In many instances tangential pressures (largely in the Caledonian orogeny) had caused enormous slices of the upper crustal rocks to fracture and be carried forward across adjoining immobile rocks, with the final result that older rocks are now left on top of newer rocks, although separated from them by a thrust fault. Such instances are described in Chapter 15, including the thrust structures of Assynt which were the first ever to be recognized. In some ways, however, the folding of the Grampians is even more complex (see Chapter 9). A final complication in any study of the geological succession in Scotland arises from the fact that the majority of Scottish rocks cannot be classified as sedimentary. It is time, therefore, to turn to an examination of the two other rock-types shown in Figure 2, igneous and metamorphic.

The formation of the igneous or 'fire-formed' rocks, which are created from different varieties of molten material, will subsequently be described in more detail, especially in Chapters 2 and 3, but it is important to notice at this point that, because of the genetic difference between sedimentary and igneous rocks, the latter do not contain fossils and must therefore be dated in other ways. In earlier years it was possible to date igneous rocks only by reference to their relationship with the surrounding sedimentary rocks, but today radiometric dating, based on measurement of the decay of radioactive minerals within the rocks, has given the geologist a much more reliable basis for the determination of absolute dating. Consequently, the absolute time-scale depicted in Figure 2 has been based on radiometric dating and, providing the datable rocks can be related to the succession, the approximate age and length of the geological periods can be established. It is interesting to note that in the Western Isles the Lewisian gneiss, the oldest rock of all, is overlain directly by the youngest deposits in the time-scale, namely the Pleistocene glacial drifts and the post-glacial (Holocene) peat bogs.

Turning now to examine the metamorphic rocks, we shall find that they differ in their mode of formation from both the igneous and the sedimentary rocks, although detailed descriptions will be reserved until Chapters 8 and 9. The majority of the rocks which form the Grampians and the Northern Highlands can be classified as metamorphic. These make up the complex structures of the Dalradian and Moinian assemblages, in which the separate lithologies are too complicated to depict in detail on the coloured geological map on the back cover. Most metamorphic rocks are rocks of sedimentary origin that have experienced prolonged pressure due to compressive forces in a mountain-building episode, metamorphism by contact with hot molten material from deep-seated magmatic sources or metasomatic metamorphism due to chemically active fluids infiltrating the rock itself. In all cases the detailed character of the constituent minerals and particles of the sedimentary rocks has been radically altered and the rocks converted to metasediments.

In Scotland two broad extremes of metamorphic change can be recognized, termed 'high-grade' and 'low-grade' metamorphism, the first implying that considerable changes have been wrought, as in the Grampians, and the second suggesting that only minor changes have taken place, as in some of the metamorphic rocks of the Southern Uplands. In the latter instance, for example, the intrusion of the Galloway granites (see Chapter 3) has caused contact or thermal metamorphism only in the immediate encircling zone of sediments around the pluton, hence the term 'metamorphic aureole'. Farther north, however, the effect of the Caledonian pressures was very much greater, so that the Dalradian and Moinian rocks were subjected to 'high-grade' regional metamorphism on a much larger scale, all the metamorphic agents tending to work together, including that of metasomatic change. Here, the sedimentary rocks of fine-grained or muddy character (argillaceous rocks) have been converted into so-called 'pelitic' rocks, whilst the coarser-grained sandstones (arenaceous rocks) have become psammitic rocks. In addition, many of the former limestones have been changed into crystalline limestones or marbles. One of the most common metamorphic changes in fine-grained sedimentaries is the formation of a slatey cleavage which cuts across the layering or bedding planes of the sedimentary rocks. Where existing igneous rocks have been influenced by metamorphism they too have been changed, according to their original composition, into epidiorites, serpentines, and so on.

Although the complete British geological succession is depicted in Figure 2, an examination will soon make it clear that many of the younger rock formations are missing in Scotland. Where certain rock formations are absent, the gap is known as an unconformity, and the first ever to be recognized was one found in Scotland in the eighteenth century by James Hutton. In a few cases it seems probable that the sedimentary rocks were never laid down, but in many cases there is evidence to suggest that some of these later rocks have been destroyed by subsequent denudation. The Scottish Mesozoic rocks mentioned in Figure 1, for example, are so small in extent that they are impossible to include at this scale, but they are nevertheless of great significance as indicators of the former limits of the Mesozoic oceans. The Jurassic and Cretaceous rocks of the Inner Hebrides imply that their counterparts may once have extended over much of the Highland zone. But the rolling outlines of the English chalklands and the Cotswold stone-belt are not repeated in Scotland, so that the Scottish scene lacks white chalk sea-cliffs and, with one interesting exception on Raasay, imposing oolitic escarpments. Nevertheless, the dearth of Jurassic freestones such as the Portland and Bath stones has not deprived Scotland of important building stone, for its wealth of granite, Old Red Sandstone and Carboniferous Sandstone has allowed a flowering of Scottish vernacular architecture. It is noticeable, however, that brick and timber, used so effectively south of the border, have rarely been utilized in Scottish buildings (except in Old Aberdeen). Here is a landscape where stone is abundant, so, despite its pleasing proportion and line, Scottish architecture is characterized by a strength and massiveness which mirrors the scenery itself.

In many instances the Mesozoic limestones and sandstones have survived in Scotland only by virtue of the fact that they have been subsequently covered by extensive lava flows of early Cainozoic age. In Chapters 12 and 13 we shall see, for example, how the basalts of Mull and Skye have virtually entombed the Mesozoic rocks, thereby protecting them from subsequent denudation. Where they do emerge, however, the younger sedimentaries have had a profound influence on the local landforms, soils and vegetation. The volcanic episode of Tertiary times also made a major contribution to the Scottish scene: many of Scotland's most spectacular landforms are found in the volcanic regions of the Inner Hebrides (Figure 1).

Between this period of volcanic upheaval and the deposition of the glacial drifts there is a major gap in the geological record, so that for all intents and purposes the deposition of the solid rocks was concluded in Eocene times. It was during this lengthy interlude of the Cainozoic that the Scottish land-surface began to assume something of the form which we see today. Amongst the variety of agents which were actively denuding the complex mosaic of rocks, running water must have been the most important, although mechanical weathering and some deep chemical weathering must also have played their parts. As the Scottish river systems established themselves on the slowly evolving land-surface, so the younger strata were gradually stripped away to reveal the more ancient rocks beneath. Some intermittent uplift and faulting probably occurred in Tertiary times, but there were no cataclysmic events once the volcanic episode had terminated. Biblical theories of terrestrial convulsions, in which all valleys were seen as cracks in the earth's surface, were discarded during the nineteenth century in favour of the Huttonian theory of earth sculpture. When Archibald Geikie set out, in his admirable *The Scenery and Geology of Scotland*, to explain the means by which the harmonious grouping of Scottish hills, valleys and watercourses had been created, it was to the ideas of Hutton he turned. By a study of Scottish scenery James Hutton had already concluded that '. . . the mountains have been formed by the hollowing out of the valleys, and the valleys have been hollowed out by the attrition of hard materials coming from the mountains.'

It remained only for the shortest, but nevertheless one of the most important, geological episodes to leave its imprint on the land-surface. Some of the most detailed modelling of Scottish landforms was achieved by ice-sheets during the Pleistocene Ice Age. Indeed, much of the attrition mentioned by Hutton probably occurred at this time, despite the fact that the overall pattern of valley systems had already been etched out. The majority of the following chapters describe the profound effects which glacial erosion has had on the upland scenery of Scotland, whilst the earlier chapters demonstrate how the ice-sheets have transported vast amounts of eroded material to the lowlands, where it forms the ubiquitous glacial drifts. These are especially important in the Midland Valley, where they form the basis of the Scottish farming scene. But not all the lowland areas benefited thus, for we shall see in the later chapters how the scouring of the glaciers and the subsequent growth of post-glacial peat bogs have created vast areas of desolate, rocky moorland in many parts of western and northern Scotland. It is in these less well-endowed tracts that the crofting system of agriculture has survived and rubble-walled cottages with sod and heather roofs have been abandoned as habitations only in the last few years.

We have now reached the point at which the shaping of the Scottish scene passes slowly from the inexorable and continuing geomorphological processes into the hands of man. It is he who now imposes the thin but fundamental cultural veneer which gradually changes the land-surface into a landscape. Landscape (or scenery) combines the topographic forms with all the complex mosaics created by human endeavour and is in reality an expression of the total environment. Landscape is, of course, dynamic: it changes through time, sometimes slowly, sometimes rapidly; it changes through natural forces and human activities. There is a dichotomy in the Scottish landscape, for it appears that great tracts of the Highlands have scarcely felt the impact of man, while the natural forms of the lowland coalfields and some of the deepwater estuaries have been all but obliterated by his technological mantle. But even the remoter places are now being invaded by the twin prongs of forestry and tourism, while the exploitation of North Sea oil is already beginning to influence parts of the Scottish scene. It must be remembered, however, that these developments are, in their own ways, as relevant to Scottish geology as are studies of the frost-shattered peaks of the Cuillins, the peat resources of Caithness, the farming patterns of Berwickshire or the building stones of Edinburgh. Thus, in the following chapters all the facets of Scottish scenery will be examined in an attempt to explain the changes which have finally shaped the fabled Scottish Highlands and Lowlands that we see today.

2. The Border Country

Three ancient routes lead from England into Scotland and all are determined by topographic factors. In the east, the coastal plains of Northumberland continue without pause into Berwickshire, where the modern routes hug the coast as their predecessors did. In the centre, the Cheviot uplands of the Middle March present more of a barrier, but a main road climbs through forested Redesdale to the border at Carter Bar before descending northwards to the Jed Water, parallel with a Roman road. Finally, the western lowlands of the Solway plain provide access to Annandale and thence to the Clyde, a route favoured by the Romans and now followed by the busiest arterial road into Scotland.

To understand something of the history of the Border and to appreciate the scenery of the Marcher country, a visitor would be advised to enter Scotland across the river Tweed at Coldstream or Berwick, for Flodden battlefield lies near by and one cannot fail to be impressed by the castles and fortified houses of this once-troubled countryside. It was not until 1157 that the Tweed was confirmed as the national boundary. This famous river is not a great physical obstacle because of its numerous shoals which are uncovered at low water, so it had to be protected by castles at Wark, Norham and Berwick. Although the latter town is part of England, it would be instructive to commence our journey there, for it is the true gateway to Berwickshire and the basin of the Tweed.

The Tweed Basin and the Lammermuirs

The grey, walled town of Berwick-upon-Tweed, described by Burns as '... an idle town, rudely picturesque' and by Dr Niklaus Pevsner as 'one of the most exciting towns in England', is really a Norman foundation but has long served as a focus for the Scottish road pattern of the Tweed basin.

All the main roads, radiating from Berwick, curve gradually round in a south-westerly direction as they cross the farmlands of the Merse of Berwickshire, and for a good reason. They have been constructed parallel with the 'grain' of the Merse landforms, a grain produced by ice-sheets as they moved down the valley of the Tweed. The ice was powerful enough to smooth any rock outcrops into low whale-back hills and to

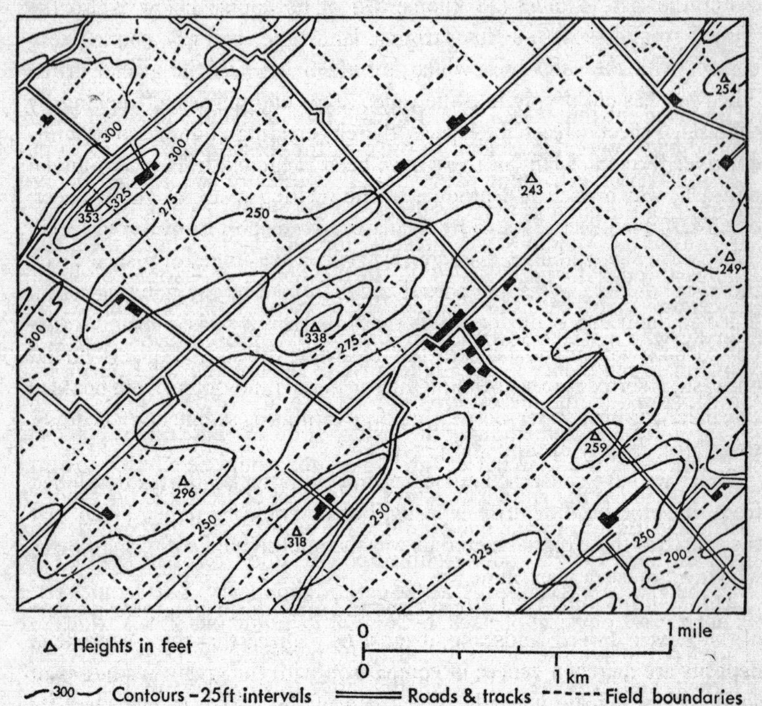

Fig. 3. *Settlement pattern in the drumlin terrain of Berwickshire.*

fashion the thick boulder clay (or 'till') into a fluted topography with streamlined hillocks, termed drumlins, whose long axes reflect the direction taken by the former ice-sheet. Since the major roads run parallel with the trend of the drumlins, the minor linking roads must cross the ridges at right angles, thus making for a tedious journey from north to south across the Merse, as one must also skirt the large rectangular fields which themselves are related to the grain of the countryside (Figure 3).

The fertile clay loams of the Merse support a flourishing agricultural economy, so that the Tweed basin is renowned for its large mixed farms – areas of over 500 acres (200 hectares) are not uncommon – on some of Scotland's richest soils. The landscape is a prosperous one, with neat hedgerows of beech and thorn dividing the extensive cornfields from the sheep- and cattle-crowded pastures.

For the last 30 miles (48 kilometres) of its course below Kelso the Tweed meanders across this farming landscape, scarcely coming into contact with the solid rock which is hidden beneath the glacial drifts. The river has cut deeply into the clays, sands and gravels left behind by the last ice-sheet to reach these parts, thereby creating a number of prominent river-terraces. Many of these have been fashioned from glacio-fluvial outwash, which itself is a product of the melting phase of an ice-sheet, when vast quantities of water are available to transport the waste material. Thus we are now able to distinguish between two types of glacial drift: the first, boulder clay, is non-sorted and is formed by the grinding action of ice on the bedrock, during which rocky detritus is picked up and transported both within and beneath the ice-sheet; the second, glacio-fluvial outwash, is sorted into coarser and finer materials and differs from boulder clay in having been water-borne, therefore exhibiting sedimentary bedding structures, usually of sand and gravel.

So far we have concerned ourselves only with the landforms developed from the superficial or drift deposits, but it should be remembered that the source of these glacial materials was the underlying floor of solid rocks which we must now examine.

Although the hillocks of glacial clays and sands form an impressive part of the Lower Tweed landscape, it must be realized that these superficial deposits are merely a veneer in comparison with the great thicknesses of solid rock which lie beneath. The basin of the Merse in fact owes its overall shape and character to the underlying geological structure, for if we were to compare a geological map with one depicting relief it would become clear that the Tweed basin is embowered in a horseshoe-shaped rim of hard-rock uplands. To the north, the Lammermuirs succeed in isolating the Tweed lowlands from the Midland Valley of central Scotland; to the west, a curving highland of various Palaeozoic sedimentary and volcanic rocks creates a high gathering-ground for the waters of the Tweed catchment; and, to the south, the long line of the Cheviot

Hills forms an appropriate upland barrier for the border itself to follow. These hills terminate eastwards in the granite mass of the Cheviot (2,674 feet, 815 metres), which overlooks the border towns of Coldstream and Kelso. But, as well as being a topographical basin, this is a structural basin, for the lowland of the Merse corresponds to a downfold of the rocks themselves, where the Old Red Sandstone and the overlying Carboniferous rocks form an eastward-tilted syncline which descends towards the North Sea. Thus the youngest rocks of this region, the Carboniferous Limestones, occur near the coastal margin, giving way inland to the older Calciferous Sandstones of Lower Carboniferous age, equivalent in character to the Fell Sandstones and Cementstones of Northumberland.

These calcareous sandstones, of creamy-brown colour, are deeply buried beneath the drumlins of the Merse, but the tilt or dip of their bedding-planes raises them high enough to appear at the surface around the flanks of the Tweed basin. Here they have been sporadically quarried for building stone, since they provide a splendid freestone which has been used in many local buildings, including those of Duns, Coldstream and Kelso. The latter market town, located strategically at the junction of the Teviot and the Tweed, is an ideal centre from which to explore the area, but its unspoiled architecture and its great cobbled square (paved mainly with Cheviot volcanics and granite setts), together with its ruined Norman abbey and yellow sandstone bridge, combine to make it an attraction in its own right. Before we embark on the first of our excursions, however, it is important to look at the geological history and geological structures of the region, since these help to explain the character of the landscapes we are setting out to examine.

On both western and northern flanks the geology of the central basin of the Lower Tweed changes as we move outwards from the Carboniferous Limestones and Sandstones to the older, more brightly coloured arenaceous rocks which make up the Old Red Sandstone succession of south-east Scotland. In the Lower Old Red Sandstone, between Eyemouth and Reston, conglomerates and grits are abundant, especially at the base of the succession, but these are succeeded upwards by the more widespread softish red sandstones. Among the conglomerates are pebbles of Silurian greywacke, slate, chert and jasper which give some indication of the type of landmass from which the materials were eroded during the deposition of the Old Red Sandstone. We shall see later that this period of geological time also witnessed a prolonged episode of vulcanicity, when hundreds of feet of lavas were poured out from volcanoes, some remnants of which can still be seen in the Scottish landscape.

In Middle Old Red Sandstone times the whole of Scotland was subject to intense folding and uplift during the Caledonian period of earth-movements (see p. 21). Subsequent denudation, before the continued deposition, has meant that the Upper Old Red Sandstone not only rests unconformably on Lower Old Red but also transgresses onto the folded and eroded older rocks, such as the Silurian and Ordovician sedimentaries which occupy so much of southern Scotland. The most famous example of this particular unconformity can be seen at Siccar Point (see p. 34), but before visiting this historic spot we must delve a little farther back into geological time in order to understand the structure and character of the older Palaeozoic rocks which build the upland rim to the west and north of the Tweed basin. The oldest rocks are the Ordovician slates and shales which crop out in a narrow band along the northern flanks of the Lammermuir Hills, but by far the most extensive are the Silurian shales, grits and flagstones which stretch from Peebles to Hawick. All these rocks were formed in the Lower Palaeozoic geosyncline (see Glossary) but have subsequently been intensely folded and uplifted by the Caledonian mountain-building episode. Countless minor crumples and faults have affected these older rocks in detail, but in general two primary folds can be distinguished: in the north a downfold affects the Ordovician rocks, but this gives way southwards, in the Silurian area, to a complex upfold. Since these primary folds are composed of numerous parallel minor flexures it is preferable to refer to them as a synclinorium and an anticlinorium respectively (Figure 4), instead of using the terms syncline and anticline.

The complex folding of the older rocks is best seen in the coastal cliffs around St Abb's Head, which some regard as the finest stretch of coastal scenery in southern Scotland. But southwards towards Berwick, where the Carboniferous rocks of the Tweed basin reach the coast, the high, rugged cliffs of contorted older rocks near Eyemouth give way to the gently dipping Lamberton Limestone which creates a rather featureless coastline. The complex structures of the older rocks may also be examined inland, at Grantshouse, where two interesting sections can be seen near the main road. Facing the railway bridge a bare, quarried dip face of Silurian greywacke exhibits good joint patterns, but it is in the adjoining quarry that the most rewarding exposure occurs, for here is a perfect anticline of Silurian greywackes and shales, which demonstrates many structural and sedimentological principles. But our way lies northward, back to the coast near Cockburnspath, for no excursion hereabouts would be complete without a visit to nearby Siccar Point.

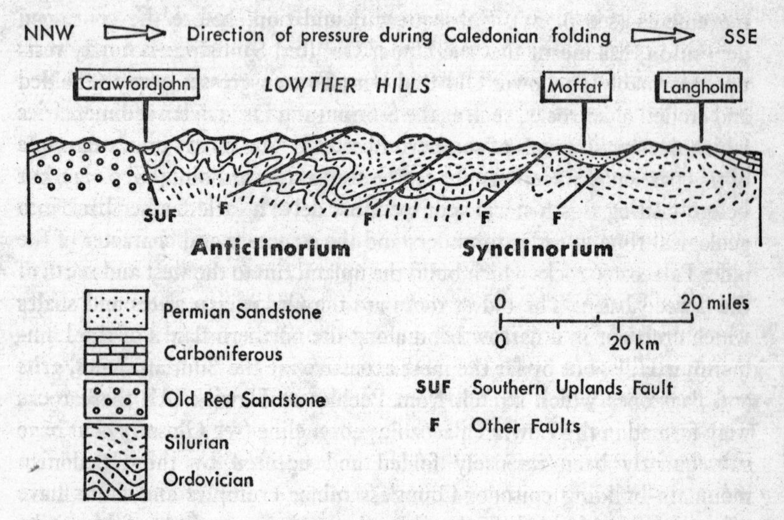

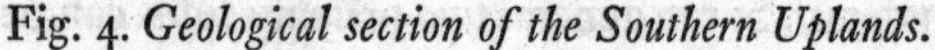
Fig. 4. *Geological section of the Southern Uplands.*

So far as historical geology is concerned, Siccar Point must take its place as one of the most important sites of scientific interest in the British Isles, for it was here in about 1790 that Dr James Hutton became the first man to grasp the concept of a geological unconformity. In his famous treatise, *Theory of the Earth* (1795), Hutton tells us: 'At Siccar Point we found a beautiful picture of this junction [between the underlying Silurian and overlying Old Red Sandstone] washed bare by the sea. The sandstone strata are partly washed away and partly remaining upon the ends of the vertical schistus; and in many places points of the schistus are seen stand-up through among the sandstone, the greatest part of which is worn away.' (See Plate 1.)

It seems true to say that from this world-famous section, and from similar ones in the banks of the Jed at Jedburgh and in the Isle of Arran,

Hutton was able to construct the principles of the cycle of erosion and sedimentation which are fundamental to the study of earth science. In recognizing the orderly system by which the earth's surface was worn down and the material then transported to the ocean floor, there to be stratified and finally uplifted to form a new landmass, Hutton was challenging the entire philosophy of the so-called Catastrophists, who saw all geological phenomena in terms of the catastrophic biblical Flood.

From the sea cliffs at Siccar Point there are splendid views north-westwards towards Dunbar. Near at hand, at Pease Bay, the Upper Old Red Sandstone can be seen grading upwards into the greyish sandstones of the Lower Carboniferous series. Beyond stands the finger of Barns Ness lighthouse on its platform of Carboniferous Limestone, while in the distance the volcanic plugs of North Berwick Law and the Bass Rock faintly punctuate the horizon.

The Lammermuirs slope steeply seawards hereabouts, but the coastline is generally low-lying and is dominated by the gigantic cement works quarrying the Carboniferous Limestone to the south of Dunbar. The Southern Uplands Fault crosses the coastline at this point and marks the northern boundary of the Southern Uplands and of the Border Country. To the north lies the Midland Valley of Scotland with its urbanized and industrialized landscapes cut off from the borderlands by the steep northern face of the Lammermuir Hills. This abrupt termination of the Lammermuirs, between Dunbar and the Soutra Pass, represents a fault-line scarp where the Southern Uplands Fault has brought the harder Ordovician rocks against the relatively softer Upper Palaeozoic sandstones of the Midland Valley. This juxtaposition has allowed the sandstones to be worn down to a greater degree to the north of the fault, leaving the Ordovician and Silurian rocks upstanding as an escarpment. The adjoining Moorfoot Hills, to the west of the Soutra Pass, are the only other clear example of a fault-line scarp in association with the Southern Uplands Fault, for elsewhere along its length much tougher conglomerates, grits and volcanics lie directly against the hard rocks of the Southern Uplands. Farther south-westwards, therefore, in Lanarkshire, the northern edge of the Southern Uplands is not so clearly defined.

Turning southwards, across the Soutra Pass, the rolling summits of the Lammermuirs can be seen stretching away to the eastern horizon, their tawny-coloured grasslands unbroken by major rock outcrops and diversified only by the dark lines of forestry plantations. The Leader river, flowing south to join the Tweed, has here taken advantage of a narrow tongue of Old Red Sandstone to carve out the attractive valley of Lauderdale. Farther down the valley, as it widens, the distinctive red soils of the ploughed land gradually replace the untamed moorlands even on the steep slopes up to 1,000 feet (over 300 metres). The improved grasslands of the valley floor, supporting sleek herds of cattle and flocks of sheep, already suggest that we are returning to the mixed farmlands of the Tweed basin. The rich, red, sandy soils and the mature woodlands of the hillslopes in this prosperous landscape of Lauderdale give more than a hint of Devon, where the Old Red Sandstone again prevails. Nevertheless, the little town of Lauder, with its tolbooth and its Scottish harling (a type of stucco) on the kirk walls, soon removes all doubt as to our real location, whilst in a few more miles we return to the beautiful scenery of Tweeddale at Melrose.

Where the middle reaches of the Tweed, and its tributaries the Gala, Yarrow, Ettrick and Teviot, have cut deep trenches into the Silurian rocks, they have allowed ribbons of farmland from the Berwickshire lowlands to infiltrate deeply into the Southern Uplands. Above the woodlands of the valley sides some of the ploughed fields extend even to elevations of 1,100 to 1,200 feet (about 350 metres) – the highest improved land in all Scotland. Within this well-wooded and cultivated foothill zone, between the uplands and the Merse, the sheltered valleys of the Middle Tweed have provided a focus for both communications and settlement in the Border Country. In the north, the towns of Selkirk, Galashiels, Melrose and St Boswells cling to the steep valley sides above the tree-lined water-courses.

Most of these towns have developed as part of the tweed and knitwear manufacturing complex that has brought fame to the Scottish borders, utilizing the well-known Cheviot hill sheep and the availability of water power for the earliest looms. The stone-built houses, a mixture of dark Silurian greywackes and ruddy Old Red Sandstone, reflect the location of these towns astride a geological boundary, although the brick mill chimneys bring a note of Victorian harshness into the townscapes of mellow Scottish stone. Long before woollen spinning and weaving had reached commercial levels, however, these valleys were famed as ecclesiastical centres, as their picturesque abbey ruins so clearly demonstrate.

Although Dryburgh Abbey is the most complete, it is the Gothic splendour of Melrose which has been the greatest source of inspiration to generations of poets, writers and artists. Built from large blocks of Old Red Sandstone, which have weathered to a delicate rose colour, this ancient abbey is the highlight of the attractive town which nestles at the foot of the Eildon Hills.

From almost any viewpoint in the Middle Tweed basin the triple summits of the Eildon Hills dominate the skyline, and they remain as aesthetically stimulating today as they must have been to the Iron Age peoples (who built a hill-fort on the northern summit) and to the Romans (who placed the aptly named fort of *Trimontium* at their feet). But it was Sir Walter Scott who brought most fame to these conical eminences, for the graceful lines of his 'delectable mountains' stimulated some of his greatest writings and it is not by chance that his home, the famous Abbotsford, 'that Romance in stone and lime', was built only a short distance away. The hills are best seen from the east, from the so-called Scott's View near Melrose, where their shapely, heather-clad summits rise above the neatly ordered fields and woodlands of the Tweed (Plate 2). In our wanderings through the border counties none of the uplands has shown such remarkably steep slopes nor such strikingly isolated summits, for neither the Lower Palaeozoic nor the Upper Palaeozoic sedimentary rocks, which we have so far considered, have succeeded in producing such prominent landforms. The answer lies in the fact that the Eildon Hills were carved from rock-types which we have not yet encountered: hard igneous rocks which bear no relation to the sedimentary succession described above.

We saw in Chapter 1 how igneous rocks differ in their mode of formation from sedimentary rocks and how it is possible to classify different types of igneous rocks on a genetic basis. Although we shall have an opportunity to consider examples of extrusive lavas and other volcanic rocks in Chapters 12 and 13, and we shall visit granitic terrains in Galloway and elsewhere, the Eildon Hills are one of the best localities for investigating the intricacies of intrusive igneous rocks. The Eildons are regarded as the remnant of an enormous composite laccolith, made up of several sheets of mainly acidic lava which have invaded the bedding-planes of the sedimentary Old Red Sandstone (Figure 5). Fed by volcanic vents from deep-seated magma chambers, the lavas must have caused the horizontal sandstones to become updomed, but without enabling the fluid

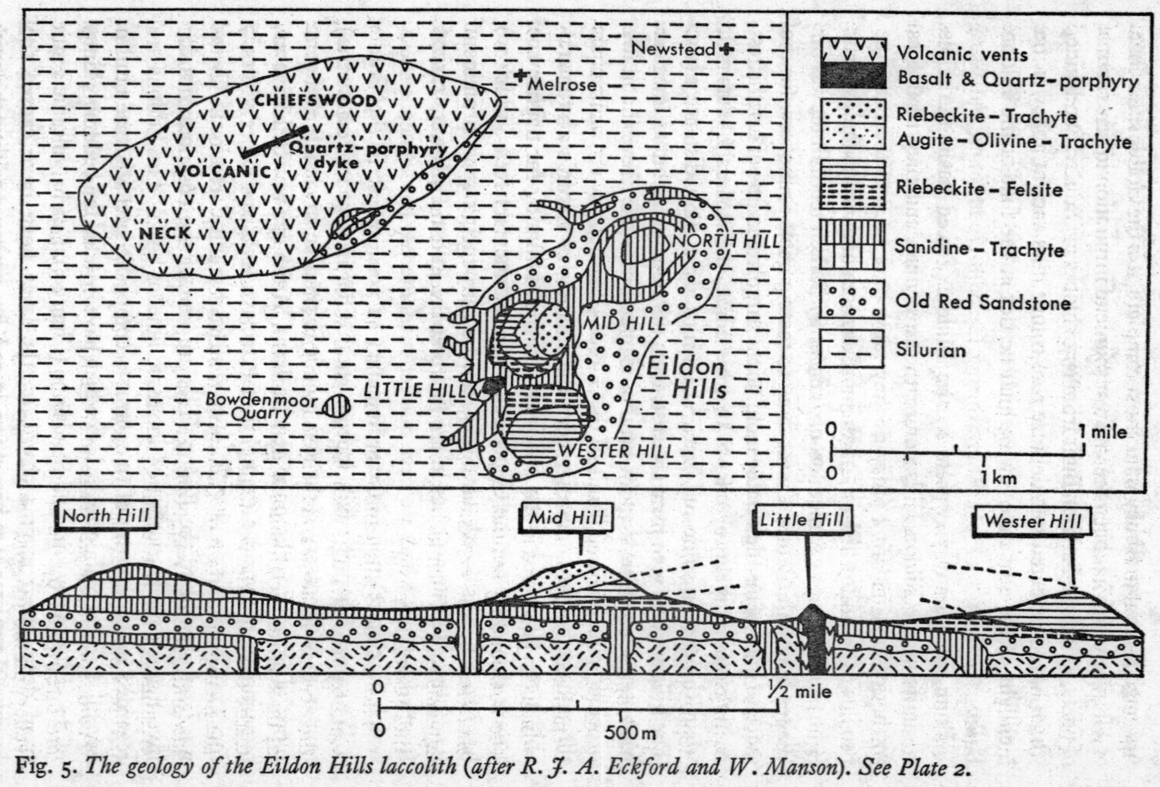

Fig. 5. *The geology of the Eildon Hills laccolith (after R. J. A. Eckford and W. Manson). See Plate 2.*

igneous material to break out at the surface as a sub-aerial lava flow. Once the intrusive material had cooled and hardened into individual sheets of igneous rocks, similar to the sills which we shall examine later, it remained interbedded with the sedimentary rocks until denudation destroyed the overlying cover and exposed the ancient laccolith. It is known that the igneous activity took place here during Carboniferous times, but many millions of years elapsed before erosion succeeded in lowering the less resistant sedimentary rocks, leaving the harder igneous material to form the spectacular Eildon Hills.

Sills, or sheets of intrusive igneous rock, are common around Melrose and have helped to form such eminences as Black Hill, White Hill and Bemersyde Hill. Before leaving the interesting geological phenomena of Melrose, mention should be made of the large volcanic neck of Chiefswood, which lies between the town and Abbotsford. Although of no great topographical significance, this oval igneous mass, almost 2 miles (over 3 kilometres) long, represents an ancient volcanic vent, but it is no use looking for a great Vesuvian cone in the landscape, for the ash and lava cone (if it ever existed) has long since disappeared. Only the walls of the vent have survived, together with the infilling of volcanic agglomerate which can be viewed in the workings at Quarry Hill. Angular fragments of the surrounding sedimentary rocks (Silurian and Old Red Sandstone), shattered by the explosive forces, can be seen mingling with pieces of igneous rocks which fell back into the former vent (chiefly basalt, trachyte and quartz-porphyry).

The Eildon Hills cannot be equalled as a viewpoint in the Border Country for, although their highest point is only 1,385 feet (422 metres), their central location and their isolation give them an advantage over some of their loftier neighbours. Apart from the Merse of Berwickshire, away to the east, and the red-floored trough of Lauderdale in the north, only the picturesque strath of Teviotdale remains to be explored in the Tweed lowlands, for in every other direction the eye is carried upwards to the crestlines of the border hills.

The largest town of the Borders, Hawick, is located near the head of Teviotdale and, as with its associates farther north, its sandstone buildings are interrupted by the alien brick chimneys of woollen mills. In a side valley, on the Jed, the historic town of Jedburgh is, however, a good deal more romantic, with its fine abbey ruin of grey, red and yellow sandstone surrounded by tall stone houses with steep slate roofs. Excellent exposures of the Old Red Sandstone can be seen in the river cliffs at Jedburgh, including one of Hutton's famous unconformities (see p. 34). But it is not the red rocks and soils of this area which take the eye, nor the attractive village of Denholm around its rectangular green, for scattered at regular intervals along the flanks of Teviotdale are prominent conical hills. As one might expect from the foregoing description of the Eildon Hills, these steep-sided summits have been carved from a variety of igneous rocks, formed during Carboniferous times. The former volcanic vents which form the Minto Hills, Troneyhill and Ancrumcraig were filled only with agglomerate, but the high summits of Rubers Law, Black Law and Lanton Hill have vents infilled with both agglomerate and basalt. Finally, Dunion Hill and the oddly named Fatlips Crags have only plugs of basalt beneath their prominent summits.

The Border Hills

Throughout our journeys in the Tweed lowlands we have constantly been aware of the Cheviot Hills which dominate the southern horizon. Although they form more of a physical obstacle than the Tweed itself, the border runs almost arbitrarily through the high, rolling hills and plateaux of the virtually featureless Cheviots. Washington Irving gazed with disappointment over a '. . . mere succession of grey waving hills, line beyond line . . . monotonous in their aspect . . .', although the eminent geologist, Sir Archibald Geikie, defended these border hills from criticism: '. . . . nowhere else in Scotland can the exquisite modelling of flowing curves in hill-forms be so conspicuously seen.'

The flowing curves and the lack of sharp peaks, arêtes and spectacular corries suggested to some early writers that the Cheviots might either have remained unglaciated throughout the Pleistocene or have remained deeply buried beneath the Southern Uplands ice-cap. Recent research by Dr C. Clapperton has shown, however, that the Cheviot had its own ice-cap, with glaciers emanating from such hollows as the Hen Hole and the Bizzle. Nevertheless, the lack of major rock outcrops on the summits (except for a few tors) has provided few landmarks to demarcate the border, except for the so-called Hanging Stone, which marks the junction of the Eastern and Middle Marches. Otherwise, the Cheviot is little more than a peat-covered dome and it is difficult to understand why Daniel Defoe, on his visit there, ventured upwards with trepidation, afraid that there would not be sufficient room for himself and his guides on the summit!

If the summits of the Cheviots cannot claim much of geological or scenic interest, the same cannot be said of their slopes, for here there is much to intrigue the student of glacial geomorphology or river development. One of their most striking features is the way many of their hill-slopes, spurs and drainage divides are indiscriminately cut through by deep, virtually streamless channels. These obviously bear little relationship to the present drainage pattern, and we are forced to turn to the Pleistocene Ice Age to seek an explanation of their form. Although earlier writers have invoked ice-impounded lakes to explain them as glacial 'spillways' carved by overflowing lake-waters, more recent research has suggested that these channels were formed entirely sub-glacially. Thus, glacial meltwaters, running under enormous pressures beneath a down-wasting ice-sheet, had sufficient energy to carve this branching and twisting channel network. The streams reworked much of the englacial detritus, ultimately leaving it as bedded sands and gravels along the valley floors and hill slopes. The ice-sheet responsible for the majority of the channels and the glacio-fluvial landforms was generated from the Southern Uplands, so the ice moved across the northern slopes of the Cheviots and down the Tweed basin.

The associated glacio-fluvial sand and gravel formed kames and eskers (see p. 225) in a complex system which can be traced down Teviotdale to Eckford and thence up the Kale Water valley, past Morebattle and into the meltwater-modified valley of the Yetholms. Here, river diversion has left an old abandoned channel of the Bowmont Water, now occupied by the marshy hollow of The Stank and Yetholm Loch. The attractive villages of Town Yetholm and Kirk Yetholm are built of a mixture of Carboniferous Sandstone and black Cheviot lavas but, as if to demonstrate the proximity of England, several of the cottages have retained their thatch. The half-timbered inn and the chestnuts and sycamores on the village green of Kirk Yetholm are also reminiscent of England, but the shiny black pitchstone and andesite walls of the kirk are singularly Scottish.

The central part of the border hills is in reality the main massif of the Southern Uplands, which separate the Solway plains of England from the

Midland Valley of Scotland. Their major geological structures, of complex Caledonian folds with an orientation from NE to SW, remain in general the same as those described above (p. 33), but here in the Lowther and Tweedsmuir Hills the elevations are somewhat greater. Between the towns of Peebles and Moffat, for example, more than a dozen summits approach or overtop 2,500 feet (over 750 metres) in the Tweedsmuir Hills. These high summits are remnants of a once higher surface that has been almost totally destroyed, leaving them standing above the general level of the dissected surface of the main plateau, the gently rolling level of which ranges between 1,500 and 2,000 feet (450–600 metres). The residual hills stand above the plateau partly because they coincide with outcrops of massive grits, which occur within the complex succession of folded Lower Palaeozoic sediments. But even these higher summits exhibit the flowing curves remarked upon by Geikie, and only in a few places can these uplands be compared with the grandiose scenery of the Lake District or North Wales, built from sedimentary rocks of similar age. The reason seems to lie in the fact that the central Southern Uplands lack the hard igneous intrusions and volcanic rocks which help to form the rugged peaks of Cumbria and Snowdonia. Instead they are composed of Lower Palaeozoic greywackes, flags and mudstones which, because of their thin bedding, usually occur as steeply inclined narrow outcrops, thereby excluding major structurally controlled features. Only where glacial overdeepening has intervened do the landforms take on the stature of their Lake District neighbours.

One such valley can be found where the Yarrow Water meanders through the uplands known as the Ettrick Forest. Upstream from the picturesquely wooded stretches around Selkirk, the Yarrow valley penetrates deeply into the highest hills. These are mirrored in St Mary's Loch and Loch of the Lowes, once a single valley lake but now divided by a delta on which stands Tibbie Shiel's Inn. The lakes occupy a true rock basin, where the valley floor has been glacially overdeepened by the same ice-sheet which breached the main pre-glacial watershed of the Southern Uplands at the head of the valley. Thus the main road now climbs easily out of the valley of the Yarrow Water and across the high col at Birkhill, before descending south-westwards into Moffatdale. Immediately the character of the scenery changes and we could be in a typical Lakeland or Welsh glacial valley.

At the upper end is Dobb's Linn, where a waterfall plunges over the vertical Silurian grits in a series of steps. It was from the exposed rocks in the cliffs of the gorge that Charles Lapworth, in the late nineteenth century, worked out the complex stratigraphical succession and the structures of this region, whilst employed as the local schoolmaster. It was here also that Lapworth, later to become an eminent professor of geology, identified many of the fossils known as graptolites, which lived in the Silurian seas some 400 million years ago. Farther down Moffatdale, where the valley takes on the characteristic U-shape of glacial terrain, we find one of the most spectacular scenes in southern Scotland at the Grey Mare's Tail, where a superb waterfall cascades from a text-book example of a hanging valley (Plate 3). The stream, falling some 700 feet (over 200 metres) to the valley floor, is known as the Tail Burn and drains the moraine-impounded Loch Skene, which lies in a corrie-girt upland basin. The large glacier of Moffatdale succeeded in lowering the main valley more than the tributary valleys, which have as a result been left 'hanging' at their point of confluence. Such glacial overdeepening was assisted in this instance by the presence of a major fault, which can be traced for 40 miles (64 kilometres) past St Mary's Loch to Moffat. The shatter-belt associated with the faulting has not only assisted erosion but also accounts for the remarkable straightness of Moffatdale itself.

To the north of Moffatdale the northern valleys of the Tweedsmuir Hills are drained by the headwaters of the Tweed, which first flows north-eastwards along the axes of the Caledonian folding before turning eastwards at Peebles. Thence, for the next 20 miles (32 kilometres) of its course, the Tweed cuts discordantly across the major structures of the Southern Uplands. One interesting effect of these contrasting reaches of the Tweed is the way in which the character of the valley changes: where it lies parallel with the strike it is often broad, with spur ends truncated, but in its discordant reaches the valley is irregular and constricted. At Drumelzier the northern hills are broken by the Biggar gap, a remarkably wide, flat valley that may once have carried the waters of the upper Clyde eastwards to the Tweed (see p. 103). Its strategic importance as the only major east–west routeway in the Southern Uplands is demonstrated by the abundance of hill-forts and pele towers in the district. The slated roofs of Peebles, and indeed those of old Edinburgh and of many border towns, were almost certainly constructed from the Silurian slates of nearby

Stobo, which once possessed the largest slate quarry in southern Scotland.

The Peebles area is more significant, however, for its large number of glacio-fluvial landforms. Since the Eddleston valley was one of the first sites in the British Isles where subglacial, englacial and marginal drainage channels were recognized (by Dr J. B. Sissons), it is worthy of examination. Today the Eddleston Water runs *southwards* to join the Tweed at Peebles, but the pattern of meltwater channels has shown that at one period in the Pleistocene the meltwaters of the Southern Uplands ice-sheet escaped towards the Midland Valley near the northern end of the Eddleston valley, at an altitude of almost 900 feet (294 metres), because the Tweed valley was buried beneath the ice. The outwash from the ice-sheet created a number of kame terraces which can be traced *northwards* to the former outlet, although the modern drainage of the Eddleston Water flows in the opposite direction. Subsequent work in the upper Tweed valley by Dr R. J. Price has shown that hundreds of glacial meltwater channels exist in this area, making Peebles an important centre for the study of phenomena associated with the downwasting of ice-sheets.

The Western Dales

It has become obvious from our examination of the Border Country that the primary drainage of the Eastern and Middle Marches has been integrated by the Tweed, which carries the rivers eastwards to the North Sea. To the west of the high hills around Loch Skene, however, the drainage is related to the Solway Firth, so that here the valleys run from north to south and have long been used as important routes across the Southern Uplands. The influence of the major south-flowing rivers (the Nith, the Annan and the Esk) on the scenery of the western borderlands is of such paramount importance that we must examine these valleys in some detail in order to explain their somewhat anomalous courses. First, however, a word is necessary about the geological structures hereabouts.

The complex folding of the Ordovician and Silurian sedimentaries, which form the central massif of the Southern Uplands, is here diversified by the presence of basins of Coal Measures and of New Red Sandstone, whilst in the south-east Carboniferous Sandstones and Limestones reappear. The influence of these relatively newer rocks on the evolution of the drainage pattern will be duly noted. This is the first time we have encountered the New Red Sandstone, which brings its distinctive colouring to the soils and building-stones of Dumfriesshire. An extension of the rocks which floor the neighbouring Vale of Eden and the Solway plain, the New Red Sandstone is represented in Scotland largely by the Permian system rather than the Trias. The Permian rocks are made up of dune-bedded sandstones and breccias which indicate that they were formed under desert conditions. The breccias, sometimes referred to as 'brockrams', are in reality fossil screes of greywacke, basalt and sandstone which formed on the surrounding slopes of the desert basins during Permian times. The basins must have been carved in the underlying Lower Palaeozoics; we can visualize how these narrow desert valleys, similar to the modern Death Valley of California, were occasionally swept by torrents, when the screes were carried out from the hillslopes as talus fans, eventually to form the basal breccias. Ultimately, as the dune sands accumulated in the arid basins, the topographic depressions would become infilled by sandstones, finally to be re-excavated (some 200 million years later) by the river systems to which we must now turn.

The major watershed between the Solway drainage and that of the Midland Valley lies along the faulted northern edge of the Southern Uplands. Indeed, the headwaters of the Nith rise to the north of the Southern Uplands Fault and look as if they belong to the Ayrshire rivers rather than to the Solway system. Professor T. N. George has demonstrated, in fact, that the Nith drainage is a composite one, having captured some of the headwaters of both the Lugar Water and the Clyde (Figure 6). It becomes clear that the Nith headstream once flowed northward through the New Cumnock gap, but was later captured by the reach which had adjusted itself to the synclinal structure of the Sanquhar coal basin. Nevertheless it is equally clear, from a study of the river pattern in the lead-mining country around Wanlockhead, that the Sanquhar basin itself was once drained by the headwaters of the Clyde (Figure 6). The elbow of capture and the marshy wind-gap of Crawick Moss bear mute testimony to the former river course. T. N. George summarizes thus: '... the unbroken continuity of Nithsdale is deceptive in its simplicity. Lower Nithsdale is mature because of its ancient establishment ... and its Permian floor. Mid Nithsdale is mature because of its rapid adjustment to the Sanquhar basin. Upper Nithsdale is mature having been abstracted from another system. The integration of the three segments has come

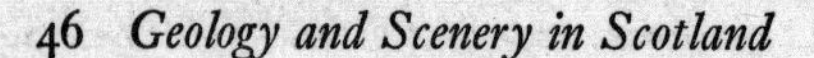

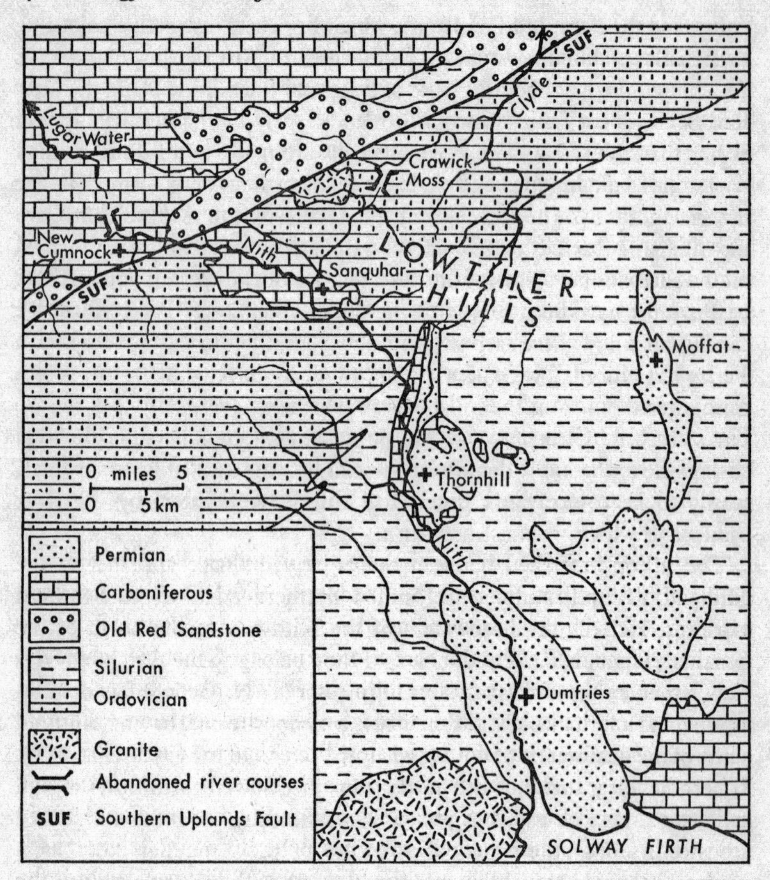

Fig. 6. *The drainage of the river Nith in relation to the geology of the Southern Uplands. The 'wind gaps' at Crawick Moss and New Cumnock are thought to be abandoned river courses.*

about by repeated piracy that radically disrupted the primary drainage of the area.'

Similarly, we can trace a general correspondence between the alignment of Annandale and the structural basins of newer rocks. But, while the Nith took advantage of both Carboniferous and Permian outliers, Annandale is related only to the Permian sandstone basins of Moffat and

Lochmaben. At the head of the Annan, a few miles above the town of Moffat, a most remarkable trough, known as the Devil's Beef Tub, sunders the uplands between the Lowther and Tweedsmuir Hills (Plate 4). It is best viewed from the A701 main road, and must have impressed Scott, since he introduced a vivid description into *Redgauntlet*: 'A deep, black, blackguard-looking hole of an abyss . . . at the bottom there is a small bit of a brook, that you would think could hardly find its way out from the hills that are so closely jammed around it.' But find its way out it did, and in so doing was partly responsible for the trough itself. Although glacial erosion has played no small part in its excavation, the Beef Tub represents the ancient floor of the Pre-Permian valley which has been re-excavated by the headwaters of the Annan, necessitating the partial removal of the Permian sandstones which still occupy the lower reaches of the valley. The same bright red sandstones have been extensively used in the local buildings, especially in the attractive town of Moffat, through which we must now pass to explore the last of the western dales: Eskdale and Liddesdale.

The valley of the Esk, lacking the more easily eroded Permian rocks, is different in form from those of the Annan and Nith. The river has been more constricted in its journey across the Silurian rocks, so that its valley remains narrow right up to the border-crossing near Canonbie, where the New Red Sandstone reappears to form the Solway plain. Because of its narrowness and its lack of major through-routes and settlement, Eskdale has remained something of a backwater, little changed since 1803 when Robert Southey spoke of its 'quiet, sober character, a somewhat scenic melancholy kind of beauty . . . its green hills . . . of a mountainous sweep and swell; green pastures where man has done little . . .' It is perhaps ironical that this quiet valley was the birthplace of Thomas Telford, the great civil engineer, who, more than any Scotsman (other than Macadam), has helped to open up the dales and hills of Scotland to modern motor traffic. Unlike the more fertile New Red Sandstone soils of the plains, the soils of this western marchland are thinner and stonier on the hillslopes, despite the thick drift infilling of the valleys. The differences in soil quality and depth are reflected in the contrasting vegetation of these southern hills. Bracken flourishes on the lower, steeper slopes where the soils are more than one foot (0·3 metres) deep. On the better drained land, known as 'white land', extensive grasslands flourish, where sheep's fescue, sweet vernal grass and bent grow. But where drainage conditions are poor, especially on the plateaux and spurs, sedges and rushes occur in association with the deep layers of peat that mantle the hilltops. Where heather has carpeted the rapidly eroding peat-hags, the colour of the hills takes on a darker hue: this has been given the name 'black land'.

Along the border itself runs Liddesdale, different from Eskdale in its geology and landforms, and hence in its character. Near its head stands the grim castle of Hermitage to remind us that we are back in the troubled borderland, but here also are the extensive conifer forests which appear to have spilled over the border from Redesdale and Kielder. The Forestry Commission started its Scottish plantings here in 1920, and the ubiquitous sheep are gradually being ousted by spruce on these lonely hills. Nevertheless, some natural woodlands have survived in the valleys, where trim beech hedges replace the stone field boundaries of the uplands. At the planned town of Newcastleton the appearance of both Carboniferous Sandstone and Limestone in buildings reminds us that we have left the Silurian uplands, while the abandoned lime kilns and small limestone quarries show how farmers once utilized the local limestone to offset the acidity of these upland soils.

By the time we have reached the Canonbie district the river is cutting once more into red sandstones, but these are the barren red Coal Measures which herald the tiny Canonbie coalfield, where production has now ceased. The valley has now opened out onto the broad Solway plain, and the red sandstone farmsteads and the pele tower of Kirkandrews reflect not only the reappearance of the New Red Sandstone but also the fact that the English border is upon us. As if to demonstrate the impending change of nationality the Permian rocks give way to the Trias within the New Red Sandstone succession between Annan and Gretna. The Trias was once used as a building stone here, although it was soon discovered that its weathering qualities were greatly inferior to those of the Permian sandstones. The proximity of the brick clays of the Solway plain, however, has meant that brick now appears in some of the buildings of this lowland fringe, as if to mirror the English brickwork of neighbouring Carlisle. As a final reminder that twentieth century technology is invading the Scottish rural scene, the nuclear power station of Chapelcross, near Annan, stands implacably on the Solway shore.

3. Galloway

Jutting westwards into the Irish Sea, the peninsula of Galloway is the most clearly defined of all the regions of Lowland Scotland. It is bounded on three sides by the maritime waters of the Solway Firth, the Firth of Clyde and the North Channel, the last of which separates it from the neighbouring coast of Ulster – a mere 20 miles (32 kilometres) away. Its two counties of Wigtown (The Shire) and Kirkcudbright (The Stewartry) are cut off from the Border Country by the broad lowland of Nithsdale, which has long been taken as the eastward margin of Galloway, though in the north the Carrick division of south Ayrshire is generally included within this distinctive region.

The landscape of Galloway exhibits a contrast between the neat orderliness of its rolling farmlands and the wilderness of its forests and moorlands – between the bright green pasturelands of the lowlands, one of the richest dairying regions in Britain, and the dark green conifer forests and rock-strewn summits of the mountainlands. Galloway has a mountainous heartland, with scenery as rugged and as attractive as in many parts of Highland Scotland. But the coastal fringes lack the fjords of the Highlands, for the erosive powers of the ice-sheets were not as intensive nor as prolonged as they were farther north and Galloway instead exhibits extensive spreads of glacial detritus. Thus, bare, ice-scrubbed rocks are uncommon in the Galloway lowlands, and the true crofting landscapes of northern Scotland, with their pocket-handkerchief plots, are absent here. Instead, there is a lowland scene of pastoral prosperity, patterned with well-timbered hedgerows and neat white farms, against a backdrop of dark, forest-clad mountainland. Nevertheless, the rock basins, peaty hollows, lochs and glacial valleys of The Merrick (2,764 feet, 843 metres), the highest peak of the Southern Uplands, are still reminiscent of the Western Highlands.

A comparison can be drawn, therefore, between these two peripheral areas of western Scotland which are known to have been peopled during the Roman occupation by Gaelic-speaking Pictish tribes, separated from each other by the Britons of Strathclyde. It is also significant that the high Border Country east of Nithsdale, with smooth, rolling plateaux incised by deep, flat-floored valleys opening out into the strath-like Berwickshire Merse, exhibits a marked scenic affinity with both the Eastern Highlands and their Aberdeenshire coastlands. Consequently, throughout this chapter reference will be made to the contrasts in landscape which exist between Galloway and the Border Country – contrasts which in part reflect the differences in geology and geomorphology between the western and eastern tracts of southern Scotland.

The Solway Coast

The town of Dumfries, the largest in the Southern Uplands, is the traditional gateway to Galloway and controls an important crossing-point of the Nith. We have already seen that the course of the Nith has been largely determined by the presence of basins of younger rocks, especially the Permian sandstones. Dumfries is located in the centre of the largest of these Permian basins, and the warm red sandstone from the neighbouring Locharbriggs quarry has left its imprint not only on the architecture of this town, known as the 'Queen of the South', but also on Glasgow and other urban settlements of the Lowlands, for Locharbriggs once provided about half the freestone used for building in Scotland. The Permian sandstones, like those of the Trias in England, are also important for their water-bearing qualities, and in earlier years some of the Dumfries manufacturing industries were based on the high-quality artesian water of the Permian basin.

The basin is not only a structural feature; it is, like the neighbouring Lochmaben basin, also a topographic hollow. Its southern end passes beneath the waters of the Solway Firth, and it has been suggested that the Solway coastline was determined very largely by the drowning of similar basins of relatively softer rocks. It is true that between the Nith and Abbey Head the coastline coincides almost exactly with the northern limit of a submerged trough of Carboniferous and New Red Sandstone, so that the older, harder rocks of Criffell and Bengairn rise steeply from the low coastal fringe. Furthermore, to the west, Loch Ryan, the Stranraer lowland and much of Luce Bay have manifestly been carved from a partly submerged basin of Permian sandstone; Wigtown Bay may have a similar relationship. Thus we begin to see a picture emerging of old, hard-rock headlands separated by bays which probably owe their origin to the presence of softer rocks, now largely inundated by the sea. In detail, however, the character of the Solway coast owes much to the variety of superficial deposits, which range from those of glacial derivation, through the suite of raised beaches, to the extensive post-glacial peat mosses and estuarine deposits.

The Solway shores are perhaps most famous for their marshes, whose wide, lonely expanses provide a quiet solitude quite unlike that of the mountainous heartland of Galloway. The sinuous channel networks of the salt marshes, uncovered during every low tide, interlace the gleaming bronze mudflats treasured by the ornithologist and marine biologist but disliked by tourists in search of firm sandy beaches. These coastal flats, with their characteristic flora of glasswort (*Salicornia stricta*), sea manna-grass (*Puccinellia maritima*), sea pink (*Armeria maritima*) and sea aster (*Aster tripolium*), are in fact more sandy than the majority of British coastal marshes, being made up almost entirely of fine-grained marine sands.

The most extensive of the coastal marshes, Lochar Moss, lies just across the Nith estuary from Galloway but is worthy of our attention, if only for its peat deposits and the romantic setting of Caerlaverock Castle. This stronghold of New Red Sandstone blocks, reflected in the once tide-filled moat, marks the zenith of feudal power and is a constant reminder of the proximity of England across the waters of the Firth. The broad, flat bog of Lochar Moss, reclaimed only at its fringes, was once the scene of commercially mechanized peat-cutting for industrial and domestic fuel. Although peat has horticultural value, however, its mechanical extraction is no longer economically competitive, whilst even hand-dug peat is rarely burned on the Solway coast today.

At both Lochar Moss and Moss of Cree, near Wigtown, the peat bogs have accumulated on layers of marine clays and sands which belong to the well-marked post-glacial raised beach that fringes the Solway coastline. In a few places the raised beach deposits themselves rest on buried peaty material and on a few tree stumps (appearing at low tide in a position of growth), which demonstrates that an old post-glacial landsurface has been subsequently inundated by a rise in sea-level. This post-glacial rise was a world-wide phenomenon, as water returned to the oceans from the melting ice-sheets. But why is it that the raised post-glacial shorelines of the Solway Firth stand some 25 feet (7·6) metres) above present Ordnance Datum? It must be remembered that when the Scottish ice-sheet was present in this region the land became locally depressed by the excessive weight of the ice. (A depression of the earth's crust of this type is called 'isostatic downwarping', and in the British Isles it reached its greatest magnitude in western Scotland, where the ice-sheets were thickest.) With the disappearance of the ice, however, the land began to recoil and return to its former level, as the excess weight was removed. Nevertheless, the rate of land recovery in north-west Britain is known to have been outpaced for a short period of time, some 8,000 years ago, when the post-glacial sea-level rose even more rapidly. Thus, for several centuries a post-glacial marine transgression of western Scotland inundated the early post-glacial forests and flooded a few miles inland, where the waves cut cliffs and left their marine clays and sands on the newly emerged land-surface. But this transgression was short-lived and by 3500 BP (Before Present) the rate of recovery of the land was greater than the rise in sea-level, so that the sea receded from the clifflines, leaving the post-glacial marine clays and raised beaches high and dry. The importance of the post-glacial raised beach in the Galloway landscape lies in its effect on coastal settlement and land use, for the raised shoreline has added many square miles of low marshland to the coastal fringe, especially in the large bays and estuaries of the Solway coast, which we must now examine.

Opposite the bleak peat bogs of Lochar Moss, the western shore of the Nith estuary is also fringed by a wide expanse of raised-beach deposits which stretch from Dumfries, past Southerness Point, to Southwick Water. For the most part these have remained as open marshlands dotted with occasional wind-sheared trees, but where mounds of glacial drift rise above the saline soils of the creeks and the raised-beach clays the land has been improved, so that fertile farmlands stand out amidst the sombre-coloured rough-grazing lands. The main coast road and the older settlements stand on the 'old land', some distance landward of the raised-beach shoreline. Here the most renowned settlement is New Abbey, with its magnificent ruin known as Sweetheart Abbey. Built largely from warm-toned New Red Sandstone, this picturesque building, like the cottages of the village itself, also exhibits the use of another local stone, the Criffell Granite (Plate 5). Immediately behind New Abbey looms the brooding hump of Criffell (1,886 feet, 575 metres), its rocky foot-slopes patterned with stone walls constructed largely from the numerous erratic boulders of granite dumped indiscriminately by former ice-sheets but now cleared from the fields.

The Criffell Granite represents only one, albeit the largest, of the Caledonian igneous intrusions which contribute significantly to the relief of Galloway. Although we shall examine the Loch Doon and the Cairnsmore Granites later in this chapter, it would be convenient to describe the mechanisms of granite formation at this point.

For an understanding of the granitic emplacements in the Southern Uplands, one must first take into account the sequence of events in Scotland during the Caledonian orogeny, and especially the Caledonian igneous activity. During the prolonged earth-movements of Lower Palaeozoic times igneous activity was widespread in the British Isles, but especially in Scotland. At the outset we must make a distinction between the so-called 'metamorphic' Caledonian belt of Highland Scotland and the 'non-metamorphic' Caledonian belt of the Scottish Lowlands, south of the Highland Border Fault. Five major groups of Caledonian intrusions (in approximate order of age) have been recognized: first came the migmatites (see Glossary) of the Highlands (sometimes termed the Older Granites), which were intimately connected with the folding and metamorphism of the Moinian rocks; second, the post-Cambrian alkaline intrusions of Assynt, in the Northern Highlands; third, the basic intrusions (especially gabbro) of north-east Scotland, which post-date the deformation of the Dalradian rocks of the Highlands; fourth, the so-called Newer Granites, the only group well represented in the less metamorphosed rocks of the Lowlands (including all the Southern Uplands granites); finally came the last group of intrusions, known as ring-complexes (see Glossary), which post-date the Caledonian folding and are best seen around Ben Nevis, although the Cheviot Granite is probably of the same age.

The Galloway granites were not formed from ring complexes, nor were they associated with metamorphic processes as were the Highland migmatites. Instead they were formed relatively simply by magmatic injection from below into the existing sedimentary sequence, referred to as the 'country-rock'. Since it was intruded at depth as a batholith, the molten magmatic material cooled slowly to form quite large crystals, compared with, for example, the finely crystalline texture of extrusive lavas. The varying proportions of the constituent minerals of quartz, feldspar, and mica, biotite or hornblende, result in different types of granite, although the true granite (in its strictest sense) is very acid, with a high percentage of quartz and a small proportion of biotite.

The Criffel Granite forms an elevated tract of land some 15 miles (24 kilometres) in length and is composed of three granodiorites and a quartz-diorite (Figure 7). It varies in texture from a coarse-grained quartz-porphyry at the centre to a fine-grained variety at Auchencairn, although the main summits of Criffell and Bainloch Hill are formed from a medium-grained granodiorite. At Bengairn, however, it is associated with a grey quartz-diorite which is very similar to granite.

Although Criffell stands out as an upland because the granite has apparently resisted forces of denudation more successfully than the surrounding rocks, the upland is not by any means coincident with the limits of the granitic outcrop. Not only is there granite beneath some of the lowlands, but the metamorphic aureole (see Glossary) has not played any part in the relief hereabouts, in contrast with that of the Loch Doon Granite (p. 58).

The coast road swings along the southern slopes of the granitic hills, which have largely been planted by the Forestry Commission, and across ice-scrubbed rock-exposures before reaching the light grey granite town of Dalbeattie. On Craignair Hill, to the west of the town, the quarries form a conspicuous scar. They have been worked since 1824, to build not only the Liverpool docks, but also those at Birkenhead, Newport and Swansea. Easily exported by boat on the Urr Water, this coarse, grey granite with white to pale pink feldspars was also used in the King George V Clyde Bridge at Glasgow. The stone walls, the smaller fields, the white cottages, the rocky knolls, the gorse and the marshy hollows of this granite terrain are more reminiscent of the Western Highlands, although the 'bright, rose-bowered, garden-circled' village of Auchencairn reminds us of our southerly location.

When the granite country is left behind, the landscape appears to soften in texture, partly by virtue of the great spreads of glacio-fluvial deposits in the lower Dee valley and partly because of the change in the agricultural

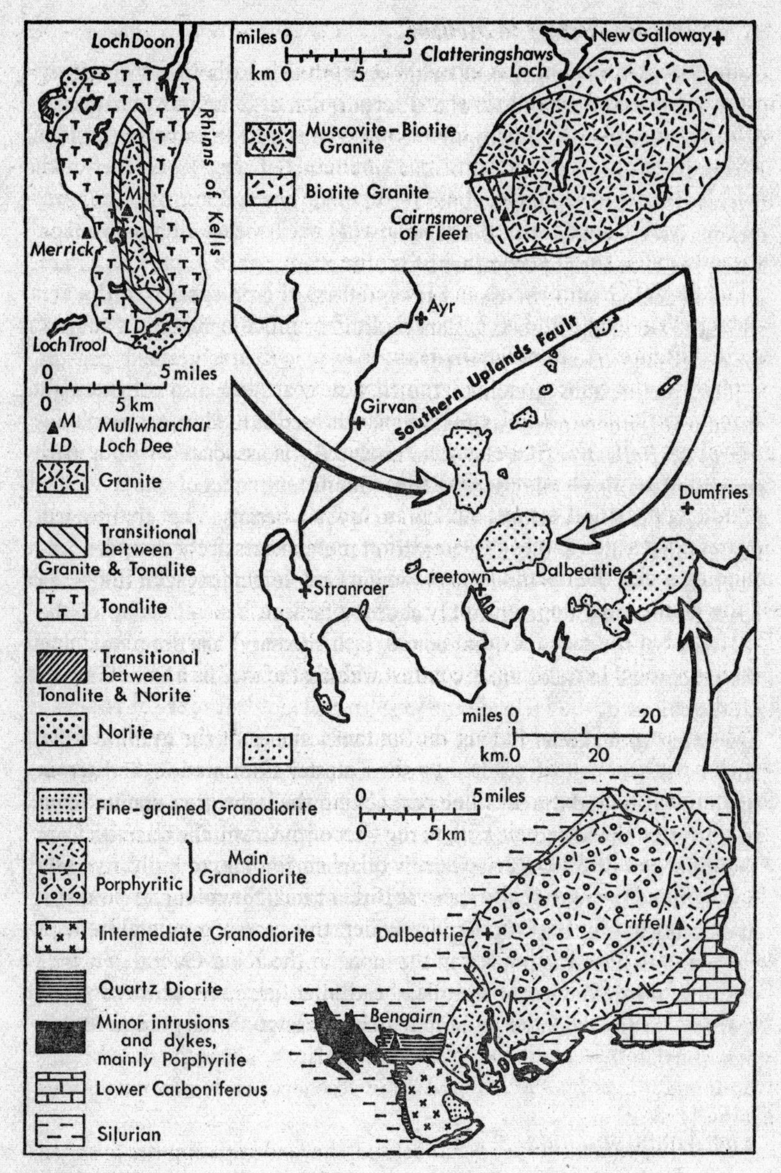

Fig. 7. *The Galloway granites (based on material prepared by the Institute of Geological Sciences).*

scene. Here 50 to 75 per cent of the improved land is under grass; the thicker drifts and the mild climate support some of the most productive dairy farms in Scotland. Cropping has never played an important role in Kirkcudbright, but there has been a significant change from beef-cattle rearing to dairying since modern transport has helped the Galloway creameries to supply the dense industrial settlements of the Midland Valley. Kirkcudbright town, with its attractive terrace houses and warehouses backing onto the river banks, reflects the prosperity of the area while the wooded shores and islands of its bay provide numerous subjects for a well-known colony of artists.

This picturesque stretch of the Solway coast has also attracted the attention of famous novelists, for Borgue was used in R. L. Stevenson's *The Master of Ballantrae* (not the town on the Ayrshire coast, as is popularly supposed), while Gatehouse-of-Fleet, with its long street of colour-washed houses, is the Kippletringan of Scott's *Guy Mannering*. Away to the north, the granite hills of the Cairnsmore of Fleet were the setting of John Buchan's *The Thirty-nine Steps*, while Carlyle rhapsodized about the coast road from Creetown to Gatehouse-of-Fleet.

Sandy bays and rocky coves abound on this tattered coastline, and inland the patchwork of fields amongst the jumbled rocky hills and hollows has all the colour of the Hebrides or western Ireland. But tourism is already beginning to make its impact on the landscape, judging by the caravan site at Auchenlane, while the industrial clutter associated with quarrying has long disfigured the coastline near Creetown. It is surprising, therefore, to learn that the attractive silver-grey Creetown Granite, well displayed in the buildings of Newton Stewart, is quarried from a narrow vein a mere 200 yards (183 metres) in width. Because of its light colour, good quality and excellent jointing it is in constant demand as an ornamental stone. Its tiny outcrop has little impact on the topography, however, and in order to examine the influence of the main Galloway granite intrusions on the scenery of this region we must now turn inland to explore the mountainous heartland.

The Mountainlands

The dichotomy of the Galloway countryside has already been mentioned. To some the true Galloway is the mountainous interior – the grey Galloway of misty hills, '. . . of brown bent and red heather, of green knowe and grey gnarled thorn' (S. R. Crockett). These are the 'High Moors' of Galloway, but they are mainly 'grass moors', in contrast with the heather moors of the Highlands. It is partly because of the dominance of grassland that sheep-rearing has been practised so extensively in both Galloway and the Border Country. There is, however, evidence that the Galloway hills were not always grasslands and moorlands: in earlier post-glacial times most of the lower hillslopes carried forests of oak, birch or pine, remnants of which are now entombed beneath the peat bogs. Sheep-grazing on the southern 'grass moors' has itself been an important factor in both the partial suppression of the heather and the disappearance of this original forest cover, though climatic change was even more important.

In a comparative study of the upland vegetation of Galloway and of the Border Country, Dr Joy Tivy has shown that the higher rainfall and humidity of the west, together with its greater cloudiness and exposure, are more conducive to peat growth than conditions in the hills farther to the east. In the Galloway mountains, therefore, we shall find that peat bogs are common, even on the valley floors (where they are known as 'flows'), that peat is still burned in the hill farms, and that the peat deposits of Galloway are exceeded in magnitude only by those of Caithness and Sutherland. The eastern hills of the Border Country lie in something of a rain-shadow, so that on their drier grasslands purple moor grass (*Molinia caerulea*), bog myrtle (*Myrica gale*) and deer sedge (*Scirpus caespitosus*) are not as common as they are on the wetter Galloway mountains. It has already been shown that rock outcrops were rare in the smooth scenery of the Border hills, but in Galloway the erosive powers of the ice-sheets have left many rocky knolls to break up the vegetation cover.

These contrasting environments mean that there are land-use differences in Galloway: improved land extends only to a height of 500 feet (150 metres); the productive sheep-runs of the Border hills – 2–3 acres (1 hectare) per ewe here – are restricted by rock outcrops so that 5 acres (2 hectares) per ewe are needed; much more forestry has been introduced on these wetter peaty soils than on the more easterly of the Southern Uplands. Thus, the high mountains of The Merrick and the Rhinns of Kells stand with their feet in the forests, most of which have been planted since 1920. The Glen Trool Forest Park, with over 100,000 acres (around 40,500 hectares) of plantations, is the second largest forest in Scotland.

But by no means all the mountains are clothed with dark conifers, for the high peaks have been left as a great natural wilderness, whilst in the valleys, alongside the rivers and lochs, many of the native broadleaved woodlands of oak and ash have survived and are being carefully managed. Of the coniferous trees, Norway spruce prefers the more fertile, sheltered sites while Sitka spruce will tolerate the poorer soils and the more exposed locations.

A mere glance at a geology map (Figure 7) will suggest that the Galloway mountains are all associated with granitic outcrops, and in the case of the Cairnsmore of Fleet (2,331 feet, 711 metres) it is true to a certain extent. In this upland, all the major summits around the lonely glacial trough of Loch Grannoch are found within the granite margins. In much the same way, however, as the Southwick Water has hollowed out the centre of the Criffell Granite upland, so the Big Water of Fleet has lowered the central outcrop of the Cairnsmore of Fleet Granite. It seems unlikely that the slightly different mineralogical composition of the innermost granite has been the sole cause in the fashioning of the central amphitheatre, for ice erosion has clearly played a significant part in the shaping of these landforms.

When we turn northwards to look at the Loch Doon Granite, however, we find that the petrological differences between the granite margins and the surrounding country rock have clearly played a major part in the formation of this central massif. Here, the hour-glass-shaped intrusion of the granite occupies an area extending from Loch Doon to Loch Dee, a distance of almost 12 miles (19 kilometres). (See Figure 7.) After our experience at Criffell and Cairnsmore of Fleet it comes as something of a surprise to find that few of the highest summits coincide with the granite outcrop. Only in the ridge of Mullwharchar (2,270 feet, 692 metres) does the white central granite play a major part in the relief of the area. The surrounding mountains of Shalloch on Minnoch (2,520 feet, 768 metres), The Merrick (2,764 feet, 843 metres), Lamachan Hill (2,349 feet, 716 metres) and the Rhinns of Kells (2,668 feet, 813 metres) are all found not on the granite but on or just outside its metamorphic aureole (Figure 8). The latter, produced by the intense heat and chemical activity of the granite batholith, must once have roofed the massive intrusion, but these altered country rocks have now been worn away from the central intrusion, surviving only on the flanks, where they create a prominent ring of uplands. The sedimentary rocks of Lower Palaeozoic greywackes, shales and flagstones have been changed into tough quartzitic schists and mica schists by contact metamorphism, as with all the Galloway granites, but only around the Loch Doon Granite does the resistance of the aureole create such a conspicuous ring of highlands. The central granite ridge of

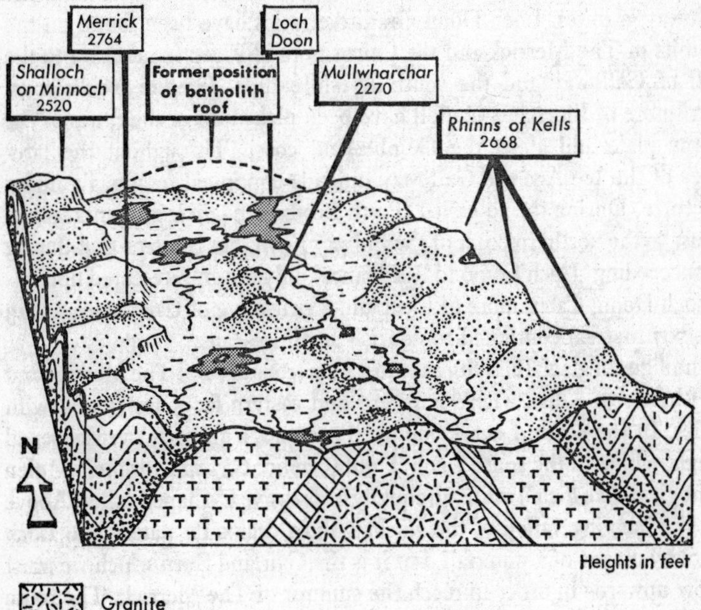

Fig. 8. *The structure of the country around Loch Doon.*

Mullwharchar is flanked on all sides by a circular corridor (about 1,000 feet, or 300 metres, O.D.) now occupied by a chain of lakes or major river systems which isolate the central ridge from its perimeter highlands. These lakes and water courses coincide everywhere with the concentric outcrop of a tonalite granite, a more basic rock with plagioclase feldspar, which must be less resistant to denudation than either of its adjoining rocks (Figure 8).

A great deal of the denudation must have been the work of Pleistocene ice-sheets, which appear to have overridden completely all of the Galloway Highlands. One of the ways in which such overriding can be proved is to note the location of the distinctive erratic blocks carried by ice from the Galloway granites. Loch Doon Granite erratics have been found on the summits of The Merrick and the Cairnsmore of Fleet, in addition to the Mull of Galloway and the south Ayrshire hills. Granite blocks from Cairnsmore of Fleet and Criffell have been picked up in the drifts of the Solway plain and along the Cumberland coast. Throughout the later stages of the Ice Age the Galloway uplands continued to act as a major ice-centre. During the so-called 'Perth Readvance' their glaciers reached almost to the northern coast of the Solway Firth, but it seems that during the succeeding 'Loch Lomond Readvance' only the granite-cored uplands of Loch Doon, Cairnsmore of Fleet and Cairnsmore of Carsphairn (2,613 feet, 797 metres) continued to nourish small ice-caps.

The signs of glacial activity are most numerous around The Merrick and the Rhinns of Kells, whose northern and eastern faces are fretted with corries. In the north, Loch Doon itself occupies a glacially overdeepened trough, whilst in the south Loch Trool (Gaelic: *Gleann t'struthail* – 'glen of the river-like loch') infills an equally impressive U-shaped valley. Above this the hanging valley of the Buchan Burn drains its waters into Loch Trool by means of a waterfall, but it is the Gairland Burn which we must follow upwards in order to reach the summit of The Merrick. The route leads past Loch Valley and over recessional moraines to the oddly shaped Loch Neldricken, which is noted both for its legendary, rush-fringed Murder Hole and for its silver granitic sand, which was once collected for sharpening scythes. Beyond this it is but a short step to the solitary Loch Enoch, cradled between the crags of The Merrick and the hump of Mullwharchar, and thence to the summit of The Merrick, the highest peak in southern Scotland. In addition to its importance as a viewpoint for the coasts of Galloway and Ayrshire, beyond which the view may embrace Ben Lomond, Goat Fell (Arran), the Irish Mournes and the Lakeland peaks, The Merrick gives us an opportunity to examine the drainage pattern of Galloway.

Despite the enormous amount of glacial modification, it has been suggested by Dr W. G. Jardine that the present drainage system of Galloway originated in mid-Tertiary times, since when drainage modifications have been the result of adjustment to structure and a certain amount of river capture. Although the initial watershed between the Midland Valley and the Solway Firth was along the northern anticlinorium of Ordovician strata, local centres of radial drainage, such as that on the Loch Doon Granite, were also operative. Here, near to Loch Doon, it appears that the headstreams of Water of Deugh and Carsphairn Lane once flowed northwards through the valley now occupied by this narrow loch, but were subsequently turned southwards past Carsphairn by river capture around Lamford Hill. A similar diversion appears to have taken place farther down the valleys of the Ken and Dee, where a former south-easterly course past Castle Douglas to Orchardton Bay on the Solway has been replaced by a south-westerly flow to Kirkcudbright Bay.

More recently the river flow of the Galloway uplands has been considerably modified by man: although Loch Doon normally discharged its water northwards to the Ayrshire coast, it is now diverted through a mile-long tunnel to the river Dee, and thence by a succession of power stations southwards to the artificially deepened Loch Ken and the Solway. The majority of the mountain lake-levels have also been raised by dams and supplemented by the large artificial lake of Clatteringshaws Loch to serve the Galloway hydro-electric scheme, the first major enterprise of this type to be established in Scotland.

Wigtownshire

In contrast with the wild moorland landscapes of the uplands of Carrick and The Stewartry, the lowlands of Wigtownshire possess a pastoral calm and a quiet orderliness. The rolling, drift-covered landscape also exhibits a uniformity of colour, for the emerald green of the permanent grasslands gives more than a hint of an Irish scene, which the numerous drumlins do nothing to dispel. Galloway is, of course, one of the nearest points to the Irish coastline, so it is not surprising to find many points of similarity between the two. Structurally, the narrow Rhinns of Galloway mirror the Ards Peninsula of County Down, whilst the New Red Sandstone basin around Stranraer has its Irish counterparts in Belfast Lough and Strangford Lough. Only the extensive Tertiary lava plateaux of Antrim are missing on the Galloway shore, but we shall see in Chapter 4 that the neighbouring Firth of Clyde witnessed Tertiary igneous activity of a different type.

Apart from the Permian basin around Stranraer (with its narrow margin of Millstone Grit) Wigtownshire is everywhere composed of Caledonian folded sedimentaries of Ordovician and Silurian age. In the north the Ordovician greywackes and shales form a broad anticlinorium which manifests itself in an area of low hills known as The Moors. In the south, Silurian rocks of very similar character underlie the thick glacial drift in the area known as The Machars. The third of the physiographic regions created by these Lower Palaeozoic rocks lies to the west of the Stranraer lowland and creates the distinctive promontories of The Rhinns.

The broad peninsula of The Machars (Gaelic for 'flat lands'; cf. 'Machair' of the Hebrides), terminating in the blunt nose of Burrow Head, is often called 'Scotland's dairy farm' and boasts one of the largest creameries in Scotland. Some 75 per cent of the improved land is under permanent grassland, partly by virtue of the dampness of these western peninsulas and partly because of the heavy clay soils of the boulder clay, which has been moulded into a southward-trending drumlin swarm. Near to Wigtown the drumlin hills are of massive proportions, their whale-back form often heightened by the hedgerows which follow the crestline of the hills, creating a scene very reminiscent of nearby County Down. Wigtown itself supports the Irish comparison, for its quiet square of rook-haunted trees and its colour-washed stuccoed terraces looking wistfully over the muddy harbour give it the qualities of an Irish town. In fact, western Galloway, cut off from Scottish affairs by the Southern Uplands, has tended to look seawards. Thus it is no surprise to find an Irish resemblance, for its cultural contacts lay to the west. St Ninian, for example, having founded the first Christian church in Scotland about the year 396 near to Burrow Head, crossed to Northern Ireland to spread the gospel. Nothing remains of St Ninian's original *Candida Casa* on the Isle of Whithorn, but the ancient arches and white harled walls of the Priory Pend of Whithorn still convey something of the antiquity and remoteness of these isolated western peninsulas of Galloway.

Even more remote are the Rhinns (Gaelic: *Roinn* – promontory), whose southernmost tip, the Mull of Galloway, is sometimes referred to as the Land's End of Scotland. The analogy is, perhaps, appropriate, since a few miles to the north a small granitic intrusion creates steep, castellated coastal cliffs near to Laggantalluch Head and Crammag Head. But here we shall find, as in the case of the Loch Doon Granite, that the granite itself has been lowered more effectively than the altered rocks of its metamorphic aureole, which now creates a ring of low hills around the granite's perimeter. Likewise, a narrow belt of igneous felsite (also of Old Red Sandstone age) has been so severely denuded that the narrow promontory of the Mull itself is almost insular. In fact, the distribution of the raised-beach deposits hereabouts suggests that during the higher sea-levels of late-glacial times the Mull of Galloway was an island. The cliff-bound western coast of this southern arm of the Rhinns is broken twice more by lowlands which extend across the peninsula to Luce Bay. Although in the first case there appears to be no structural weakness in the geological succession to account for the lowlands between Port Nessock Bay and Terally Bay, the fact remains that the raised beaches can be traced right across the isthmus. In the second case, however, less resistant Lower Palaeozoic outcrops of Birkhill and Hartfell Shales may have aided valley formation between Clanyard Bay and Kilstay Bay, which was followed by the late-glacial marine transgression.

Northwards, past the attractive white-washed town of Portpatrick, which climbs the rocky cliffs behind its tiny harbour, the coastline of the northern promontory of the Rhinns is more complex than it appears on a map. The apparent simplicty is broken by a detailed irregularity, for the frequent intercalations of the Lower Silurian shales and mudstones among the beds of massive greywackes have produced a profusion of minor bays and clefts among the headlands and ribs of harder rocks. In many places the raised-beach platform fronts the old cliff line, and in one place Ouchtriemakain Cave stands high and dry some 20 feet (six metres) above present sea-level, having been carved out by a much earlier marine advance before post-glacial uplift of the area.

Because the Permian-floored isthmus between Luce Bay and Loch Ryan is mantled with a variety of raised-beach deposits we can only infer that the whole of the Rhinns promontory was an island during late-glacial and early post-glacial times. Along the Luce Bay coast the southern end of this lowland is fringed by the most extensive sand-dune formation in Galloway, with some of the dunes more than 50 feet (15 metres) in height. Heather has begun to colonize the older, innermost dunes, but the outermost dunes remain relatively mobile, occasionally revealing the remnants of Mesolithic sites. The middens and artefacts are found at about 50 feet (15 metres) O.D., in association with one of the higher raised beaches, and may represent strand-looping activities by very primitive inhabitants who depended entirely on hunting and primitive fishing some 8,000–5,500 years ago.

In contrast with the ancient archaeological remains and lonely shores of Luce Bay, the shores of Loch Ryan display a bustle which accords with their former importance as a naval base and their present function as a cross-channel port to Ireland. The naval shipyard of Cairn Ryan now lies mouldering beneath its steep coastal cliffs, but Stranraer exhibits a vitality linked to its status as a rail terminus for the Northern Ireland crossing.

Stranraer was termed 'Stranrawer' in the sixteenth century, and is said to have taken its name from the row ('raw') of houses which lined the Strand Burn, now culverted beneath the curving, narrow streets of the town. Despite the town's apparent prosperity, it is as well to remember that the coastal trade between Galloway and Ireland has gradually declined (with the exception of that using Stranraer), leaving behind a number of decaying burghs and forlorn silted harbours along the Galloway coast.

4. The Firth of Clyde

It was shown in the last chapter that the territorial limits of Galloway extend across the northern boundaries of Wigtownshire to include the Carrick division of Ayrshire, a fact recognized in the new administrative boundaries of 1974, which exclude Carrick from the hybrid administrative region of Strathclyde. Nevertheless, the geological boundary of Galloway manifestly follows the pronounced line of the Southern Uplands Fault, which demarcates the structures of the Midland Valley from those of the Southern Uplands, of which Galloway is so clearly a part. The boundary is less obvious here, however, than elsewhere along this important fault, since Ordovician sedimentary rocks are found both north and south of the fault in the Carrick area, a distribution not found elsewhere in southern Scotland. Furthermore, these sediments have been invaded by thick layers of ultrabasic rocks, including serpentine, and also swamped by volcanic lavas, ashes and tuffs from an Ordovician volcano whose cone-like form has long since disappeared. But the igneous rocks have helped to bolster the resistance of the Ordovician sedimentaries to ensuing phases of denudation, so that today the high coastal headlands and moorlands between Ballantrae and Girvan appear to have closer affinities with the upland landscapes of Galloway than with those of the Midland Valley.

Nevertheless, to the north of Loch Ryan the traveller will first become aware of the singular linearity of Glen App, which has been carved along the line of the Southern Uplands Fault, before turning northwards into the distinctive landscapes of the Firth of Clyde. Once past the little town of Ballantrae, set in its tiny basin of New Red Sandstone amongst the sharp volcanic hills, it is only a short distance to the neat and bustling town of Girvan, which faithfully reflects the character of the Ayrshire coastal settlements in the mixture of Old Red Sandstone and brickwork in its domestic architecture. The tumbled rocks and sea stacks of the raised-beach platform near Kennedy's Pass give way northwards to the gentle curves of a drift-formed coastline at Girvan: the wild and cliff-girt shores of northern Galloway are now behind us. Ahead lies the Firth of Clyde, whose shimmering waters reflect the ever-changing light and shade of this western seaboard, where Keats saw the 'craggy ocean pyramids' of its islands starkly silhouetted against the vivid sunsets of these parts.

In good visibility, no one travelling along the Ayrshire coast can fail to be impressed by the islands which dominate the western horizon. Ailsa Craig, Arran, Holy Island, Bute and the Cumbraes are names which conjure up romantic visions to seamen and tourists alike, for their craggy hills and rocky cliffs bring a touch of the Highlands, and especially the Hebridean Highlands, deep into Lowland Scotland. Their topographic form, as we might suspect, is largely a result of their distinctive geological structures. Their height, ruggedness and steepness are primarily a reflection of their igneous character, although the age of the igneous episode is, in the main, different from that of any so far encountered.

Ailsa Craig

The smallest but most remarkable of the islands, Ailsa Craig, stands isolated '... in lone sublimity, towering above the sea and little ships' (Wordsworth), some 9 miles (14·5 kilometres) west of Girvan. Although the island is less than one mile (1·6 kilometres) in diameter, its summit attains a height of 1,114 feet (340 metres), with a grassy dome-like top surrounded by vertical cliffs up to 500 feet (150 metres) in height (Plate 6). Yet, curiously, these coastal cliffs stand beyond high-water mark (except at Stranny Point) and are fronted by a talus slope in some places and by a post-glacial raised beach along the eastern margin. This has led to the suggestion that the cliffs are fossil features which, like the caves, were most probably fashioned by waves during the post-glacial submergence. With the exception of Water Cave in the south-west corner, the majority of the caves have been carved along dolerite dykes (see Glossary) that slice through the granitic rock which builds the remainder of the island. Swine Cave, for example, is cut in a gigantic dyke which is no less than 57 feet (17 metres) in thickness, although the average dyke-width here is of the order of 6 feet (2 metres).

The main rock is a very fine-grained microgranite in which the hornblende takes the form of a mineral known as riebeckite. The fine-grained rock has a well-marked vertical joint pattern and, since denudation has picked this out, the coastal cliffs often exhibit a columnar structure which an early visitor, Dr John Macculloch, compared favourably with those found in Skye, the Shiant Isles or even Staffa (see p. 251). The speckled bluish-grey colouring of the granite has proved to be an invaluable marker of glacial erratics; its singular character has enabled geologists discovering Ailsa Craig erratics in the coastal drifts of Cumberland and Wales to estimate the former tracks of the Scottish ice-sheets during their southerly expansion.

The micrograniti boss is thought to be the basal remnant of a volcanic vent which once functioned here, although there appears to be some disagreement about the age of its formation – Tertiary is the more usually accepted date, but a Carboniferous age has also been postulated. In addition to creating the towering cliffs, which house the second largest gannetry in Scotland (St Kilda has the largest), the tough microgranite has been used in the manufacture of curling-stones for Scottish sportsmen. The quarries, on the north coast of the island, are no longer worked, however, and only the lighthouse on the raised-beach headland below the castle shows signs of life on this lonely sentinel at the entrance to the Firth of Clyde.

Arran

To the north of Ailsa Craig, and echoing on a more majestic scale the spectacular steepness of its cliffs, lies the much larger island of Arran. Although the majority of tourists receive their first glimpse of it from the Ayrshire shore or from the Isle of Bute, one of the finest views is that obtained from a Clyde-bound ship. Generations of cross-channel passengers from Ireland would agree with Sir Dudley Stamp's belief that the wild and mysterious beauty of Arran plays a large part in rendering the sea approach to the Clyde one of the finest scenic entries in the world. It is therefore surprising to find no comment from Burns about this island, even though he lived for many years on the Ayrshire coast.

The rugged grandeur of the island is very largely due to its geological structures; no other British island exhibits such geological complexity for its size. Its 165 square miles (427 square kilometres) are cut into two almost equal parts by the Highland Boundary Fault, so that the northern area exhibits the geology and scenery of the Highlands whilst the larger southern part is more typical of the Lowlands. The variety of its rocks, ranging in age from Pre-Cambrian through Palaeozoic and Mesozoic formations to the widespread Tertiary igneous phenomena, together with the diversity of its scenery, make Arran a geologist's paradise, for in a sense it represents a microcosm of Scottish geology.

The contrasts in geology and lithology between north and south Arran are fundamental to the topographic differences, which in turn have influenced the vegetation and land use of the two parts. In the north the older, harder Highland rocks, together with the massive Tertiary granitic intrusion, have been fashioned into a deeply dissected highland massif, with rugged peaks rising to almost 3,000 feet (900 metres) separated by deep glacial troughs. The southern portion of the island contains a greater variety of younger, less resistant sedimentary rocks which, despite their scattering of tough igneous intrusions, have produced a much lower undulating plateau surface, rising to 1,700 feet (500 metres) in height (Figure 9).

In north Arran the steepest slopes coincide with the edge of the granite pluton and, since this occurs within a short distance of the coast, there is space for only a narrow ribbon of coastal plain before the boulder-strewn slopes sweep steeply upwards to the highest peak, Goat Fell (2,868 feet, 874 metres; Gaelic: *Gaoth Bheinn* – Mountain of Winds). In Scotland only the isles of Rhum and Skye have similar high peaks in close proximity to the water's edge, while the comparison between Arran and the Mourne Mountains of Ireland is very obvious. It is not by chance that all these examples of high coastal mountains occur where they do, for they all coincide with massive plutonic intrusions of Tertiary age whose location straddles the north-western seas of the British Isles. The steepness of the slopes, the heavy rainfall and the acidity and thinness of the soils on the Arran granite and on its metamorphic aureole have limited cultivation and improved pastureland to the narrow coastal margin, where raised beaches overlie the fringe of sedimentary and metamorphic rocks. The remainder of the northern portion is left as rough moorlands occupied only by deer and sheep. Thus the landscape is one of large-scale grazing farms with irregular field boundaries and a few settlements clustered into the ancient pattern of the 'clachan' (see below).

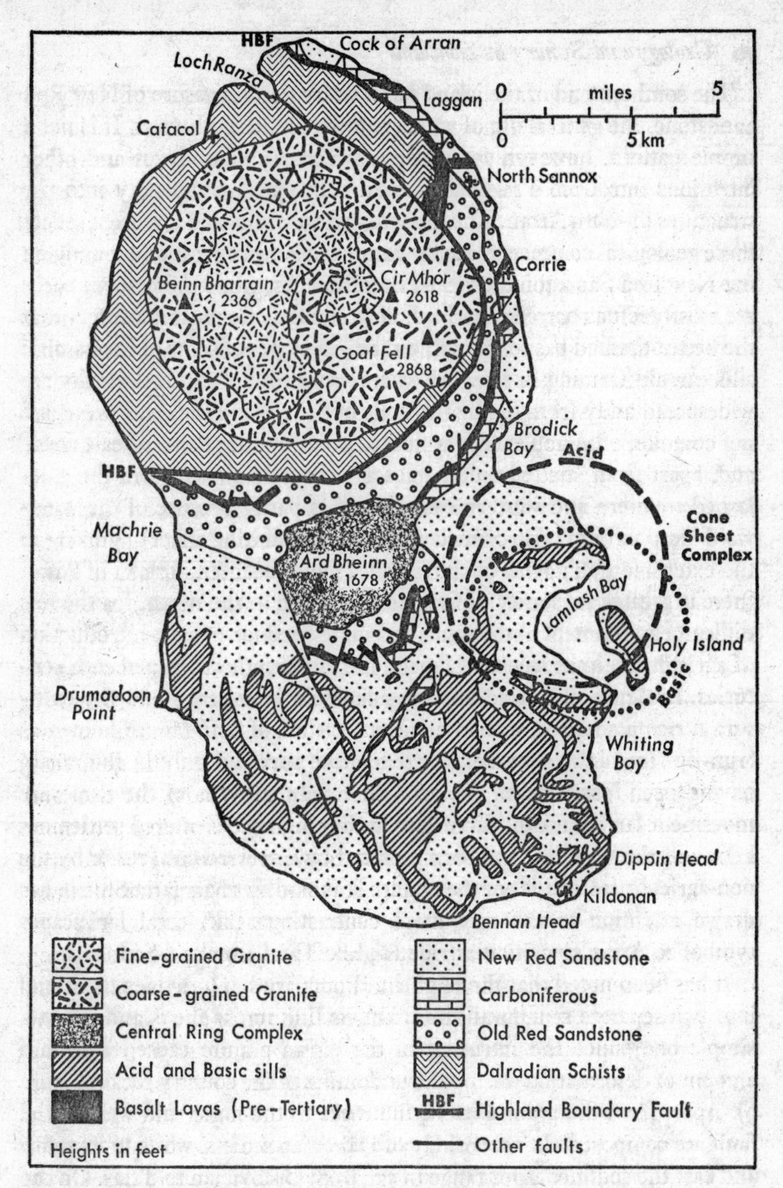

Fig. 9. *The geology of the Isle of Arran (based on material prepared by the Institute of Geological Sciences and S. I. Tomkieff).*

The southern end of the island has a widespread exposure of New Red Sandstone, the general dip of which is towards the south-west. It is not a simple pattern, however, for numerous Tertiary sills, dykes and other intrusions introduce a ruggedness and lithological complexity into the structures of south Arran. The topography, soils and land use often reflect these geological contrasts, with the improved land being found mainly on the New Red Sandstone and the raised beaches, while the igneous rocks are mostly left as barren moorland. Nevertheless, we must not forget that the better-drained base-rich soils on the dip-slopes of some of the basaltic sills can aid farming in a region where waterlogging and soil acidity are widespread away from the lighter soils of the raised beaches. Forests are not common on Arran except around Brodick on the sheltered east coast, and, apart from stunted alder, trees are virtually absent from the gale-lashed southern and western coasts. This is partly because of the windward location but is also connected with the inhibiting effect of grazing in the extensive sheep runs of the island. In the southern portion of Arran there is greater regularity of field patterns than in the north, for the less difficult environment of the south was more conducive to the introduction of agricultural improvements during the eighteenth and nineteenth centuries. In Arran, therefore, the irregular field boundaries and the cultivation rigs associated with the ancient system of land tenure known as 'run-rig' (the infield–outfield system) have survived only in the north, having been lost beneath the geometric field patterns of the enclosure movement farther south. Similarly, in the south the clustered settlement known as the 'clachan' has been replaced by dispersed cottages or by the non-agricultural villages of the eastern seaboard. Dr Margaret Storrie has drawn attention to the way these contrasting agricultural landscapes symbolize Arran's position astride Highland and Lowland Scotland.

It has been noted that the Highland Boundary Fault divides the island into two separate structural units; but its line across the island is not a simple one, since the intrusion of the Arran granite caused a certain amount of deformation during the updoming of the country rocks (Figure 9). Apart from the granite itself, the rocks to the north and west of the fault are composed almost entirely of Dalradian schists, while to its south and east the sedimentaries range in age from Ordovician to Trias. On the narrow coastal margin of north-east Arran a fringe of Old Red Sandstone and Carboniferous formations can be seen in the coastal cliffs between

Corrie and Loch Ranza. Amongst the Carboniferous strata, which have been folded into a large anticline, it is possible to recognize basal conglomerates, calciferous sandstones and a thick limestone (the Corrie Limestone), but it is the presence of Coal Measures which is perhaps the most interesting discovery, although the collieries at the Cock of Arran have long since ceased production. In geological circles, however, the coast between the abandoned mines and Loch Ranza is better known for yet another of Hutton's famous unconformities. In this instance the Upper Palaeozoic cornstones can be seen to overlie steeply dipping Dalradian schists. This unconformity was in fact the first of the three to be discovered by Hutton (see p. 34).

As these Old Red Sandstone and Carboniferous rocks are traced southwards past the frowning steeps of Goat Fell, their narrow outcrop swings westwards across the waist of the island, parallel with the edge of the granite pluton and the Highland Boundary Fault (Figure 9). The reason for the change in direction of the strike of these sedimentary rocks must be sought in the displacement caused by the granitic magma. As the granite was intruded it caused an updoming of the country rock, as is shown by the steep outward dips not only of the Palaeozoic rocks near Brodick but also of the schists on the western flanks of the gigantic intrusion (Figure 10).

South of the so-called String Road, which crosses the waist of the island between Brodick Bay and the west coast, it appears at first sight that the geological succession is becoming less complex as the Carboniferous series dip southwards to pass beneath the Permian and Triassic sandstones. This is certainly true if we examine only the coastal cliffs at Brodick Bay in the east and Machrie Bay in the west, where the sedimentary succession can be seen dipping gently towards the south. But, having already seen the character of Ailsa Craig, we find something familiar about the uplands which surround Glen Craigag in the centre of Arran. Closer inspection reveals that their cliffs and summits have been carved not from uncomplicated Palaeozoic sedimentaries but from a most remarkable jumble of igneous rocks known as a 'ring-complex'. This is the well-known Central Ring Complex of Arran, and to understand its formation we must now look briefly at the history of Tertiary igneous activity in western Scotland, though most of these igneous phenomena will be discussed in more detail in Chapters 11, 12 and 13.

The Tertiary igneous rocks of Scotland, like those of north-east Ireland, form part of the so-called Brito-Icelandic or Thulean province, which also includes the Faroes. It has been suggested that the enormous thicknesses of basaltic lavas which have survived in many parts of this province were once part of a continuous land surface that has now been fragmented by submergence and continental drift. In addition to the thick piles of basaltic lava there are numerous intrusive bodies ranging in character and

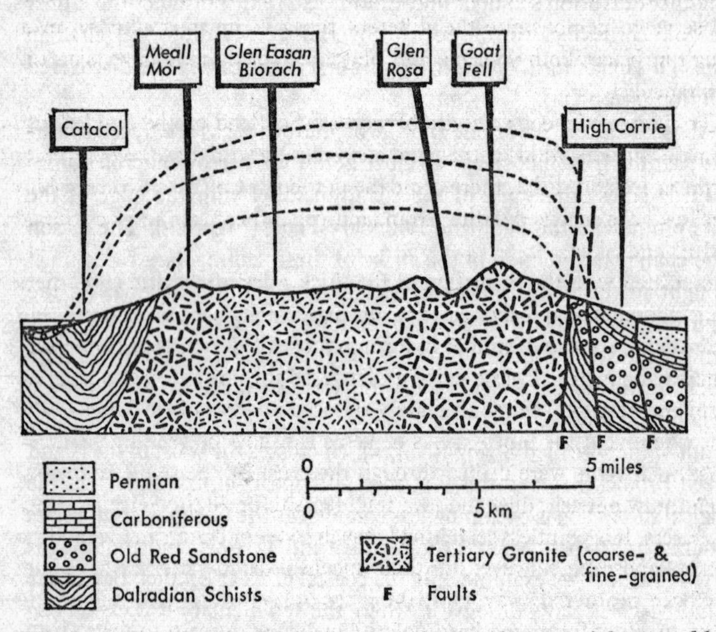

Fig. 10. *Geological section of the Arran granite (based on material prepared by the Institute of Geological Sciences).*

size from the large plutons, such as the Arran granite, through ring-complexes and sills to the more extensive dyke-swarms. The mineralogical composition of the igneous rocks also exhibits variety, with ultrabasic, basic and acid intrusives all being represented by one or other of the Scottish centres.

Although some modern geologists would dispute his views, Dr Alfred

Harker considered that when studying the Scottish Tertiary igneous rocks one could trace a sequential progression from an early volcanic phase (associated with outpourings of mainly basaltic lavas) through a plutonic phase to a final phase of minor intrusions, all this being associated with a change from basic to acid activity. Professor Gordon Craig has shown that the basic to acid sequence is misleading, for in some cases there is evidence to show that acid bodies were formed before the basic ones (for instance, in Rhum). Nevertheless, the simple sequence of events outlined by Harker will serve to demonstrate the different types of igneous activity, even though in places both volcanic and plutonic activity may have occurred simultaneously.

The Tertiary igneous activity of western Scotland can be divided into five main episodes which correspond fairly closely with those recognized in Northern Ireland. First, there came the outpouring of basaltic lavas, but, since few have survived in the Arran landscape, these can best be studied in Mull and Skye.

Associated with the formation of the thick piles of basaltic lavas there was a period of explosive activity during which gigantic central vent volcanoes were created, similar in form to those of Vesuvius or Stromboli. As magmatic material was created at depth, so the volcanic gases built up enormous pressures, being ultimately released at areas of weakness in the crust where volcanic pipes, necks or vents can now be seen in the landscape. Such vents were drilled through the crust by the same explosions which threw out ash, dust and lava fragments to build the volcanic cones and craters. It is pointless searching for such cones or craters in the modern Scottish landscape, however, for their unconsolidated materials have long since been weathered away. Only the plugs of lava which cooled slowly in the vents, often forming a hard dolerite or felsite, remain to form steep-sided hills in an otherwise rolling plateau landscape. We have already seen several examples of such landforms in the Border Country (see Chapter 2), although those igneous rocks were of Palaeozoic age. Where cooling was rapid the rocks exhibit a small or finely grained crystalline structure, as noted in the microgranite (or riebeckite) of Ailsa Craig (see p. 67). If, however, the cooling was slow, as with the rocks which solidified at great depth (hence the term 'plutonic rocks'), then the crystals are generally large or coarse, as in the porphyritic granites. Even if it is impossible to see a Vesuvian type of volcanic cone in the Arran landscape, study of the geological structures in the hills around Glen Craigag has allowed the intricate details of the Central Ring Complex to be reconstructed.

The sequence of events which led to the formation of this remarkable ring phenomenon is extremely complicated, but the reader is referred to Figure 11, where an attempt is made to illustrate the formation of a caldera and its associated volcanic vents. Shortly after the emplacement of the main Arran Granite, a local doming took place in the sedimentary rocks in the centre of the island. This uplift resulted in the creation of a circular fracture, almost 3 miles (4·8 kilometres) in diameter, which sliced right through the sedimentary rocks and the plateau basalts which

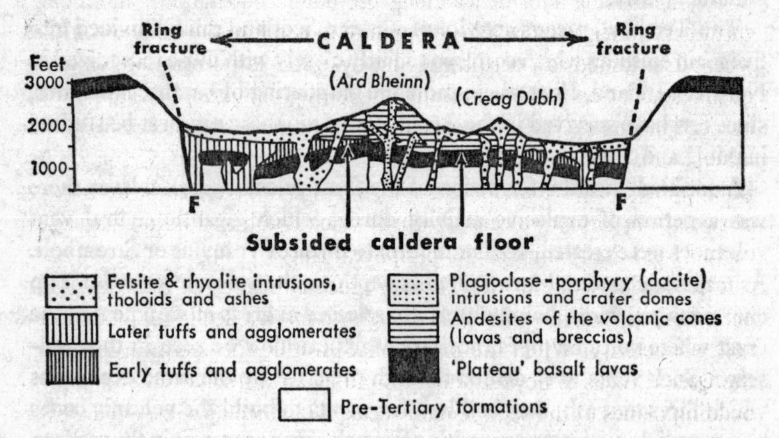

Fig. 11. *Diagrammatic section of the Central Complex of Arran before erosion (after F. H. Stewart).*

must then have covered them. Cauldron subsidence (of some 3,000 feet, over 900 metres) then followed, producing an enormous surface depression or caldera, the amphitheatre of which was to act as a receptacle for further lava extrusion from the volcanic cones which later erupted on its floor. Many tourists on safari in the Serengeti Plains in East Africa are familiar with the Ngorongoro 'Crater' (more accurately 'caldera'), which exemplifies the type of scenery that must have existed in central Arran in early Tertiary times.

One interesting point which emerges from a study of the geology of the Central Ring Complex is the presence of large masses of Rhaetic, Lias and

Cretaceous sedimentary rocks within the caldera complex, suggesting that these must have helped to build the cover of country rocks before the igneous activity occurred. Since these are no longer found on Arran, they must subsequently have been destroyed elsewhere on the island. The early basaltic lavas have also virtually disappeared, buried by great thicknesses of later volcanics which have covered the collapsed floor of the caldera. Four volcanic cones appear to have developed on the caldera floor, and these emitted a variety of both basic and acid lavas throughout their short-lived history, together with pyroclastic rocks (see Glossary) of similar composition (Figure 11). Both during and after this explosive phase, continuing subsidence along the outer ring-fracture and along a subsidiary inner fracture allowed granite to rise into the caldera complex, where it solidified as a peripheral ring-dyke and as a central body. Subsequent denudation has lowered the entire ring-complex, but some of the harder igneous rocks have resisted the lowering more successfully than others, creating the upland peaks surrounding Glenn Craigag: Ard Bheinn and Creag Dubh, for example, were formed from rocks in the vents of two of the former volcanoes, whilst Cnoc Dubh and Beinn Bhreac are carved from the granite of the peripheral ring.

The third episode in the Tertiary igneous history of Scotland is the period of sill formation which is thought to mark a declining phase in the volcanic activity. When the magma had insufficient energy to penetrate the crustal rocks it flowed underground as an intrusive sheet along the bedding planes of the sedimentary rocks, but, since its subsequent resistance to denudation has proved greater than that of these rocks, it is often left protruding as a prominent scarp face in the local scenery (Plate 7). Because the magma did not initially reach the surface, however, it sometimes cooled rather more slowly, so that despite having the chemical composition of the surface lavas its crystalline character is often coarser.

In southern Arran the Tertiary sills dominate the topography. They are interbedded with the less resistant New Red Sandstone, forming numerous scarps, terraces and 'sills' of waterfalls (hence their name). Three varieties have been recognized in south Arran. The coarsest and most massive are those formed from quartz-porphyry, well illustrated in the south-west by the conspicuous cliffs at Drumadoon Point (Plate 7), Brown Head and Bennan Head. The second group of sills is that of the finer-grained quartz-dolerite intrusions, which help to form some of the scarps in south-east Arran, such as Cnoc na Garbad (250 feet, or 76 metres, thick), Auchenhew Hill and many of the waterfall steps of Glenashdale. The final group, composed of ogabbro-like crinanite occur around Lamlash in the form of a broken circle. Since these sheets of basic igneous rocks cut across the bedding of the Permian sandstones and form a circle centred on Lamlash Bay, they have been interpreted by Professor S. I. Tomkieff as a cone-sheet complex (Figure 9). Where the northern crinanite sill cuts the coast at Claughlands Point and Hamilton Rock, its scarp forms the northern face of Claughlands Hill and provides a splendid defensive site for the ancient hill-fort of Dun Fionn. From Figure 9 it will be noted that Lamlash Bay has in part been created from the eroded centre of the sedimentary rocks within the cone-sheet complex, whilst the massive rampart of Holy Island (1,030 feet, 314 metres) is thought to have been formed from an invading ring-dyke of riebeckite–trachyte, similar in composition to that of Ailsa Craig.

The two other igneous episodes of Scottish Tertiary history are represented by the intrusion of the narrow, vertical, linear sheets known as dykes, and of the deep-seated plutonic masses of granite and gabbro. Although dykes can be seen clearly along the shore platforms of southern Arran, they can best be described in detail in Mull and Skye, where they produce more spectacular landforms. The mechanisms of plutonic activity, and in particular that of granite emplacement, have already been touched upon in previous chapters (see pp. 53–4), and so need not detain us long at this juncture. It will be sufficient at present to examine the character of the Arran Granite and of the landforms it has helped to create.

The Arran Granite is the second largest Tertiary granite outcrop in the British Isles, inferior in size only to that of the Mourne Mountains of Ireland, whose age of 58 million years it nevertheless equals. It is in fact composed of two types of granite: coarse granite makes up the outer part of the circular intrusion and helps to form the sharper and higher peaks of the eastern and western margins (Figure 9); fine-grained granite has invaded the central area of the coarser granite but has proved less resistant to denudation, thereby coinciding with the lower, rounded mountains around Loch Tanna and parts of Glen Iorsa. A detailed examination would demonstrate that the granite is grey in colour because of its mixture of white feldspar, black mica and glassy quartz. If, however we were lucky enough to discover one of the hollow 'druses', which were created from gas-filled cavities during the cooling phase, there would be a good chance of finding larger rock crystals of smoky quartz, although the minerals of beryl and topaz which are also said to be present are not of gemstone quality.

The granites are criss-crossed with a maze of dykes and veins of various sizes, which are composed of greenstone, aplite, porphyry or basalt. These invading dykes and veins either stand out as ribs or show as notches or channels in the rocky landscape, depending on their relative hardness in relation to the surrounding granite. On the slabby south face of Cir Mhór (2,618 feet, 798 metres), for example, prominent vertical grooves split the face and notch the skyline of A'Chir ridge (Plate 8). Here the jointing in the granite has also helped to create the serrated appearance of the ridges which run steeply up to the summits from Glen Rosa and Glen Sannox. But the generally rugged appearance of these northern hills is due largely to the action of ice, and it has long been realized that the glacial troughs of the glens, their 'hanging' side valleys and the arêtes and horns of the summits were all created during the Pleistocene Ice Age (Plate 8).

Sir Archibald Geikie calculated that more than 1,000 feet (over 300 metres) of rocks have been removed from the summits of these northern hills since the Tertiary dykes were intruded into the granite. Such denudation would have been sufficient not only to 'unroof' the granite dome by eroding the covering rocks, but would also have helped to fashion the so-called '1,000-foot' surface which surrounds the highest peaks. Here, a prominent plateau surface forms a plinth some 1,000–1,200 feet (300–365 metres) above sea-level, above which Goat Fell and its satellites rise. Early geologists described this as an uplifted marine-cut platform of Pliocene age, but Dr Sissons believes that such a view '... introduces into Scottish Tertiary history a great submergence for which there is no geological evidence'. An alternative explanation would be to regard the surface (which clearly truncates the geological structures) as having been formed by subaerial agencies (sometimes termed a 'peneplain' – see Glossary) and subsequently uplifted. These ideas will be discussed at greater length in Chapter 9.

Whatever the process which helped to fashion the irregularities of Arran's pre-glacial landscape, it is manifestly true that these scenic irregularities were considerably emphasized by the Pleistocene ice-sheets.

Since it is near to the high mountain areas of the south-west Grampians, it is no surprise to discover that Arran, together with the other islands in the Firth of Clyde, was overwhelmed by ice-sheets from the Scottish Highlands on several occasions. But, because of its high mountains, it also served as a local ice-centre in its own right, this being particularly significant during the last 3,000 years of the Pleistocene, when its own ice-limits can be correlated not only with those of the Scottish mainland but also with the remarkable series of late-glacial shorelines which occurs around the Firth of Clyde.

A comprehensive study of the island's late-glacial history by Dr A. M. D. Gemmell has shown that the highest of the raised shorelines (the so-called '100-foot' raised beach) occurs at 88 feet (27 metres) in south Arran and at 108 feet (33 metres) in north Arran, the tilt being a result of differential post-glacial isostatic uplift. Even more significant, however, is the fact that this highest strandline is missing from Arran's northern coast between Corrie and Catacol Bay, suggesting that Highland ice-sheets were still present in northernmost Arran at that time (c. 13,000 BP), inhibiting beach formation in this tract. An intermediate raised shoreline (the so-called '50-foot' beach) is also up-tilted from south to north (55–75 feet, 17–23 metres), but since it occurs sporadically all round the island it suggests that by about 12,500 BP the mainland ice had disappeared and the sea had access to the entire Arran coast, except for the mouths of some northern valleys which still contained local glaciers. The retreat of the valley glaciers from the coast may be equivalent in age to the recessional phase sometimes referred to as the 'Perth Readvance' elsewhere in Scotland. Dr Gemmell has shown that by 10,800 BP, after a short ice-free phase, valley glaciers had reappeared in northern Arran, leaving fresh morainic evidence in all the major glens, but being especially well-marked in Glen Iorsa and Glen Rosa. By 10,300 BP, however, this local glacial equivalent of the so-called 'Loch Lomond Readvance' had disappeared, leaving Arran free of ice and with a rising sea-level ready to fashion the post-glacial raised beaches (20–40 feet, 6–12 metres) all round the island. These raised beaches were not only to provide a foothold for the earliest Mesolithic settlers but were also later to be used for the ancient practice of 'run-rig' subsistence farming, characteristic of the western isles of Scotland. In such steep and harsh environments the raised beaches were often the only tracts of flat and fertile land available, and it is still fascinating to discover at Balliekine, in north-west Arran, survivals of these unconsolidated field strips and their sprawling clachan.

Bute and the Cumbraes

It is patently obvious to any visitor that the Isle of Bute contrasts scenically with Arran in nearly all respects, not least in its lack of high mountains. Largely as a result of having no hills over 1,000 feet (300 metres), Bute is distinguished by extensive areas of grassland and woodland, except on its northern moorlands. Yet, like Arran, Bute stands astride the Highland Line – that distinct division between Highland and Lowland Scotland which is marked by the Highland Boundary Fault. Since Bute lacks the gigantic Tertiary granite of Arran, the fault is more clearly marked here and can be traced south-westwards across the island from Rothesay, past Loch Fad, to Scalpsie Bay. The fault helps to divide Bute into two main geological tracts, each with its own distinctive scenery.

In the northern two thirds of the island the Dalradian schists, grits and phyllites (see Glossary) have weathered into poor, acid soils capable of supporting only moorlands. Nevertheless their steep, rocky northern coasts are fringed by rich woodlands which play no small part in the picturesque vistas of the famous Kyles of Bute, the 'lovely arc of sea-canal' that separates Bute from the mainland shore of Cowal. Between Kames Bay and Ettrick Bay the barren northern moorlands are broken by a belt of improved land which coincides with an isthmus now floored with raised beaches. The indentation which has produced these beautiful sandy bays appears to be related to the differential erosion of the less resistant Dunoon Phyllites, which here crop out as a wide band within the Dalradian succession.

South-east of Loch Fad the Highland rocks are replaced by the Carboniferous sandstone and Old Red Sandstone of Lowland Scotland, and the resultant change in the landscape is most marked. The moorlands give way to a tract of cultivated fields and thick woodlands, especially on the sheltered eastern coast where most of the settlement is located. Rothesay's Victorian villa-fringed promenade has long been the 'Margate' of Clydeside, being the steamer terminus for the traditional Glaswegian relaxation of cruising 'doon the watter'. South of Kilchattan Bay, however, the woodlands and fields turn to pleasant rough moorlands once more, for here the Old Red Sandstone has been buried beneath basaltic lavas of Carboniferous age, which in turn have been cut by Tertiary sills, similar to those of south Arran. Where the sandstone emerges from beneath the igneous rocks it is picked out by a scrub of elder, birch and rowan, but on the lavas and sills the ubiquitous bracken holds sway.

The scenery of this southernmost peninsula of Bute is very similar to that of Little Cumbrae Island, which lies two miles to the east. It is not surprising to find, therefore, that the mossy turf and bracken of Little Cumbrae mantle a rocky tract of Carboniferous lavas similar not only to those of south Bute but also to the extensive lava tracts of the nearby Ayrshire mainland.

While Little Cumbrae is deserted because of its thin soils and poor grazing, the same is not true of its neighbour, Great Cumbrae. This well-farmed island, composed almost entirely of Old Red Sandstone and white Carboniferous cornstones, mirrors both the improved lands of south-central Bute and the Largs–Ardrossan coastlands of Ayrshire, whose geological structures it repeats. The almost featureless plateau of Great Cumbrae, where woodlands often mark the poorer soils, is diversified only by dykes and sills of igneous rocks which form prominent ridges smothered in gorse, bracken and scrubby trees. As in Arran and Bute, the cultivated land is found mainly on the flatter and better-drained areas of raised beach. But the lowest beaches are near to the water-table (see Glossary), and their marshiness is indicated by the brilliant yellow flag irises dotted amidst the dark green rushes. Surprisingly, the marshy vegetation on the lowest raised beach of Little Cumbrae is not as well developed as this, possibly because it has been hindered in its ecological progression towards acid bog mosses by the presence of nutritive minerals flushed out from the weathered basic lavas which abound there.

5. The Midland Valley – West

After the exciting and romantic scenery of the islands in the Firth of Clyde, the mainland of Ayrshire, Renfrewshire and Lanarkshire may come as something of an anticlimax. Whilst their shores may serve as magnificent viewpoints for both the islands and the nearby Highlands, they have less intrinsic beauty of their own. But that is not to say that these landscapes are uniform and uninteresting: amid the urban bustle of Clydeside and the coalfield towns, their scenic tapestry is stitched with patches of rich farmland and extensive strips of wild moorland, and almost hemmed in by the encircling waters of the Clyde. Above all, this is the land of Burns, whose works were largely inspired by Ayrshire life and have in turn brought immortality to the towns and villages of these parts.

The 60-mile (96-kilometre) west-facing coastline between Girvan and Gourock not only traverses the length of this varied landscape but also spans the entire width of the Midland Valley of Scotland, between Galloway and the Highlands. Thus, having in earlier chapters skirted the southern and western fringes of this Scottish heartland, we must now turn briefly to examine the general geological structure and topography of the Midland Valley, before setting out to explore its fascinating scenery.

Many writers have pointed out that the term 'valley' is a misnomer when used to describe the structural unit which is contained between the Highland Boundary Fault and the Southern Uplands Fault. Apart from the fact that it includes peaks higher than many of the Pennines (for example, Ben Cleugh – 2,363 feet, 720 metres), it has a surprisingly small amount of land below 400 feet (120 metres), which is the usual upper limit of improved land hereabouts. A glance at a relief map will show how the isolated hill masses, such as the Campsies and the Ochils, break up the lowlands into isolated basins and serve to create sub-regions within the apparent unity of the Midland Valley itself. Thus we shall find that three distinct regions can be identified: the present chapter deals with the western region, where the Carboniferous basins of Ayrshire and Lanarkshire are linked by the Clyde and integrated by the overwhelming presence of the Glasgow conurbation; the eastern region, with the Midlothian and Fifeshire coalfields, is cradled around the Firth of Forth and the citadel of Edinburgh (see Chapter 6); in the north the extensive tracts of Old Red Sandstone are furrowed by the great corridors of Strathmore and the Tay, and appear to turn their backs on the rest of the Lowlands (see Chapter 7). But, while these three regions may exhibit distinctive landscapes, they are all part of the structural unit known as the Midland Valley, wherein the oldest rocks to be seen at the surface (apart from a few tiny outcrops of Silurian – see pp. 90, 107) are the basal conglomerates of Old Red Sandstone age. The reason why the Lower Palaeozoic and older rocks are virtually invisible in the Midland Valley is that they have been let down *en masse* between two major parallel fault systems to form what is termed a 'rift valley'. Consequently, except for a few minor exposures, the older rocks are now deeply buried below layers of Upper Palaeozoic sandstones, limestones, shales and coals, which now form the floor of the so-called Valley. On either flank, beyond the faults, the older rocks remain as massive upstanding blocks, from which the Southern Uplands and the Highlands have subsequently been carved (Figure 12).

The great parallel fracture systems of the Southern Uplands Fault and the Highland Boundary Fault were initiated during the instability associated with the period of Caledonian mountain-building in early Palaeozoic times. A prolonged episode of tectonic uplift was terminated when the centre of a gigantic arch of updomed crustal rocks began to crack along lines of weakness which followed the north-east to south-west Caledonian grain. As a result, a vast strip of land, some 50 miles (80 kilometres) in width, was gradually lowered to create basins in which Old Red Sandstone, Carboniferous and Permian rocks were later to be deposited. The continuing tectonic instability also manifested itself in the form of widespread vulcanicity throughout these depositional episodes. We shall see that many of the major landmarks in the Midland Valley are related to the igneous phenomena of Upper Palaeozoic age, since, together with the Old Red Sandstone grits and conglomerates, these are its most resistant rocks.

Post-Carboniferous earth-movements (of Hercynian age) have folded and faulted the thick layers of Upper Palaeozoics and have succeeded in separating the coal-bearing beds into distinct basins which have now become isolated from each other as a result of the denudation of the Carboniferous rocks from the intervening anticlines. It would be as well to remember, however, that the Scottish Carboniferous (like that of Ireland) is not as simple as the succession found in England and Wales. Of the two major coal-bearing strata in Scotland, the Lower Coals occur in the so-called Carboniferous Limestone Series, in which sandstones are

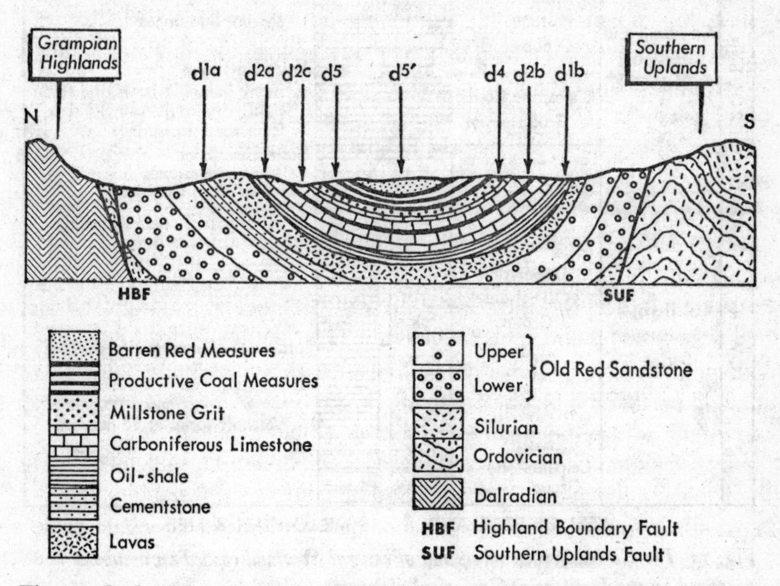

Fig. 12. *Geological section of the Midland Valley (based on material prepared by the Institute of Geological Sciences). The notation of numbered letters refers to the Carboniferous succession shown in Figure 13 and to the maps of the Geological Survey of Great Britain.*

common and limestones are few. The Upper Coals are found, appropriately enough, in the Coal Measures, but even here there is a very small proportion of coal and many shales and clays. To complete the picture of apparent confusion in the Scottish Carboniferous succession, the so-called Millstone Grit, which divides the Upper and Lower Coals, has very few gritstones but many valuable fireclays (Figure 13).

In the western part of the Midland Valley, coalfields have survived in

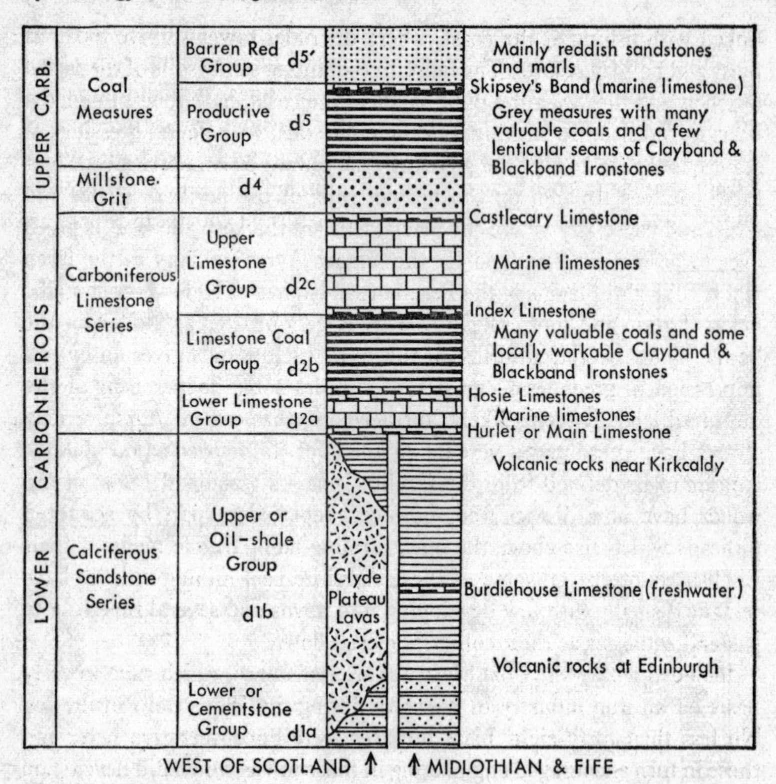

Fig. 13. *The Carboniferous succession of central Scotland* (*based on material prepared by the Institute of Geological Sciences*).

two major downfolds, namely the Lanarkshire Basin and the Ayrshire Basin. These are separated by a ridge of low hills in Renfrewshire, where an upfold of older rocks capped by Lower Carboniferous lavas has introduced a wide tract of featureless moorland between the colliery-dotted plains.

Ayrshire

Because of its eastern rim of lava plateaux, the Carboniferous basin of central Ayrshire appears to turn away from the Midland Valley to face westwards to the sea. In fact, its earliest settlements were almost certainly linked with fishing at the coast, where the older towns survive today as ports and holiday resorts. The light, free-draining sandy soils of the raised beaches, together with the dune sands of the coast, have long provided a basis for subsistence farming, but the same soils, with liberal dressings of seaweed manure, today offer a greater cash reward by supporting excellent early potatoes. Inland, on the heavier soils of the northern shales and lavas, and especially where the glacial tills are thickest, the land is given over to permanent grassland for the famous Ayrshire dairy cattle. Even where the lighter loams of the Permian sandstones of Kyle allow sporadic fields of grain and root crops, these are largely utilized as fodder for the herds. Thus the amphitheatre of the Ayrshire lowlands gives an overall impression of greenness – grassland occupies some 70 per cent of the improved land. It seems likely, furthermore, that central Ayrshire may always have presented a verdant picture, for its district name of Kyle appears to be derived from the Gaelic *Coille* – a woodland. Most of the woods have now disappeared, however, replaced in part by scattered pitheads which rise above the open farming land. Unlike many of their English counterparts, some of these Ayrshire coal-mining villages have remained small, with few developing into towns and several reverting to pastoral activities as their collieries closed down.

Included within the Coal Measures are ironstones, which gave an early basis for an iron industry in Ayrshire during the Industrial Revolution. No less than forty-eight blast furnaces were once operative here, and these in turn encouraged engineering in those towns which did develop on the coalfield. Today the large Glengarnock steelworks dominates the industrial landscapes of the Cunninghame division of northern Ayrshire, where the collieries are now largely defunct. Almost all the current coal output comes from the southern half of the coalfield in the so-called Mauchline Basin, where the productive Coal Measures are deeply buried beneath thick layers of Carboniferous Barren Red Measures, Permian sandstones and their associated basaltic lavas. It is easy to understand how the easy accessibility of both the Lower and Upper Coals, which emerged at the surface to the north of Kilmarnock, led to their early exploitation and exhaustion. Conversely, the deeper coals, in the centre of the basin to the east of Ayr, were left until modern mining methods were able to extract them from depths of more than 2,000 feet (600 metres). (See Figure 14.)

We cannot leave the Mauchline Basin without some mention of the well-known Mauchline Sandstone of Permian age, which caps the low plateau country of Kyle above the important buried coalfield (Figure 14). Because of its excellent qualities as a building stone, this bright red sandstone has been quarried for many years and widely exported through

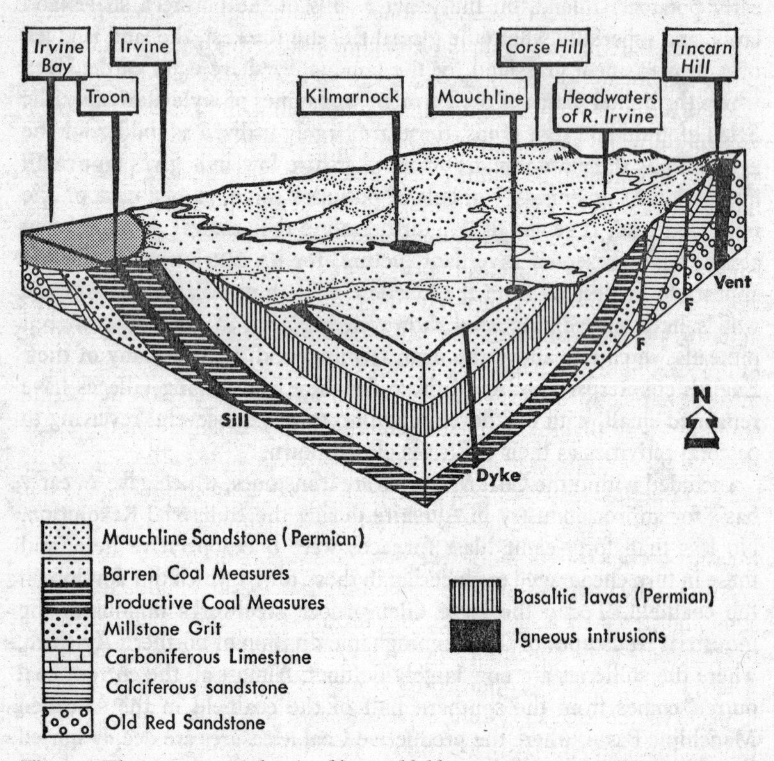

Fig. 14. *The structure of the Ayrshire coalfield.*

the Ayrshire ports. Equivalent in age to the Permian sandstones of Dumfries (see p. 50) and Penrith, this sandstone, 1,500 feet, or more than 450 metres, thick, is best displayed at the Ballochmyle Quarries at Mauchline, where its large-scale dune-bedding and wind-rounded quartz grains testify to its formation in a Permian desert environment.

Trial borings at Mauchline to ascertain the depth of the coal-bearing rocks have revealed that another interesting lithological horizon occurs within the Ayrshire Carboniferous basin. Lying below the Coal Measures there is evidence of a bauxite deposit which may underlie the entire coalfield. Its outcrop, however, is best seen farther west where a 4-foot (1·2-metre) bed of bauxitic clay can be traced from the coast at Saltcoats for 10 miles (16 kilometres), through Kilwinning to Kilmaurs, near Kilmarnock. It occurs within the Millstone Grit Series, in association with contemporaneous sandstones and lavas and a poor lateritic iron ore. Despite its 47-per-cent alumina content, however, only a small proportion was found to be suitable for conversion to aluminium and, apart from sporadic production of alum, it has been used mainly as a fireclay for lining electric steel furnaces. The geologist is able to reconstruct earlier climates from a study of the sedimentary rocks, and here the presence of bauxite and laterite indicates the different climate which must have prevailed in Carboniferous times. We find that a tropical land surface existed here after the cessation of the Middle Carboniferous vulcanicity and, following lengthy periods of deep weathering, laterites and bauxitic clays must have formed because of a concentration of residual iron hydroxide and aluminium hydroxide in the surface soils. The subsequent inundations by the brackish lagoon waters of the Coal Measure swamps involved the deposition of other Carboniferous clays (later to be used as fireclays), and it was in these and in the underlying bauxitic clay that the Coal Measure forests established their roots.

Around the Carboniferous basin a discontinuous rim of Old Red Sandstone, bolstered by lavas and igneous intrusions of similar age, creates a perimeter of moorland hills wherever it occurs. This is especially true where this semicircle of harder rocks meets the Firth of Clyde, so to understand the diversity of the Ayrshire scene it would be instructive to traverse this western coastline from south to north.

Between Girvan and Ayr the Lower Old Red Sandstone hills thrust their steep, thickly wooded slopes seawards, constricting the narrow coastal plain near Turnberry. The famous golf course here is on the dune sands which have accumulated on the lowest raised beach. Before long the coast road climbs back into the sandstone hills and past the imposing Robert Adam masterpiece of Culzean Castle. To the north, as the basaltic and andesitic lavas reach the coast, the cliff scenery becomes more rugged and the hinterland more wild, rising inland to the moorlands of Brown Carrick

Hill (943 feet, 287 metres). The stark ruin of Dunure Castle, on its sea-cliff, emphasizes the wildness of the scene, as do the 200-foot (60-metre) cliffs at the Heads of Ayr, where a volcanic agglomerate marks the site of a former vent which has drilled through the surrounding Calciferous Sandstones. These lower sandstones (or cornstones) are the basal members of the Carboniferous succession: from here northwards our journey will take us back into the coastal fringes of the Ayrshire coalfield.

The relatively simple succession of the coalfield basin, where the rocks become progressively younger towards the centre of the syncline, is complicated both by faulting and by igneous intrusions. As if to demonstrate this, the coastal headland on which Greenan Castle stands, at the southern end of the Bay of Ayr, is formed from an outcrop of tough volcanic ash. Despite this material's resistance, however, the degree of coast erosion here means that the castle is now perched precariously on the cliff edge, whereas some 200 years ago a carriage could be driven safely around it.

Less than 2 miles (3 kilometres) away, where the River Doon crosses from the sandstone hills onto the Coal Measures, stands the tiny village of Alloway, famous as the birthplace of Robert Burns. It is, perhaps, appropriate that this 'high altar of Burnsism' spans such an important geological boundary between the rural scenery of the Old Red Sandstone and the industrial scenery of the coalfield, for Burns's writings marked an important transition in Scottish social history. M. Lindsay said: 'Burns caught and fixed that old, agrarian Scotland which had persisted almost since the Middle Ages, just as it was beginning to disintegrate before the forces of industrialism.' Yet, paradoxically, since this is Burns's country, Ayrshire remains the most English-looking of the Scottish counties. Perhaps it is the regularity and orderliness of its field patterns, with their neatly trimmed hedges, that reminds us of the English Midlands. Or is it the unspoiled villages and the attractive farm cottages with their thatched roofs, a feature once common but now so rare in Scotland? Despite these similarities, however, the absence of trees in the Ayrshire hedgerows confirms that we are traversing the windswept seaboard of a Scottish county rather than the leafy richness of the English Plain.

From the attractive town of Ayr, northwards as far as Saltcoats, the coast becomes generally low-lying as we approach the centre of the basin. Here also we can find a remarkable collection of raised beaches and dunes fronted by lengthy sweeps of sandy foreshore. There are rocky dolerite sills which form headlands at Prestwick, Troon (Celtic: *Trwyn* – a promontory) and Saltcoats, but in general the bays exhibit smooth curves which are orientated at right angles to the approach-routes of the dominant marine waves. Here the coastal dunes have been carried some distance inland by the prevailing winds but are now largely fixed by marram grass and the famous golf courses of Troon and Prestwick. These, like those at Turnberry and St Andrews (see p. 128), owe their velvet sward to the light, coarse sands, which are poor in organic matter and mineral plant-nutrients but will support fine-bladed grasses on their acid soils. In raised-beach times the sea must have transgressed some distance inland to create lengthy tidal reaches on the rivers Irvine and Garnock and produce an irregular coastline, different from that of today.

The marine clays of these coastal terraces have been sporadically exploited for brick-making, while the flatness of the raised beaches has, despite their high-grade soils, attracted the recent development of Prestwick international airport and the planned 'overspill' town of Irvine. As if to emphasize further the changes in the Ayrshire scene, the nuclear power station of Hunterston intrudes its implacable presence into the coastline between Ardrossan and Largs. Farther north the coastal plain is almost squeezed out by the steep, lava-capped sandstones of the Renfrew Heights, and by the time we reach the lighthouse of Cloch Point, where the Clyde estuary turns abruptly eastwards, we are aware of a different scene.

Clydeside

On turning the 'corner' at Cloch Point there is a tendency to ignore the tableland of the Renfrew hills and let the gaze linger seawards on the Cowal shore, where the more rugged Highland backdrop of majestic hills frames the far-reaching tidal waters of Loch Long and the smaller inlets of Holy Loch and Gareloch. It is a picture of skimming cloud-shadows, mist-trailing moorlands and sun-dappled waters, often dotted with the multi-coloured spinnakers of racing yachts. Here '. . . the lochs confer on the Clyde estuary features of interest, beauty and romance possessed by no other great maritime river in Britain' (N. Munro). But once we have travelled eastwards to Greenock a vastly different landscape meets the eye: here begins the man-made urban sprawl of Clydeside and its attendant industrialization – a landscape which some would judge to be more typical of the Midland Valley than of Ayrshire. The reason may be that the Carboniferous basin of central Scotland (Lanarkshire–Stirlingshire) possessed certain advantages of industrial location denied to Ayrshire, not least the presence of the rivers Clyde and Forth. But, before turning to examine the geological and topographical factors which led to the growth of Glasgow and its industrial region, it is important to look at the encircling rim of uplands which overlooks Glasgow from the south, west and north.

Stretching north-westwards from the Southern Uplands through the Eaglesham Heights (1,232 feet, 376 metres) and the Hill of Stake (1,713 feet, 522 metres), a line of drab, moorland plateaux acts as a divide between the Ayrshire and Lanarkshire basins. In the far south near Lesmahagow the hills are made of Silurian greywackes and shales which have been exposed at the crests of two small anticlines. This is one of the few examples of Lower Palaeozoic rocks within the Midland Valley rift. In the main, however, these tablelands have been carved from an extensive occurrence of Lower Carboniferous lavas – the so-called Clyde lavas – which are more than 2,000 feet (over 600 metres) in thickness. It has been calculated that they formerly covered an area of at least 600 square miles (1,500 square kilometres), so their former limits have either been buried by subsequent sedimentary deposition or destroyed by denudation. The widespread extent of these mainly basaltic lava flows suggests that they were not produced from a single vent but from fissure eruptions, while the dearth of explosive volcanics, such as ashes and tuffs, further implies that the lava emission was largely non-violent.

The uplands of this Carboniferous lava plateau extend north-eastwards across the Clyde into Dunbartonshire and Stirlingshire, where the prominent Kilpatrick Hills (1,316 feet, 401 metres) and the Campsie Fells (1,897 feet, 578 metres) overlook the Clydeside smoke-haze from the north. The numerous flows of lava, which were extruded one upon the other, can be seen in profile along the southern flanks of the Campsies. Here, erosion along the line of the Campsie Fault has left the tiers of no less than 30 lava flows to form the cliffs and steep heathery slopes of the Kilsyth and Strathblane Hills. These provide us with a good example of a fault-line scarp, with the down-throw of the Carboniferous sediments to the south of the fault helping to explain the wooded lowlands of Strathblane. It must be remembered that a scarp of this type – that is, with the down-throw block forming the lower ground – will not always result from a fault movement of similar character. A fault-line scarp will only form when differential erosion is able to pick out differences between hard and soft rocks which have been brought together by faulting. In this instance the tough Campsie lavas on the up-throw side are in juxtaposition with the less resistant Carboniferous sandstones and limestones of the down-throw side (Figure 15a). As if to illustrate the complexity of the structure, exactly the opposite topographic phenomenon has resulted several miles to the west along the same Campsie Fault (Figure 15b). Here the lavas occur to the south of the fault, on the down-throw side, but owing to their toughness have formed the north-facing fault-line scarp of the Kilpatrick Hills.

In addition to their steep basalt escarpment, the Campsie Fells are noted for a series of prominent conical hills, especially along their northern face near Fintry (Figure 15). Such landmarks as Dumgoyne (1,402 feet, 427 metres), Garloch Hill (1,781 feet, 543 metres) and the cliffs north of Earl's Seat (1,897 feet, 578 metres) have all been carved from volcanic vents, of similar age and character to that at the Heads of Ayr (see p. 88). Within Strathblane itself, the wooded pinnacle of Dumgoyach and the smaller knob of Dunglass are equally impressive examples of volcanic necks, but we have to travel back to Clydeside to find the most famous of all. Here at Dumbarton the pinnacled rock, looking incongruous amongst the mudflats, rather like a stranded Ailsa Craig, has long provided a defensive site on the narrow waters of the Clyde (Plate 9). Claimed by some as the birthplace of St Patrick (note the adjoining Kilpatrick Hills), Dumbarton has an antiquity greater than that of Glasgow itself. Standing at the mouth of the Vale of Leven, on the Leven river so admired by Thomas Pennant and Tobias Smollett long before its power had been utilized to run the mills, Dumbarton could claim a strategic significance on Clydeside, for here the great river estuary is narrowed as it cuts through the hard basaltic rim of the Renfrew Heights and the Kilpatrick Hills.

The relationship between geology, scenery and subsequent urban growth is clearly demonstrated in this section of the Clyde valley, for the proximity of the lava hills has both helped and hindered the development of the Clydeside towns hereabouts. The rocky reefs of the Dumbuck ford long frustrated the attempts of ocean-going vessels to sail upstream to Glasgow, thus leading to the creation of such out-ports as Greenock and Port

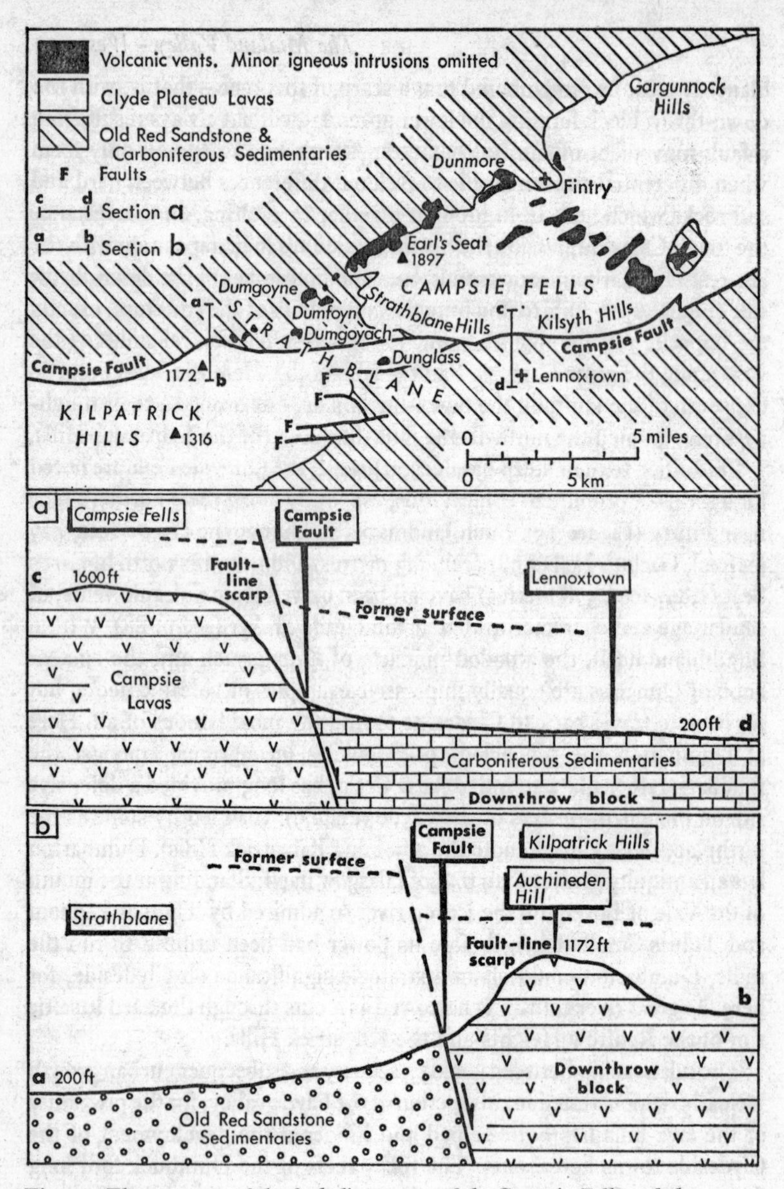

Fig. 15. *The structure of the fault-line scarps of the Campsie Fells and the Kilpatrick Hills.*

Glasgow. The same hard rocks which formed the reefs also caused a contraction of the channel downstream, resulting in deep tidal water along the southern shore. Nevertheless, the lava plateaux finally played an inhibiting part in the history of Greenock and Port Glasgow, for the narrow coastal terraces on which these towns were built are not conducive, because of their constricting hill-slopes, to the establishment of large cities or great dockyards. To find such development we have to turn upstream to Glasgow and Clydebank which, once freed of their offending reefs and shoals, were able to expand on the extensive flats of the so-called Howe of Glasgow.

To the east of the lava plateaux there is a clear relationship between the topographic amphitheatre or Howe (Howe – a hollow place) in which Glasgow has grown and the geological structure of the Lanarkshire part of the Carboniferous basin of central Scotland. Although the synclinal basin of the so-called Central Coalfield is diversified by some hard-rock ridges, generally related to the Upper Carboniferous grits and sandstones, there is a general coincidence between the down-folded limestones, shales and coals on the one hand and the low-lying basin of the Middle Clyde on the other. Owing to the thick layers of glacial drift and alluvial deposits, however, the detailed effects of solid geology on the topography are masked within the basin itself. Nowhere are such scenic relationships as clear as they are on the surrounding upland rim, where drift deposits are thin or absent. Nevertheless, by a careful study of the colliery waste tips (bings) and pit-heads it is possible to demonstrate the approximate geological boundaries: the Lower Coals in the Carboniferous Limestone Series (Figure 13) occur on the western flank of the coalfield, creating a line of mining towns from Johnstone, through Glasgow, Kirkintilloch and Kilsyth, to Denny and Stirling; the Upper Coals of the younger Productive Coal Measures crop out nearer the centre of the basin, from Hamilton and Motherwell to Armadale. Between the Upper and Lower Coals both the Upper Limestone Group (which includes several thin limestone beds and many thick layers of shale) and the Millstone Grit intervene in the lithological succession. Since neither of these contains coal seams of any significance, their line of outcrop in the Carboniferous basin has created a 'barren' zone in the coalfield, where pit-heads and spoil-heaps are absent from the landscape.

It took many millions of years of subsequent denudation to destroy any post-Carboniferous rocks which might have been deposited in the Midland Valley and to hollow out the basin of Strathclyde. The final touches must have been given by the Pleistocene ice-sheets, which ultimately lowered the less resistant rocks during several periods of selective glacial erosion. We can reconstruct the direction of former ice-movement not only from the drumlins of the Glasgow district but also from the distribution of the distinctive erratics from the Glen Fyne Granite, which is located near the northern end of Loch Lomond. Thus we have evidence that Highland ice-sheets advanced south-eastwards across the Howe of Glasgow before moving *up* the Clyde valley. In the process the ice left thick layers of till (or boulder clay) to obliterate the former drainage pattern and infill the pre-glacial rock basins.

In the early nineteenth century a Glasgow clergyman, Dr Waddell, in the belief that the sea had once reached much farther inland, perhaps during the Biblical Flood, pointed to the place-names between Glasgow and Bothwell as having maritime derivations. At about the same time some of the early Scottish geologists began to announce the discovery of marine shells in the glacial drifts of Strathclyde, thereby lending a measure of support to the then popular ideas of an early marine transgression. Such shelly 'tills' were also found in Ayrshire, where one local geologist, John Smith, continued to argue strongly for the marine origin of glacial 'drifts' long after the majority of geologists had abandoned the idea. Such rejection was based, among many other things, upon a close examination of the marine shells, which showed that they had lost their original markings and had often been scrubbed and even striated by the ice-sheets which had transported them from the former sea-bed of the Firth of Clyde to a resting place many miles inland. Nevertheless, the existence of marine shells in the moraines which are found at the southern end of Loch Lomond suggests that during an earlier phase of the Pleistocene an arm of the sea must have extended far into the Lomond valley, which must then have looked similar to the present-day sea inlets of near-by Loch Long and Loch Fye.

The glacial tills virtually obliterated the pre-glacial valley of the Clyde, which borings have revealed as a buried rock-cut strath now lost beneath some 300 feet (90 metres) of glacial infill. Consequently the post-glacial river, having been unable to rediscover its former valley (except in a few places), has now cut a new course in the easily eroded superficial deposits.

Where the Clyde and its tributaries have succeeded in downcutting to the old pre-glacial land surface they have carved out spectacular rock gorges. The most famous of these is near Lanark, where the Clyde follows a narrow gorge 100 feet (30 metres) deep, below its well-known series of falls (see p. 100).

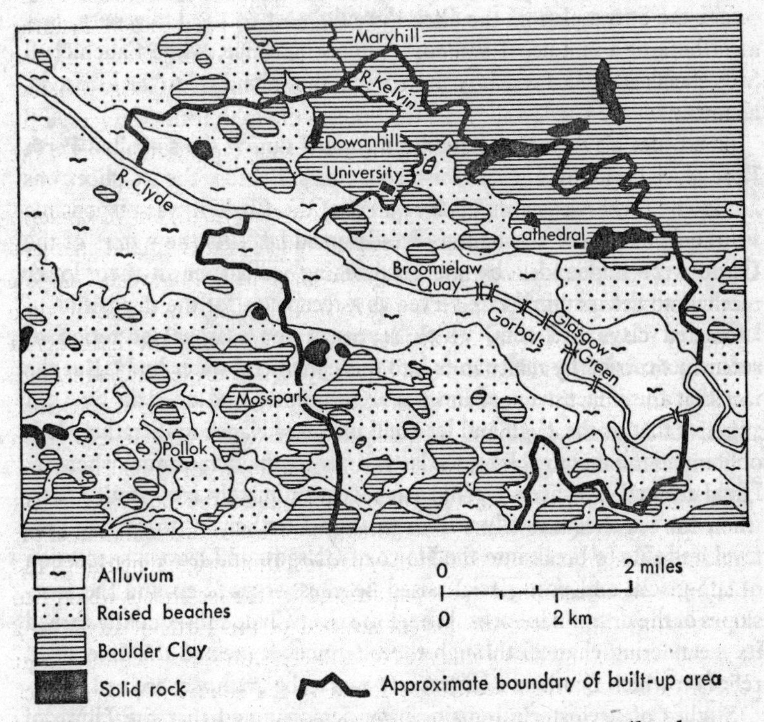

Fig. 16. *Distribution of the superficial deposits in the Glasgow area (based on material prepared by the Institute of Geological Sciences). Note the boulder-clay drumlins. See Plate 10.*

Glasgow itself is built on a cluster of drumlins, with many of the famous buildings, including the University, located on their crests (Figure 16). The majority of these streamlined hills are about 100 feet (30 metres) in height and half a mile (0·8 kilometre) in length, and all exhibit blunter, steeper ends towards the west-north-west (that is, the direction from which the ice-sheet came) and more gentle tapering tails in the opposite direction. Not only have the drumlins affected the local drainage, as in the case of the meandering Kelvin river (Figure 16), but they have also influenced the layout of the city streets. The curving streets of Maryhill and Dowanhill, to the north of the Clyde, and those of Pollok and Mosspark in the southern suburbs reflect the natural curves of the drumlin slopes. In the city centre the street-plan of the Park drumlin, east of the University, is a most striking example of this topographic influence, while Sauchiehall Street makes use of a low col between the drumlins to maintain its alignment.

The drumlins appear to have been formed during the so-called Perth Readvance period of Upper Pleistocene times. Because the ice-sheet was advancing south-eastwards up the valley of the Clyde it was responsible for the formation of a temporary ice-dammed lake. As the waters of the Clyde were impounded by the encroaching ice-barrier in their lower reaches, so the marginal lake served as a receptacle for the deposition of laminated clays and silts, which represent the seasonal accretion of sediments carried by meltwaters into this ephemeral water body. But the fact that these lacustrine sediments are crumpled and overlain by a till suggests that as the Highland ice continued to advance up Strathclyde it obliterated the marginal lake, the latter's final extinction coming when the Highland and Southern Upland ice-sheets ultimately coalesced. Later, when the ice-sheets finally withdrew from Strathclyde, the rising sea-level was able to break into the Howe of Glasgow and leave a succession of late-glacial and post-glacial raised beaches wrapped around the foot-slopes of the drumlin hillocks. Where the river Clyde subsequently carved its meandering channel through these terraces it created the diversified relief on which Scotland's largest city was to be gradually erected.

Studies of prehistoric remains have demonstrated that the Howe of Glasgow remained devoid of settlement for several millenia in early post-glacial times. Despite the presence of Mesolithic hunters and fishermen on the neighbouring shores of the Firth of Clyde, the thickly wooded clay soils of the inland basin were apparently unattractive to these early settlers. The same situation seems to have prevailed during the advent of the Neolithic agriculturalists in about 4000 BP, for they also settled along the western coasts, utilizing the lighter soils around the Firth of Clyde and leaving their enormous megalithic tombs of so-called Clyde–Carlingford type. Similarly the succeeding Bronze Age settlers, known as the Beaker Folk, settled in the western coastlands, penetrating no farther inland than Dumbarton Rock. Even the Iron Age settlements avoided the thickly wooded mid-Clyde valley and the Howe remained a virtual no-man's-land until Roman times, when the area was probably reclaimed and cleared during the building of the neighbouring Antonine Wall. This northern bulwark of the Roman Empire was only briefly held against the marauding Highlanders, the Midland Valley finally being given up in A.D. 196 when the legions retreated to Hadrian's Wall.

Eventually, as the post-Roman kingdom of Strathclyde evolved, the Howe became something of a cross-roads, where the Clydesdale routes to the Firth met those from Ayrshire and the Lothians. There appear to have been many sandy islets in the river hereabouts, and these provided both a bridging-point and a ford at the centre of the Howe, later to become the focal point of Glasgow. From here the Stirling road led northwards across the floodplain and the water meadows to the back of the raised-beach terrace, where the Cathedral was ultimately to be founded on a steeper and higher bluff of boulder clay. Crossing this main axis of the High Street, the Edinburgh–Dumbarton road was later to become the Gallowgate and Trongate of the medieval town, whose Tolbooth and Mercat Cross were located at the major cross-roads. On the south side of the river, down on the lowest of the Clyde terraces, the ford and bridgehead settlements of Hutcheson Town and Gorbals grew up.

The advantages of Glasgow's nodal location in the Strathclyde basin were, however, partly offset by the vagaries of the river itself. By the mid seventeenth century Glasgow's harbour was silted up and the Clyde held only 3½ feet (1 metre) of water at high tide. Thus we find that during the next century the out-ports of Greenock and Port Glasgow usurped much of Glasgow's trade because of their deeper-water facilities. But determined efforts by the city council to dredge and embank the channel finally achieved success in 1773 when the hard igneous rock barrier of the Dumbuck ford was destroyed. Henceforth the Clyde became navigable as far as Glasgow, where the combination of local coal and blackband iron ore aided the city in achieving rapid fame as a centre of shipbuilding and engineering during the Industrial Revolution. It is often remarked that Glasgow was built upon coal: this is both literally and metaphorically true, for the readily accessible outcrop of the seams within the city boundary gave it an early advantage. Now that these Upper Coal seams have been virtually exhausted, however, Glasgow has been left at something of a disadvantage. The main coal output of Scotland has switched to Ayrshire and Fifeshire, where the Lower Coals have greater reserves.

Clydesdale

As we leave Glasgow to follow the river Clyde upstream there is no immediate change in the cultural landscape, for our journey takes us through the southern tracts of the Central Coalfield. The scene remains one of nineteenth-century industrial wasteland – factories, collieries and steelworks – now overlaid by a twentieth-century veneer of new housing estates and motorway networks. Although the collieries of Hamilton, Motherwell and Airdrie have declined, the same is not true of the sprawling steel mills and their associated industries, which have been stimulated by government finances in an area of economic recession. Despite the fact that the iron industry was one of the earliest industrial enterprises in early-nineteenth-century Lanarkshire, the local ores proved too acid for use in the steel-making which subsequently developed here. The ironstones, composed of soluble ferrous carbonate and an insoluble clay matrix, can be found in many of the Carboniferous formations. Their earliest Scottish exploitation was in the shallow surface workings of the Central Coalfield, where the iron nodules could be easily extracted from the accompanying shales. Those deposits which had sufficient combustible carbonaceous material to render them self-calcining were known as the blackband ironstones. These, at first rejected by the coal miners as 'wild coal', were discovered near Airdrie in 1801, but it was not until about 1830 that the invention of the blast furnace allowed them to be fully utilized. Prior to this, iron ores were imported from Cumberland and Lancashire to be smelted on the shores of Loch Etive and Loch Fyne, where wood-charcoal could easily be provided from the Highland forests. Before the discovery of the blackband ores the clayband ironstones had been smelted by using local coking coal, first used in Scotland at the famous Carron furnaces in 1759 (see p. 118).

Today, iron-ore extraction in the Midland Valley has ceased, but the demand for steel in Clydeside engineering, shipbuilding and car manufacturing has allowed the gigantic Ravenscraig steel mills at Motherwell to thrive, using imported iron ore and coking coal and limestone flux mostly brought in from northern England. Thus, all the local geological factors which helped to locate the Lanarkshire iron industry are now virtually unimportant.

Throughout this Scottish 'Black Country' the Clyde cuts a swathe of open space, allowing the rural greenery to invade the urban and industrial sprawl, and, since Clydesdale has long been historically important, it is not surprising to find stately homes, ruined castles and pit-mounds cheek by jowl. For instance, the famous Bothwell Castle, '. . . a large and grand pile of red freestone, harmonizing perfectly with the rocks of the river, from which, no doubt, it has been hewn' (Dorothy Wordsworth), overlooks abandoned coal tips. William Lithgow spoke of Lanarkshire in 1640 as a land of '. . . orchards, castles, towns and woods planted side by side', and even by Cobbett's time Clydesdale was still a pleasant countryside. As we travel upstream towards Lanark we can begin to appreciate what impressed these writers, for here the Clyde has cut a narrow gorge in the thick layers of glacial drift, creating river bluffs and cliffs now clothed in thick woodlands. In the Lanarkshire basin one passes rapidly from industrial blight to unspoiled farmlands above Hamilton and Wishaw, and because of the sheltered nature of the Clyde valley in this area it has come to be known as the Orchard Country. Mentioned as the 'appleyards of Lanark' by Bede in the eighth century, these orchards can best be seen from Crossford Bridge or the bridge at Garrion, where they rise in tiers above the loitering and twisting Clyde. Although the valley is narrow and often wooded, the low river terraces are flanked by thick deposits of glacio-fluvial material which nourish the strawberry plants, raspberry canes and plum, pear and apple trees which cling even to the steepest banks in their tiny plots. Acres of greenhouses can also be seen climbing up the valley sides, for here too is an important location for tomato-growing.

The tributary streams, often flowing in rock gorges such as those of the Nethan and the Lee Burn in the Carboniferous limestones and sandstones near Crossford, join the Clyde by steep gradients, as if to herald the famous Clyde Falls, which we must now examine. At Lanark the Clyde flows across a broad outcrop of Old Red Sandstone which serves to separate the reach of the Upper Clyde from that of the Lower Clyde. So far we have followed the slow-flowing waters of the Lower Clyde, but as the valley narrows upstream towards Lanark we suddenly find the Clyde tumbling some 30 feet (9 metres) over the Stoneybyres Falls. As we progress 2 miles (3 kilometres) upstream the gorge becomes narrower until the majestic 90-foot (27-metre) stepped falls at Corra Linn are reached, and a little farther up the chasm the smaller Bonnington Linn completes the sequence. All the gorges and waterfalls have been carved from gently dipping greywackes of Lower Old Red Sandstone age, and they combine with the ancient oak and ash woods and the 'time-cemented tower' of Corehouse Castle on the cliff edge to produce a romantic scene celebrated alike by Wordsworth's pen and Turner's brush. Corra Linn is the best-known landmark because of its high fall of water, the power from which was utilized early in the Industrial Revolution by Richard Arkwright and David Dale. By running the turbulent waters in a subterranean aqueduct through the greywackes, they were able to harness the power of the Clyde to their cotton-spinning mill, set up in 1783 at New Lanark and at that time the largest in Britain. Later this important stone-built mill was incorporated into the famous industrial village planned by Robert Owen as an early experiment in practical socialism and set amidst this 'landscape of woods and rocks worthy of the hand of Poussin' (Thomas Gray).

At this point, however, we must ask ourselves why the waterfalls have formed at Lanark, where they separate the gently flowing, graded reaches of the Upper and Lower Clyde from each other. To seek an explanation it is necessary to examine both the tectonic and the erosional history of the Upper Clyde, for the headwaters, which rise high up in a mossy hollow on Clyde Law in the Southern Uplands, have been not only regionally uplifted but also subjected to major river capture. Let us first examine the effects of regional uplift on the drainage network.

Upstream from Bonnington Linn the Upper Clyde occupies a wide, fertile strath as far as Lamington, where it leaves the inner recesses of the Lowther Hills. Throughout this entire reach the river is a sluggish, meandering watercourse with little change of gradient, and it appears to have been adjusted to a base-level of some 560 feet (171 metres) above present sea-level. We have already noted, however, that near Lanark the character of the valley changes as the river tumbles into a series of gorges before entering the Lower Clyde section of its course, which is adjusted to sea-level near Glasgow. Such a change in river character is a result of rejuvenation by tectonic uplift of the river system. This drastic interference with the river's normal equilibrium when, theoretically, it is just able to transport its load, means that the river is given increased energy. In turn this manifests itself in rapid downcutting of the channel into its former valley in an attempt to restore the smooth, long profile of equilibrium, theoretically hyperbolic in form (Figure 17a). Thus the Upper Clyde reflects the former profile of the river, before the regional uplift of some 560 feet (171 metres), whilst the Lower Clyde represents the episode of downcutting to a new base-level due to increased energy after rejuvenation (Figure 17b). We can see from Figure 17b that the sudden change of gradient between the two stretches is known as a knick point. In this instance the Lanark Falls mark the change from the old to the new profile. Since the river is constantly working to smooth out its profile and regain equilibrium (Figure 17a) by removing irregularities in its course, knick points tend to move upstream. Such headward recession is relatively rapid in the more easily eroded rocks, but once the waterfalls reach a hard-rock band there is a slowing-down of their upstream retreat: hence the Lanark Falls are now 'held' on the hard outcrop of Old Red Sandstone.

Having established the reasons for the different heights of the Clyde Valley above and below the Lanark Falls, it now remains to explain the somewhat anomalous drainage pattern of the Clyde above Lanark where, more than a century ago, Sir Archibald Geikie pointed to a pre-glacial connection between the Clyde and the Tweed. The finest viewpoint in Upper Clydesdale is undoubtedly the summit of Tinto Hill (2,320 feet, 707 metres), from which the remarkably circuitous route of the Clyde from Lamington to Lanark can be viewed. Within this 20-mile (32-kilometre) reach the apparently haphazard changes of river direction have suggested, as D. L. Linton says, that '. . . this portion of Clydesdale must be of composite rather than of simple origin, and is in fact an assemblage of portions of various valleys, now linked into a continuous whole, but formerly quite distinct.'

The view eastwards from Tinto Hill leaves little doubt that the wide, flat through-valley of the so-called Biggar Gap must have played a significant part in the evolutionary history of the Upper Clyde, as was pointed out by Geikie, who concluded that all the Clyde headwaters once followed this remarkable gap to the Tweed, '. . . thus entering the sea at Berwick instead of at Dumbarton'. Professor Linton has traced the mechanisms of this major river capture, which is outlined in Figure 17c. He believed that the

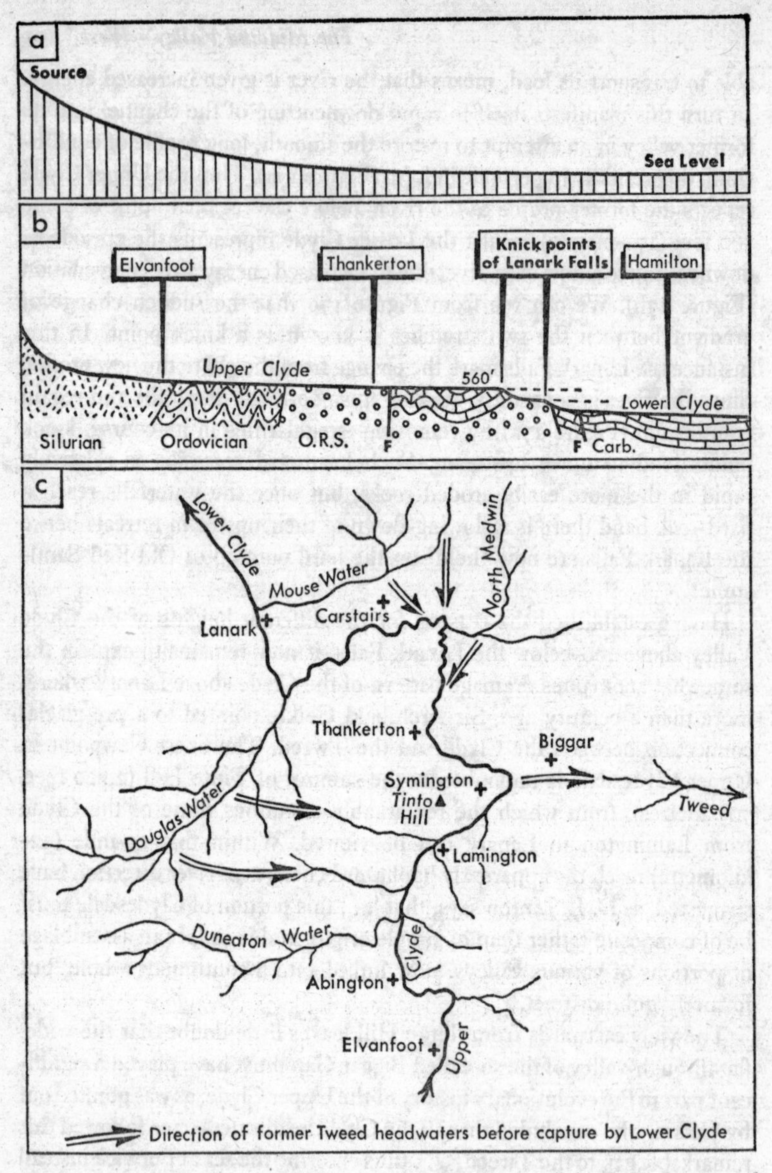

Fig. 17. *The evolution of the Clyde drainage (based on work by T. N. George and D. L. Linton): (a) The hypothetical curve of the long profile before rejuvenation. (b) The present river profile related to geology. (c) A hypothetical reconstruction of former drainage directions in the basin of the Upper Clyde.*

former watershed between Clyde and Tweed lay near Lanark, but that the Lower Clyde, working along the less resistant strata of the Lanarkshire Carboniferous basin, also had a shorter route to the sea. These two advantages allowed it to capture the headwaters of the east-flowing Tweed drainage one by one (Figure 17c).

First, what was to be the river Douglas, utilizing a structural syncline, cut off any of the Tweed headstreams which might have risen to the west of the Tinto Hills. Later, the Medwin tributary of the Clyde cut back eastwards into the outcrops of Calciferous Sandstone, thus beheading a series of south-flowing Tweed tributaries in the vicinity of Carstairs and Carnwath. These include the Mouse Water, the Dippool Water, the North Medwin and the Tarth, all of which now turn through a right angle and flow to the Lower Clyde. Finally, the Upper Clyde itself was captured near Symington, leaving the former trunk route of the Biggar Gap virtually streamless and the Tweed robbed of some 150 square miles (388 square kilometres) of its former drainage basin. During the Ice Age the Biggar Gap was followed at various times by glaciers and glacial meltwaters, and possibly even by overflowing waters from the impounded pro-glacial lake of the Lower Clyde (see p. 96). It is true that its valley floor is now choked with drift, and that the drainage hereabouts has been slightly modified by glacial interference, especially by the well-known Kame of Carstairs. But there is little doubt that river capture was already completed before the onset of the ice-sheets, and that this is not an example of glacial breaching of a major watershed.

South of Abington the Clyde is little more than a mountain torrent twisting through the Southern Uplands, for we have already crossed the boundary fault of the Midland Valley. Here we are back amongst the place-names of the Border Country, where the 'Laws', 'Gills', 'Riggs' and 'Dodds' are reminiscent of Cumberland and Northumberland. It is time to retrace our steps to the Midland Valley and explore the valley of the Forth.

6. The Midland Valley – East

A traveller journeying eastwards from the Clyde to the Forth would probably cross the divide between these two major river basins without realizing it, for in this central section of the Midland Valley, the very heartland of the Scottish nation, the watersheds are generally low and ill-defined. Nevertheless, the contrasts between west and east soon become apparent in the landscape – contrasts that seem to be epitomized in the differences of character between the cities of Glasgow and Edinburgh. In contrast with the long, narrow estuary of the Clyde, the Firth of Forth widens rapidly seawards, bringing the presence of the North Sea and its persistent 'haar' (coastal fog) deep into the central valley. The Clyde's overwhelming backdrop of Highland fantasies is also missing here, for the bordering hills of the Lammermuirs and the Ochils stand discreetly back. Yet the very flatness of the Forth carselands and the coastal tracts of Fife and the Lothians serves only to magnify the small but craggy igneous outcrops which dot the eastern plains. The importance of these isolated crags as defensive sites will be emphasized when we examine the location of Edinburgh itself, sited at the only place where the upland hills approach the Firth, thus restricting the Lothian coastal plain. Farther upstream, Stirling too was founded at a strategic gap where the Forth breaks through the hard-rock barrier of the Ochils and the Gargunnock Hills.

Despite the coalfields of the Lothians and Fifeshire, these eastern landscapes do not bear the imprint of industrialism quite so heavily as those of Ayrshire and Lanarkshire, perhaps because the farmlands are here more widespread and permeate the scattered industrial areas more easily than they do the massive conurbation along the Clyde. But it is also by geological, topographical and climatic contrasts that we are able to define the differences between the eastern and western regions of the

Midland Valley. We have seen, for example, how the extensive lava plateaux of the west create not only unproductive rush-choked uplands but also poor, waterlogged soils. Such plateaux are less widespread in the east, where broad basins of mainly Carboniferous rocks have helped to produce the more fertile soils which characterize the wider coastal plains. Here, too, the glacial drifts are more widespread, as are the extensive marine clays associated with the raised shorelines, and where these have been drained some excellent farmlands have been created. The lower rainfall of these eastern tracts has led to a greater emphasis on arable farming, so that the endless grassland and fodder crops of Ayrshire's dairying region are here replaced by a more satisfying chequer-board pattern of yellow grain-fields interspersed among various shades of green. But the westerly prevailing winds are still channelled through the corridors of the Midland Valley, so even in the Lothians shelter-belts of trees along the field boundaries are a common sight.

The geological and topographical differences noted above are also evident within the region itself, enabling us to distinguish three separate divisions: the Lothians to the south of the Forth; the area of the Middle Forth (including the Ochils); and the peninsula of Fife and Kinross, between Forth and Tay.

The Lothians

The geographical unity of the Lothians has long been recognized, for this ancient province of Lothian is neatly cradled between the Southern Uplands and the Firth of Forth. Although the southern limits of the three counties (West Lothian, East Lothian and Midlothian) extend upwards into the Lammermuirs and Moorfoots, these hill masses lie beyond the Southern Uplands Fault and have been treated elsewhere (see Chapter 2). The remainder of the Lothians is not all lowland, however, for the high ridge of the Pentland Hills thrusts its shoulders into the fringes of Edinburgh to remind us that the term Midland 'Valley' is something of a misnomer. As in other parts of the Central Lowlands, such isolated hill masses divide the drift-covered plains into a number of basins; in this case the Pentlands act as a substantial barrier between East and West Lothian. The hill masses themselves are often structurally controlled, with the Moorfoots and the Pentlands both exhibiting fault-line scarps.

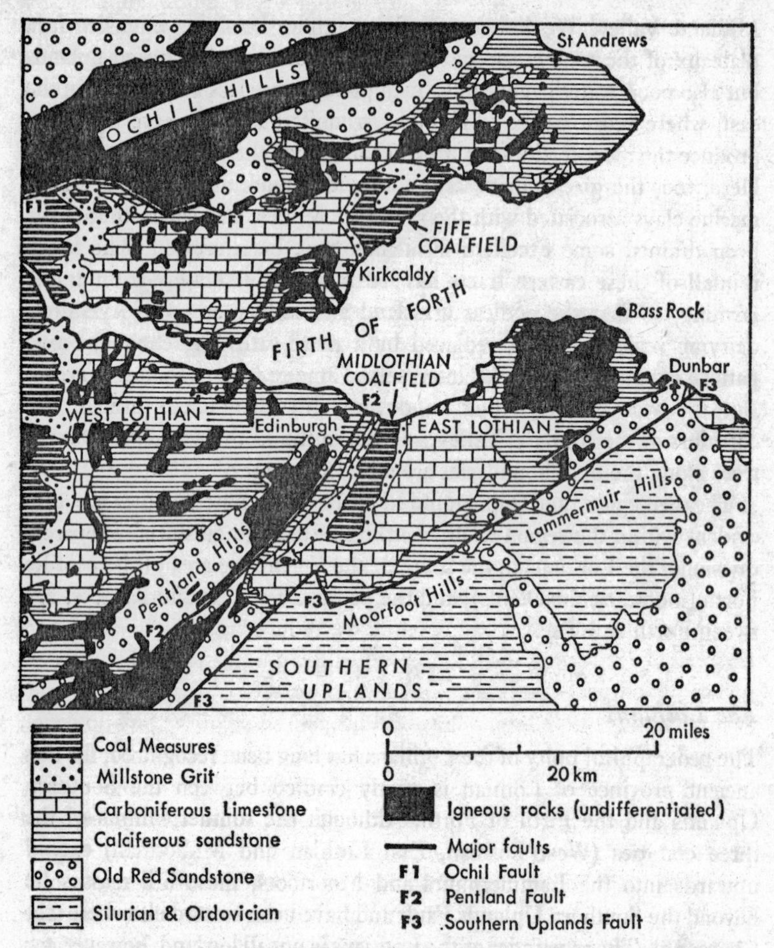

Fig. 18. *The geology around the Firth of Forth (based on material prepared by the Institute of Geological Sciences).*

Not all the topographic contrasts are related to faulting, however: a study of the geological map (Figure 18) will show that, while the lowlands generally coincide with rocks of Carboniferous age, the hills are carved from a variety of pre-Carboniferous rocks which appear to have resisted denudation more successfully than the Carboniferous sedimentaries.

The Pentland Hills dominate the southern outskirts of Edinburgh, their long, grassy slopes rising steeply from the suburban housing estates to a series of isolated summits (notably Scald Law, 1,898 feet, 579 metres). This 'old huddle of grey hills' was described by Robert Louis Stevenson, who lived for many years at the hill-foot hamlet of Swanston, overlooking his beloved old city of Edinburgh, which appeared to him '... like an island in the smoke, cragged, spired and turretted'. Generally speaking the thin, acid soils of many Scottish hills are too deficient in phosphate to support anything but peat and heather moor. But, because of the basalts and basic tuffs which crop out near their northern end, this part of the Pentlands possesses soils with a higher content of mineral salts than usual and is thus conducive to the growth of such grasses as *Agrostis* and *Festuca ovina.* The resulting sward is kept closely grazed by the Blackface sheep which wander across the treeless slopes; no woodland could survive such longstanding grazing pressure. Nevertheless, remains of birch and Scots pine in Boghall Glen suggest that forests once flourished on parts of the Pentlands, perhaps on the more acid soils of the Old Red Sandstone which forms their north-western and south-western flanks. It is recorded that juniper and heather also reached far down the slopes in earlier centuries, but burning and grazing have reduced such vegetation to a fraction of its former extent. Consequently the present flora is by no means the natural climax, and in fact many traces of ancient ridge-and-furrow patterns have been recognized beneath some of the present rough grazings of the lower slopes. This suggests that the Pentlands long provided a livelihood for early settlers, probably before the marshy 'muirs' and 'mosses' of the lowlands had been drained and cleared of their extensive oakwoods (see p. 109).

The structure of the Pentlands is that of a denuded anticline, in which the stripping of the former Carboniferous strata has revealed a complex mixture of Old Red Sandstone sedimentaries and lavas and a few highly folded Silurian rocks. The latter form the core of the anticline and represent one of the few exposures of Lower Palaeozoic rocks within the Midland Valley proper. Among the volcanic rocks, which attain a thickness of some 6,000 feet (1,800 metres) at the northern end of the hills, no less than ten distinct groups of lava flows have been recognized. These range from the lower basalts and andesites of Warklaw Hill, up through the trachytes between Carnethy and Scald Law summits, to the uppermost basalts and andesites, including the attractive porphyrite of Carnethy. The conspicuous dome of Black Hill (1,636 feet, 499 metres), near the northern end of the range, is composed not of extrusive lavas but of an intrusive laccolith of felsite, contrasting markedly with the neighbouring craggy lavas. Buried beneath the southern end of the volcanic tract are great thicknesses of conglomerates and grits of Lower Old Red Sandstone, but the pink Upper Old Red Sandstone is far more extensive, forming much of the southern end of the Pentlands, south-west of the Cairn Hills. As it is traced northwards, the Upper Old Red Sandstone thins considerably before disappearing beneath the Basal Carboniferous Beds in the southern suburbs of Edinburgh. Here, in Craigmillar Quarry, the Old Red Sandstone has been widely exploited for its fine-quality freestone, which has contributed greatly to many of the buildings in the Edinburgh townscape. In addition, the same formation has served as a source of pure water-supply for the city's brewing and paper-making industries.

Seen in profile, the Pentlands appear to be steeper along their eastern than on their western flanks. This is because on the eastern side the Pentland Fault has brought the relatively harder Old Red lavas directly against the more easily denuded Carboniferous rocks of the Midlothian coalfield without the intervention of the Old Red Sandstone. The fault-line scarp thus produced can be traced north-eastwards through Carlops to Straiton, with the fault itself continuing beneath the city to the Forth at Portobello. Such a direction is typical of the NE–SW Caledonian 'grain' which dominates the structural trend of the Southern Uplands, but since other faults and fold axes of the Midland Valley do not always follow this direction we must attempt to explain something of the apparent anomalies within the Midland Valley structures.

At first glance the alignment of the Pentland Hills is seen to conform with the Caledonian trend-lines (of Middle Old Red Sandstone age). Taken in association with the neighbouring syncline of the Midlothian coal-basin, however, the direction of folding is seen to swing round into a more northerly alignment as it is traced across the Forth into Fifeshire (Figure 18). Since the Carboniferous rocks themselves are also folded, we must be dealing not with an episode of Caledonian folding but with one of Hercynian age which came later, in Carbo-Permian times.

We have seen in earlier chapters how the structural framework of the

Midland Valley is Caledonian (as demonstrated by its boundary faults), but within this framework it can be shown that many of the important folds and faults were created by a phase of Hercynian compression. During this ensuing period of mountain-building the pressure was exerted from a southerly direction, and this has manifested itself in two sets of Hercynian structures: first, those with a 'normal' east–west trend, such as we have already seen in the Ayrshire Coalfield, the Glasgow–Airdrie syncline and the Campsie Fault (see p. 92); we shall see below that the Ochil Fault is another striking example of this east–west trend (see p. 123); second, those in which the underlying Caledonian structures have influenced the folding and faulting, causing them to take on a 'Caledonoid grain'. The best examples of the latter type are found in the Midlothian syncline and in the Pumpherston anticline of West Lothian, which affects the oil-shale workings (see p. 116). In fact, the complexity of both the geology and the scenery of the Midland Valley can be explained very largely in terms of the intersecting 'latticework' of the Caledonian and Hercynian structures. Nowhere is this better illustrated than in the Central Coalfield itself, which has an east–west axis near Glasgow but exhibits a north-north-east alignment farther north between Falkirk and Alloa, where it crosses the Forth into Clackmannanshire.

East of the Pentland Hills the Lothian plains stretch uninterruptedly to the North Sea at Dunbar. Here is some of the finest arable land in Scotland, where crops of barley and wheat flourish on the thick glacial drifts – wheat preferring the heavier soils of the boulder clay (mainly derived from Carboniferous rocks), while barley is grown more successfully on the equally basic but lighter soils of the glacio-fluvial drifts. The large, regular, square fields with their scattered settlements and individual holdings create a landscape similar to that of Berwickshire, where the 'rationalized' field pattern has led to equally efficient farming. This, then, is a typical landscape of 'improvement', where the traditional Scottish custom of 'run-rig' was abolished early as the mosses were drained and the ancient oak woods cleared. Only at Roslin Glen and in the park of Dalkeith House have sufficient oak woods survived to remind us that, prior to the eighteenth century, the view eastwards from the Pentlands would have presented a very different picture. We now have to climb to the foot of the Lammermuirs to find remnants of the old fields, the twisting road patterns and the ancient villages which stand on the poorer, more acid soils of the upland fringe, overlooking the geometric farmlands and shelter-belts of the plains. It is no surprise to discover, therefore, that these attractive hill-foot villages – Gifford with its beechwoods, the sandstone and pantiled cottages of Garvald, the grass-fringed street of Stenton – are part of a Conservation Area. From these vantage points, however, it is not the gold and green chequer-board pattern of the lowland farmlands that attracts our attention as we look north across the Lothian plain, for the dark, isolated igneous hill masses disrupt the scene so brusquely that they cannot fail to arouse our curiosity.

The southernmost of the hills, Traprain Law (724 feet, 221 metres), rises abruptly above the valley of the Tyne river, which follows a gorge overdeepened by glacial meltwaters across the hard rock-band of Carboniferous basalts, some of which have been quarried for use as a building stone at East Linton. The prominent cliffs and dome-like summit of the Law demonstrate very clearly the shape of the phonolitic igneous rock which was intruded into the Calciferous Sandstones in the form of a laccolith. The sedimentary rocks, formerly up-arched by the intrusion, have now been eroded from the dome, leaving the fine-grained, well-jointed igneous rock clearly exposed. On the summit the remains of a hill-fort have led to a suggestion that there was an important defensive settlement here between A.D. 100 and 400.

Traprain Law now provides us with a magnificent viewpoint for the other Lothian igneous hills: to the west are the Garleton Hills (590 feet, 180 metres), formed largely from Carboniferous tuffs and trachytic lavas; immediately to the north is the wooded Pencraig Hill, also a laccolith but more subdued than Traprain Law, while away in the distance the prominent crags of North Berwick Law and the Bass Rock dominate the northern coastline. North Berwick Law (613 feet, 187 metres) is a particularly precipitous crag, its red-mottled trachytic intrusion rising incongruously above the flat farmlands which occupy the Carboniferous sandstones of the coastal plain. Several of the near-by sea cliffs have been carved from associated volcanic rocks, including agglomerates occupying vents which have been drilled through the older rocks. The most interesting of the vents is that upon which the picturesque Tantallon Castle stands, for here we can see not only the so-called 'bread-crust' type of volcanic bombs, bedded dark-green basaltic tuffs and a variety of pyroclastic rocks, but also a large mass of sandstone which appears to have been deposited in a hollow within the temporarily quiescent Carboniferous volcano. The famous Bass Rock, a few miles off shore, has a similar structure to that of North Berwick Law, its fortress-crowned cliffs having been carved from an intrusive plug of trachyte which, like North Berwick Law, formerly supplied a volcanic vent.

Although Traprain Law and North Berwick Law were intruded into the Carboniferous strata in somewhat different ways, their subsequent history in the Ice Age is strikingly similar, for both exhibit the landform known as a crag-and-tail. Nowhere else in Scotland are there such superb examples of this phenomenon as in the Lothians: in addition to the two mentioned above, crag-and-tail features may be seen at Dechmont Law (West Lothian), Blackford and Craiglockhart Hills (Edinburgh) and – classic examples – at Calton Hill and Castle Rock, right in the city centre (see p. 114). It is known that ice-sheets, moving in an easterly direction down the Forth, helped to create these singular landforms, with the igneous 'crag' being overridden by ice but helping to protect the leeward 'tail' from such intense erosion. Thus the western faces of these crags are usually precipitous, while their eastern slopes taper off more gently into the tail itself. There is a mistaken belief that the tail is composed of glacial drift, despite the correct observation made by Sir Archibald Geikie more than a century ago that the tail is composed very largely of solid rock, often with only a veneer of drift. Thus, it is an erosional and not a depositional feature.

Glacial erosion has also helped to smooth off the irregularities of the sedimentary rocks in the Lothians, so that the landscape of today, despite its covering of glacial drifts, is very largely an erosional one glacially moulded into tapering streamlined ridges. The 'fluted' drift landscape of East Lothian, made up of kames, drumlins and kame terraces, combines with the ice-moulded ridges and depressions to create a regular east-north-east alignment which the bare igneous hill masses do little to disrupt. The effect of this alignment on the drainage pattern is very clear, especially in the case of the Tyne, but even more striking is the way in which many of the Lothian streams occupy valleys that were almost entirely excavated by glacial meltwaters. Dr Sissons cites the case of the North Esk in Midlothian, which flows for a dozen miles in a deep melt-water channel which it has inherited from glacial times. Parts of the Upper Tyne and Biel Water exhibit similar characteristics.

Because of the presence of the Midlothian coalfield and the industrial activities of Leith, Musselburgh and Prestonpans, much of the Lothian coastline to the east of Edinburgh has been developed. This has included the construction of the major Cockenzie power station, which is supplied by coal from new mines sunk in the neighbouring coalfield. The two new

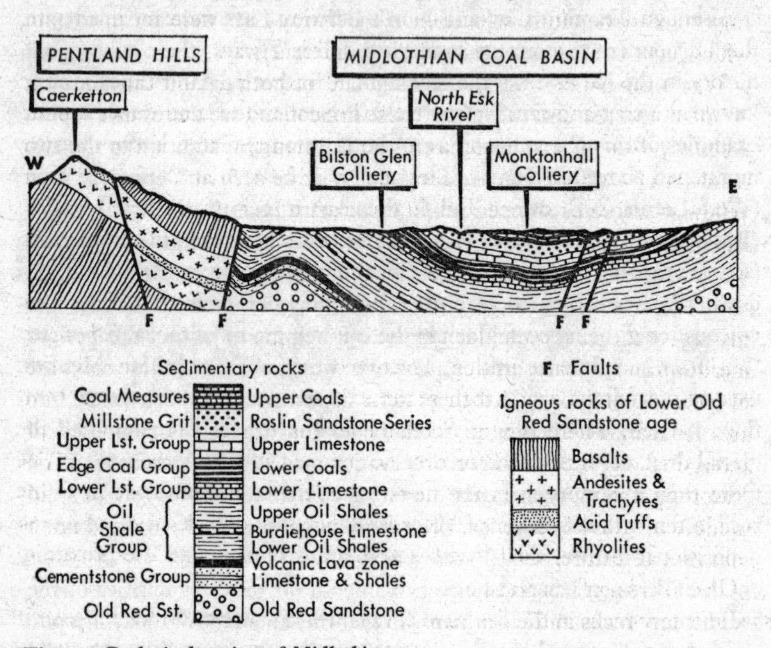

Fig. 19. *Geological section of Midlothian.*

collieries of Bilston Glen and Monktonhall have been designed to exploit the very deep reserves of the Lower Coals (Carboniferous Limestone Series) which occur at the base of the Midlothian Carboniferous syncline. Earlier collieries mined only the easily accessible Upper Coals and the steeply dipping linear outcrops of the Lower Coals (the Edge Coals), where these occur on the fringes of the basin (Figure 19). Records show that monks from Holyrood and Newbattle were mining surface outcrops more than 750 years ago, but most of the reserves in the Upper Coals and the Edge Coals were exploited during the Industrial Revolution and were worked out by the mid twentieth century. The coastal fringes of the Forth have long been known for their salt manufacture, especially those around Prestonpans, which derived its name from the coastal salt pans. Sea-water provided the major raw material, and although peat and timber were originally used in the evaporating process the abundant coal outcrops, both here and in Fifeshire and West Lothian, became more important in the ultimate location of the Scottish salt industry in a region which lacks the brine deposits of the English Trias.

When compared with the mining and industrial landscapes around Tranent and the Esk valley, the coastal marshes and sand dunes around Aberlady Bay present a picture of rural serenity. Here is the classical mansion of Gosford House, designed by Robert Adam and built from yellowish freestone shipped round the coast from a quarry near Dundee. Aberlady village has a more rustic charm, with its mixture of red pantiles, blue slates and whitewashed, grey or red stone cottages. Around it is a nature reserve where the grey-green hummocks of sea buckthorn and the marram of the dunes complement the brighter greens of the neighbouring meadows and orchards in this attractive corner of East Lothian. Mention of orchards reminds us that these same coastlands are the most important area for market gardening in Scotland, accounting for more than half the acreage of vegetables and for over 60 per cent of their production. The reason for the location of the market gardens is the combination of the sandy and gravelly soils of the raised beaches (which have lime-rich mixtures of marine shells) with a favourable climate and the proximity of the Edinburgh markets.

Hill-top towns and cities have always held an attraction for the visitor because of their aesthetically pleasing morphology (hence the attempt to re-create the style of an Italian hill town at Cumbernauld), and Edinburgh is no exception. Generations of writers and artists have tried to recapture the qualities of this attractive city, often referred to as the 'Athens of the North'.

The profusion of igneous hill masses within the city boundary is the main reason for Edinburgh's distinctive character: the abrupt changes of level give splendid vistas and viewpoints at every turn (Plate 11). Two other factors have also played a part in the creation of this urban masterpiece. First, Edinburgh is built almost entirely of the creamy local sandstone, which gives it an air of congruity rare in most British towns away from the Cotswold stone-belt. It has been said that this Calciferous

Sandstone from Craigleith Quarry is to Edinburgh what Portland Stone is to London, Pentelic marble to Athens and Pietra Serena to Florence. Secondly, the city was fortunate in its architects and planners, who built with vision, grace and good sense. Much of Edinburgh is of classical design, and the Georgian architecture of the New Town stands comparison with that of Bath. The Parthenon-style buildings of Calton Hill bring more than a touch of the Acropolis to the city, although Gothic Revival buildings also add their own character – the Scott Monument,

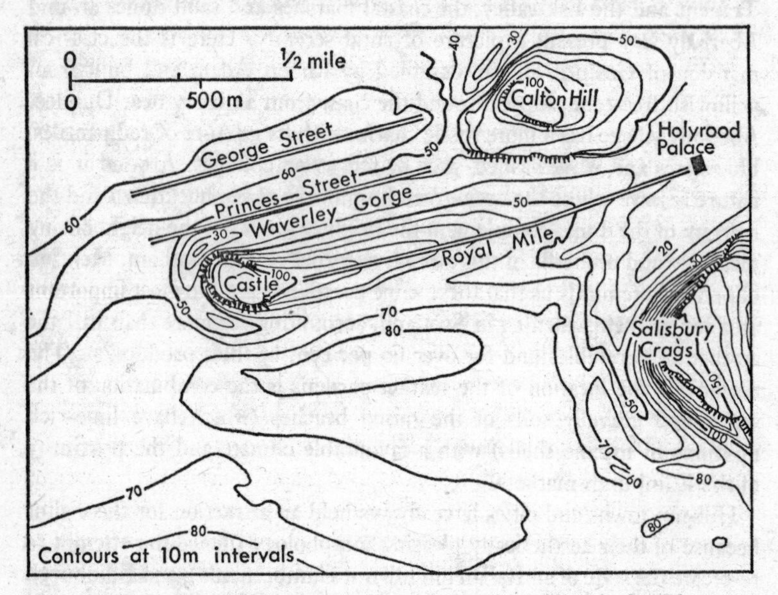

Fig. 20. *The landforms of central Edinburgh (after J. B. Sissons). See Plate 11.*

Pugin's Tolbooth Church and Gilbert Scott's St Mary's Cathedral, to name but a few. But it is to the ancient 'wynds' and 'pends' of the Old Town that the tourist will keep returning, where the names Grassmarket, Lawnmarket and Cowgate are redolent of the Middle Ages, and where the street canyons of the High Street are an early example of high density urban housing.

The Castle Hill and High Street axis continues eastwards to Holyrood House as the Royal Mile and is, perhaps, the best known example of a crag-and-tail in geological literature (Figure 20). The basalt plug on which the Castle is built, which can be seen emerging through the floor of the War Memorial Chapel, has been greatly oversteepened by ice-sheets which were then deflected along the northern and southern faces. To the north the ice bulldozed the softer sedimentaries to form the Waverley gorge, later to become a marshy, lake-filled hollow, and then drained and reclaimed for the Princes Street gardens and the railway tracks. South of the Castle Rock the ice similarly excavated the depression now occupied by Cowgate and the Grassmarket. Between these hollows stands the tail of the Royal Mile, '... whose ridgy back heaves to the sky' (Scott), inspiring the analogy of spine and ribs much used in descriptions of the Old Town street pattern. Near the Portcullis gate of the Castle the actual junction of the basalt plug of the crag with the grey Carboniferous marls of the tail may be seen, whilst the north-western cliffs of the crag exhibit glacial grooving and striae. But the crag-and-tail of Blackford Hill, some 2 miles (3·2 kilometres) to the south, is even more significant in this respect, for it was here in 1840 that the eminent Swiss geologist, Louis Agassiz, recognized the grooving in the overhanging andesite cliff as the work of a former ice-sheet – the first such recognition in Scotland.

It is little wonder that Edinburgh has been the home of many notable geologists, considering the wealth of geological phenomena both in and around the city. The well-known landmark of the Salisbury Crags sill (Plate 11), for example, provided Hutton with irrefutable evidence to support his hypothesis of magmatic intrusion during the prolonged Neptunist–Plutonist arguments of the eighteenth century. Immediately to the east of this prominent escarpment the remnants of the long extinct Carboniferous volcano of Arthur's Seat (822 feet, 251 metres) dominates the city (Plate 11). The complex structure of the volcano includes no less than thirteen lava flows, extruded variously from five vents: the Lion's Head (forming the summit crag); the extensive Lion's Haunch (composite basalt and vent agglomerate); the smaller Crags Vent (agglomerate); and the basalt-filled vents of Pulpit Rock and Castle Rock. Remnants of the former ash/lava cone can be seen at Whinny Hill and Calton Hill, from which exposures it was possible to elucidate the geological history of the volcano. It appears that the initial eruptions took place in the shallow waters in which the Carboniferous Cementstones had accumulated, and there is evidence to show that throughout most of its life the cone remained partly submerged, being finally buried by the waterlaid sediments of the Carboniferous Oil Shales.

To the west of Edinburgh the new arterial road network sweeps the traveller rapidly to the Forth Bridge or into the rolling plain of West Lothian, where one crosses what is virtually the complete Carboniferous succession *en route* to Stirling. Immediately to the west of the Pentlands lies a wide expanse of Lower Carboniferous Sandstone which includes the Oil-Shale Group. This tract was for a short period the most important oil producing area in the world, as witnessed by the conspicuous truncated cones of spent shales – the well-known red shale 'bings' (*beinn* – hill) of West Lothian – which dominate the skyline around the New Town of Livingston (Plate 12).

Six years before the first American oil well was drilled in 1859 the West Lothian oilfield was in production. James Young, finding that the West Lothian coal miners were lighting their homes by burning 'cannel' (candle) coal, discovered that the Bathgate coal was practically oozing with oil. The thick brown shales found in association with the coal never offered as good a yield as the cannel coal but, in contrast with the limited coal reserves, the extent of the oil shales was sufficient to produce 25 million gallons (112 million litres) of crude oil per year in the late nineteenth century. Thus the towns of Mid Calder, West Calder, Broxburn and Pumpherston grew in size, while the Bathgate works also produced ammonium sulphate, paraffin and candles. During the twentieth century, however, imported liquid oil became cheaper to produce than shale oil, leading to a final closure of the West Lothian oil-shale mines in 1962. The near-by Grangemouth refinery, located there originally because of the proximity of the shale oil, has grown into a massive petro-chemical complex on the banks of the Forth, now dependent on ocean-going tankers for its supplies. Ironically, much of its expansion was made possible by bricks manufactured from the shale 'bings'.

The oil shales belong to the Calciferous Sandstone Series of the Lower Carboniferous and appear to have been deposited by a special type of alternating lagoonal and estuarine sedimentation in a restricted basin. Decaying plant and animal remains in the lagoons supplied the carbonaceous matter which ultimately impregnated the shallow-water mud-flats. Sun-cracks suggest that dessication was intense, but occasional marine bands indicate periodic incursions of the sea. A freshwater limestone, the

Burdiehouse Limestone, is also present, a further indication of the rapidly changing environmental conditions of the time.

The sedimentary rocks have been invaded by numerous dolerite sills and dykes which, like the Bathgate lava hills, produce prominent ridges in the otherwise featureless drift-covered landscape. Because of the thick boulder clays, often moulded into drumlins which contrast in shape with the steep-sided 'bings', the soils are generally heavier and more poorly drained than in East Lothian. This has been aggravated not only by the ubiquitous marls and shales of the bedrock but also by the widespread hollows caused by underground mining subsidence. These factors, together with the higher rainfall, mean that there is a greater proportion of permanent grassland in the West Lothian landscape (some 50 per cent), and the spread of rushes remains a perennial problem for the dairy farmer. More than a century ago these pastures were won from the mossy wastes by new drainage techniques and by chemical improvement of the soil with lime produced by burning the Burdiehouse Limestone with local cannel coal in newly erected limekilns, now left standing forlornly amongst the hayfields.

One variation from the widespread clay soils is to be found around Linlithgow, where a ridge of glacio-fluvial deposits creates light sandy soils well suited to barley. The quiet, grey old town of Linlithgow, with its great square-towered castle beside a tiny loch, was a royal stronghold in Norman times, once related to Edinburgh very much as Windsor is to London. But the Calciferous Sandstones here descend westwards beneath the Carboniferous Limestone and the coal-bearing rocks of the Upper Carboniferous Central Coalfield. Thus the rural scene soon gives way to the industrial sprawl of the Carron valley and the extensive carselands of the Middle Forth.

The Middle Forth

No other area in Scotland has a nodality to equal that of the narrow corridor of lowland which separates the volcanic heights of the Ochils from those of the Gargunnock Hills. The town of Stirling has been founded at the veritable cross-roads of Scotland, where the Forth leaves the Old Red Sandstone of the Menteith basin to follow the narrow gap to the Firth. The Middle Forth is, therefore, something of a transition zone at the very centre of the Midland Valley: its coal mines and blast furnaces echo the landscapes of the Lanarkshire coalfield, of which this is the northern extension; its agriculture continues the trend already noted in West Lothian, where arable farming becomes increasingly subordinate to dairying as we move towards the more humid west; and its architecture is already beginning to include the typical pantiles and Low Countries building style which dominates the Fifeshire towns and villages farther east. Only the river Forth gives unity to this narrow zone which spans parts of the old counties of Stirlingshire, Clackmannanshire, Perthshire and Fife – a unity now recognized in the creation of the new Central administrative region of Scotland.

The industrial district around Falkirk is notable as the cradle of the Industrial Revolution in Scotland and its landscape still bears the scars of this long-term exploitation of the local resources. It all started when the blast furnaces at Carron began to utilize the neighbouring 'splint' coals and clay-band iron ores of the Coal Measures to produce the first Scottish pig-iron in 1759. The scarcity of coking coal meant that the 'splint' coals were fed raw into the furnaces, replacing charcoal and the water-power of the river Carron, and building up a local industry which was producing some 27 per cent of British pig-iron by 1850. Scottish iron-ore mining has now ceased, so the Scottish steel industry, which has moved south to Motherwell, is now dependent on Furness and Cleveland ores. But Falkirk has retained some important metal industries, including the manufacture of iron castings and aluminium sheeting. Coal-mining has also survived, especially in Clackmannanshire, where the scattered new collieries among the conifers of Devilla Forest now supply most of the needs of the gigantic thermal power stations of Kincardine and Longannet. Both of these are situated on the banks of the Forth because of their voracious demand for enormous quantities of cooling water. It is for the same reason that Scotland's nuclear power stations are located on the coast.

Above Kincardine the Forth narrows considerably, and until the building of the Forth Road Bridge this was the lowest bridging-point on the river. It is in this section, upstream as far as the Lake of Menteith, that the carselands of the Forth may best be studied. Today the scene is one of pastoral serenity, with the river winding leisurely through crops of hay and corn and the widespread permanent pastures which characterize the heavy soils of these claylands. It is difficult to realize, therefore, as one looks down on this riparian landscape from Stirling Castle or Wallace's Monument (Plate 13), that the formation of the Forth glacial deposits, the raised shorelines and their overlying peat mosses represents some of the most complex chapters in the Quaternary history of Scotland that we have so far encountered.

The traditional view of Scottish raised beaches sees them as having three simple shorelines – the so-called 100-foot (30-metre), 50-foot (15-metre) and 25-foot (8-metre) beaches – with the two highest belonging to the late-glacial period and the lowest being of post-glacial age. Painstaking work by Dr Sissons has demonstrated that such a viewpoint can no longer be upheld, and in its place he and his colleagues have erected a complex picture of fluctuating sea-levels, ice-front advances and retreats and alternating periods of glacio-fluvial, estuarine and marine sedimentation, together with concluding episodes of post-glacial peat growth. The story is a long and complicated one and we are able to look only at the bare outlines in the following summary of events, which took place not only in the valley of the Forth but also in that of the neighbouring Tay.

As the Scottish ice-sheets withdrew westwards from the North Sea coastlands, rising sea-levels left the oldest late-glacial beaches in East Lothian and East Fife. More important, so far as the Stirling area is concerned, however, was the subsequent glacial advance (or still-stand during the general retreat) known as the Perth Readvance, for its associated ice-limit can be traced from Perth around the west flank of the Ochils, and as a lobe through the Stirling gap (Figure 21). Vast quantities of outwash can be traced from the former ice-front, and this helped to build the most distinctive late-glacial raised beach of the Forth Valley – the Main Perth Raised Shoreline. Forming broad, flat terraces at 120 feet (36 metres) near Bannockburn, this feature can be traced down to Stenhousemuir (105 feet, 32 metres) and Falkirk (100 feet, 30 metres), with an eastward decline in altitude of about 2·25 feet per mile (2·3 metres per kilometre). Such a gradient is sufficient to bring it down to sea-level near the mouth of the Firth, the tilting being a result of upwarping of the land due to former glacial loading (see p. 52). Other terraces of approximately this age (13,500–13,000 BP) are associated with the Main Perth Beach and these can best be seen around Falkirk along the Carron valley, where they have been extensively built over. As we cross the line of the

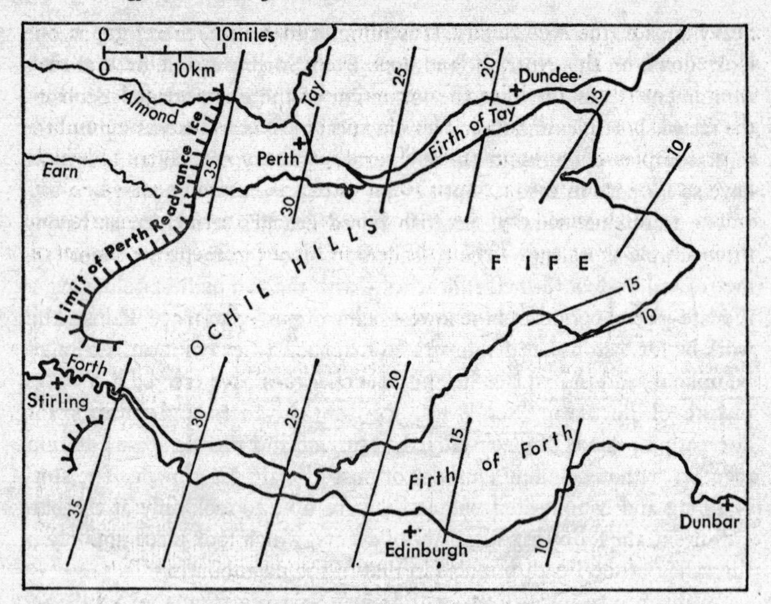

Fig. 21. *The limits of the Perth Readvance ice-sheet and the elevation of the Main Late-glacial Shoreline in the Firth of Forth and Firth of Tay (after J. B. Sissons). Heights in metres.*

Perth Readvance and move westwards into the Carse of Stirling these terraces disappear, to be replaced by a wide plain of flat carse clays and peat mosses some forty square miles (102 square kilometres) in extent. It is now known, however, that these innocuous meadowlands hide an even more complex sequence of sediments related to buried channels, and buried beaches and peats, discovered only from borehole data (Figure 22).

The so-called High, Main and Low Buried Beaches of Figure 22 were in turn created partly from the glacio-fluvial outwash of a later ice advance (the Loch Lomond Readvance) which built the Menteith Moraine near Arnprior some 10,300 years ago. The Lake of Menteith is in fact impounded behind the kames of this morainic belt, occupying a gigantic kettle-hole left after the melting of an enormous ice-block. The intermittently falling sea-level of these times was responsible for the formation of the three buried raised beaches 9,500 to 8,800 years ago, and we can envisage a long narrow estuary stretching through the Stirling gap as far as Menteith at that time. It was upon these abandoned shorelines that forests of alder and birch began to grow, indicating the passage from late-glacial into post-glacial times. The remains of the trees became entombed in peat deposits, the appearance of reeds in the upper layers of which suggests that there was a return to an estuarine environment when the forests were drowned by the major post-glacial marine transgression around 8,500 years ago. This is the reason for the widespread blanket of

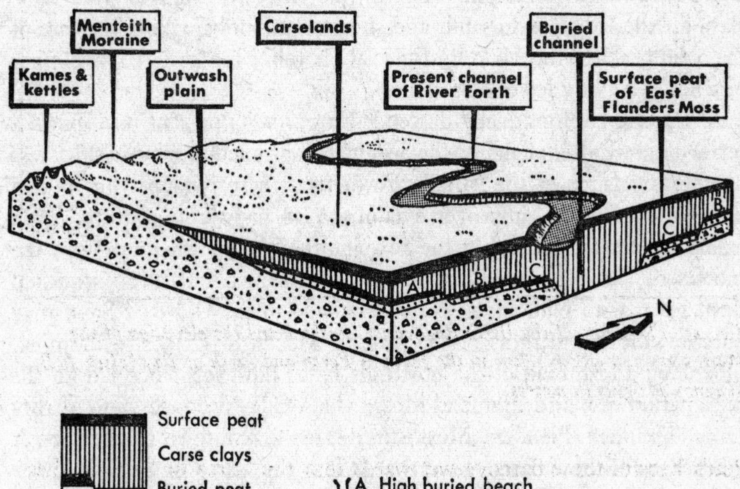

Fig. 22. *Superficial deposits of the Carse of Stirling (after J. B. Sissons).*

stiff blue carse clays, which are nothing more than uplifted fossil mud-flats formed at the edge of the so-called Flandrian Sea in much the same way as the present tidal flats have formed at the edge of the modern shorelines.

Dr Sissons has reconstructed a vivid picture of the Carse of Stirling in 8,500 BP, with the Menteith morainic ridges and the raised bogs of Flanders Moss standing like reed-encircled islands above the slowly creeping tides. As he describes it: '... the surface of the mudflats was interrupted by channels, particularly that of the Forth, and it is likely that many of the present stream courses, including the peculiar meanders of the Forth itself west of its junction with the Teith, are the direct descendants of these mudflat channels.' The surface of this main post-glacial raised shoreline can be seen to descend gradually downstream across the Carse of Stirling from a height of 49 feet (15 metres) at Gartmore (West Flanders Moss) to 42 feet (13 metres) at Stirling – and this is the so-called '25-foot' (8-metre) post-glacial beach! As it is traced down the Firth (at Leith it is about 30 feet, or 9 metres), its character changes from carse clays at Aberlady Bay to sands and shingles at Dunbar. As in the case of the older beaches, the Main Post-glacial Shoreline is associated with other beaches at slightly lower elevations.

As the sea-level gradually descended towards its present position in a series of intermittent halts, so the peat bogs continued to grow in thickness on the carselands of the Middle Forth, and these remained undrained until the agricultural improvements of the late eighteenth century. Apart from peat-cutting for fuel by the inhabitants of the peripheral villages, the mosses of Flanders and Blairdrummond remained virtually untouched until 1767, with ploughland being restricted to the Old Red Sandstone outcrops on the margins of the Vale of Menteith. Extensive draining, ploughing and liming of the peatlands have, however, since led to the reclamation of Blairdrummond Moss, which today produces high quality crops, although Flanders Moss survives as a reminder of the former landscapes of these parts.

The town of Stirling, with its fourteenth-century bridge, has long been regarded as the most strategic location in the Midland Valley, controlling the central access to the Highlands at a point where many decisive battles were fought between the ninth and fourteenth centuries. Its brooding castle, high on a dolerite crag, symbolizes the military significance of the Stirling gap. Although, as a royal palace, it became the first British building to be influenced by Italian Renaissance designs, the narrow cobbled wynds and crow-stepped gables of the town itself are Scottish enough. Across the silvery meanders of the Forth the conspicuous Wallace Monument rises from another eminence of intrusive dolerite (Plate 13), whilst the wooded foothills above the Bridge of Allen contain the old copper mines which once supplied the royal mint at Stirling. Our journey has now taken us across the Forth, past the ruins of Cambuskenneth Abbey, where the medieval monks once tilled the rich alluvium of the carselands. Above us stand the Ochils, whose moors and glens bring a touch of the Highlands to the Midland Valley scene.

Stretching more than 25 miles (40 kilometres) from south-west to north-east, the Ochil Hills rise to considerable elevations (for instance, Ben Cleugh, 2,363 feet, 720 metres), their plateau-like crests seamed with deeply cut valleys, including the remarkable glacial trough of Glen Eagles. From any southerly viewpoint their most striking feature is the conspicuous scarp which runs for some 20 miles (32 kilometres) from Blairlogie eastwards to beyond Dollar (Plate 13). This straight mountain-front rises steeply from the Clackmannan plain, and it is no surprise to find that here in the west the abrupt change of slope coincides with the line of the Ochil Fault. This is a major feature in the structure of the Midland Valley, running for over 55 miles (nearly 90 kilometres) from the Clyde to central Fifeshire and exhibiting a southerly down-throw of some 10,000 feet (3,000 metres). (See Figure 18.) Such a displacement has brought resistant Lower Old Red Sandstone andesitic lavas into juxtaposition with Carboniferous sediments. The latter now form the lowland of the Clackmannan coal basin, whilst the lavas build the 1,700-foot (518-metre) fault-line scarp of the Ochils (Plate 13). But, as if to demonstrate that there are exceptions to the broad association of uplands with the more resistant rocks, the relatively low plain of Kinross, to the west of Loch Leven, has been levelled across identical andesitic lavas. Furthermore, the Ochil Fault is here found at the Cleish Hills and is therefore not responsible for the continuation of the Ochils' southern scarp-face to the east of Dollar.

It has been suggested that many of the river valleys of the Ochils have been carved along lines of structural weakness, possibly related to a Hercynian fault-mosaic which followed earlier trends of Caledonian (NE–SW) and even Charnian (NW–SE) structures. But some of the streams appear not to have been influenced either by structure or by glaciation. (Of these, the east-flowing headwaters of the Devon are the most important.) To explain this and similar anomalies the process of superimposition from a former cover of sedimentary rocks has been invoked. A small area of Carboniferous rocks, lying isolated as an outlier to the north of the Ochils near Bridge of Earn, was regarded by Geikie as evidence of this former cover, which, if it ever existed, has now been stripped away from the summits of the Ochils. If the streams had once flowed across the Carboniferous cover rocks, the drainage pattern would have been lowered as these were gradually destroyed until it finally came into contact with the older structure lines of the underlying lavas. In some cases the streams may have become so well established in their courses that they did not become 'adjusted to structure' but instead retained their 'superimposed' courses.

Finally, a word about the location of the so-called 'hill-foot towns' of the western Ochils: here swift streams on the fault-line scarp have provided a source of power for the long-established textile industries of Tillicoutry, Alva and Menstrie.

Fifeshire

Anyone travelling across the county of Fife could not fail to be impressed by the contrasting landscapes of this eastern peninsula. The surface features closely reflect the underlying geology and structure, so that the predominantly WSW–ENE grain of the Carboniferous, the Old Red Sandstone and the volcanic and intrusive rocks has produced a relatively simple pattern of landforms. The resulting alternation of upland and lowland zones allows the following tracts to be recognized: first, in the south, the Carboniferous coastlands along the Forth, including the Fife coalfield; second, the line of prominent igneous hills which forms the boundary between Fife and Kinross; third, the Old Red Sandstone corridor known as the Howe of Fife, drained by the river Eden; fourth, the low Carboniferous Sandstone plateau of the East Neuk, with its fringe of picturesque fishing villages; finally, the northern ridge of volcanic hills which cuts off the major part of Fifeshire from the Firth of Tay.

The Fifeshire scenery therefore exhibits a remarkable diversity: its rich, rolling farmlands are interspersed with coalfield clutter, the wooded parks of the great estates, picturesque lochs and heathery hills, all encircled by coastal dunes and rocky cliffs, with pretty fishing villages reminiscent of those in Cornwall. Because its soils are fertile clays and loams, developed mainly from sandstones and limestones, and because the climate of the coastal plateaux is as favourable as that of the neighbouring Lothians, Fife has the highest proportion of arable land of all the Scottish counties. Since much of this is devoted to wheat, especially in the coastlands, and since the igneous hills of the interior have remained as patches of moorland, it is easy to understand the traditional description of Fifeshire as 'a beggar's mantle fringed with gold'. In earlier centuries such a description would have been even more appropriate than today, for the centrally located Howe of Fife was once a badly drained basin of forest and marshland. Recent reclamation, however, has made its potentially fertile soils some of the most productive in Scotland.

Despite the attractiveness of its agricultural landscapes, much of the charm of Fifeshire is to be found in its towns and villages, where the influence of Dutch and Flemish architecture is manifest in curved gables and high-pitched roofs. Indeed, many of the Scottish tolbooths around the Forth echo the character of the seventeenth-century Dutch *rathaus*, while the famous Townhouse of Culross has often been compared with Leiden town hall. Culross, with its walls and roofs rising in tiers above the banks of the Forth, is as picturesque as any North Devon village. The outside stairs on the steep, narrow streets are especially notable, with the rosy pantiles of the whitewashed houses bringing a pleasant relief from the more usual slates. Dunfermline, by contrast, is a city of grey stone, with its massive Norman abbey echoing the dignity of Durham cathedral. This ancient royal city replaced Iona as a burial place for Scottish kings when Malcolm III built the original Dun (fort) on the lip of a gorge. But the same stream which initially gave security eventually supplied water-power for milling, so that today Dunfermline is dominated by its linen mills and silk factories.

Kirkcaldy too is an industrial town, but its growth has depended on linoleum as well as linen. Its initial development was restricted by the narrow raised beaches, so its nick-name 'the Lang Toon' was more understandable in the past than it is now that its suburbs have stretched away from the coastal margin. The contrasting character of the Kirkcaldy collieries reflects the changing history of the Fifeshire coalfield. To the east of the town, the older Frances Colliery has only touched the fringes of the coal reserves, having retained methods traditional since the seventeenth century. To the west, however, the massive Seafield Colliery, using modern technology, has been able to exploit the considerable reserves which exist beneath the Firth of Forth. Here, at depths of almost 2,000 feet (over 600 metres), no less than twenty seams are now being worked far out beneath the sea, so that Fifeshire production is gradually replacing that of the declining Lanarkshire pits. Inland, the colliery winding-gear and bings can be seen dotting a countryside which has been badly scarred by decades of coal-mining and surface subsidence, especially in the Cowdenbeath–Lochgelly area, where the sprawling, unattractive mining villages do nothing to mitigate this impression. But rehabilitation schemes are now changing the countryside as the bings are levelled and their waste used to infill some of the marshy hollows. The larger subsidence 'lochs' are also being landscaped and planted with trees.

Cutting through the Carboniferous sedimentary rocks are numerous igneous intrusions, including many basic sills. These fine-grained, olivine-rich rocks are responsible for the majority of the smaller ridges of southern Fifeshire; they also create the well-known islands of Inchcolm and the Isle of May at opposite ends of the Firth. The constituent rock of the islands is teschenite, the same rock-type which forms Salisbury Crags in Edinburgh. The highest hills of Fife have, however, been carved from thick sheets of quartz-dolerite, locally pierced by volcanic necks and agglomerate-filled volcanic vents. Thus, the central ridges of dark-coloured uplands which overlook Loch Leven from the south and east are all of predominantly igneous composition. In the west are the largely forested Cleish Hills (1,243 feet, 379 metres), whilst Benarty Hill (1,131 feet, 345 metres) stands isolated in the centre. But it is the Lomond Hills, farther east, which are most prominent, with their shapely crests (West Lomond, 1,713 feet, 522 metres; East Lomond, 1,394 feet, 425 metres) rising dramatically above the flat farmlands of the Howe of Fife. Both the Lomond summits owe their character to the resistance of the volcanic necks which have been drilled through the surrounding sills and sedimentary rocks alike.

Between the steep northern slopes of the Lomond Hills and the volcanic uplands of northern Fife the grits and conglomerates of Upper Old Red Sandstone age have been picked out by denudation to form a narrow strath known as the Howe of Fife, which continues westwards as the Loch Leven basin. Considering its large size, it is remarkable that Loch Leven does not occupy a glacially eroded rock basin, but investigation has shown that, like the Lake of Menteith, it rests in a hollow within the glacio-fluvial deposits. By comparison with the widespread boulder clay of southern Fife, this Old Red Sandstone vale is thickly infilled with glacial outwash – the corridor must have acted as a major outlet for glacial melt-waters during the waning of the ice-sheets. Like the Loch Leven basin, the Howe of Fife once boasted a loch at its centre, but this was finally drained in 1745, with only the alluvial flats reminding us of its former presence. Today the basin is largely under arable farming, the light, sandy soils favouring sugar-beet, which is refined at near-by Cupar.

An exception to the farming scene occurs in the centre of the Howe, where extensive conifer forests have been planted on the thick gravels of Edensmuir. Forests have flourished here for centuries, for this was an ancient royal forest, the ruined palace at Falkland testifying to its earlier importance. The former forests and marshes at the centre of the Howe have caused the village and town settlements to be restricted to the perimeter, generally on encircling bluffs of boulder clay. Of these, Auchtermuchty is by far the most interesting, for it retained its thatched roofs much longer than most other Scottish towns. Although only a handful of thatched houses survive today, it is interesting to note that reeds (usually *Glyceria maxima*) from near-by Lindores Loch were often used in this fast-disappearing rural craft.

The coastal plateaux of the East Neuk are composed largely of Calciferous Sandstone, although Carboniferous Limestone and Millstone Grit occur along the western margins. But the sedimentary rocks are thickly mantled with glacial drift so that the only occurrences of solid rock which can be seen, apart from the coastal cliffs and platforms, are the conical hills of the Laws. Like their Lothian counterparts, such eminences as Largo Law (952 feet, 290 metres), Kellie Law (597 feet, 182 metres) and the oddly named Bungs of Cassingray (694 feet, 211 metres) are all carved from igneous rocks, generally volcanic necks and vent agglomerates. These basic volcanics and the Lower Carboniferous rocks through which they have been injected have combined to produce fertile loamy soils, with a high proportion of phosphate of lime owing to the presence of the mineral apatite, derived from the underlying igneous rocks. Thus, all but the highest craggy hills have been farmed, making the East Neuk one of the most prolific areas of mixed farming in Scotland. Cattle and sheep are fattened on the rich grasslands of the interior, where beech- and ash-fringed lanes lead to attractive stone-built farmsteads.

The coastal margin is even more attractive, for here we can find a collection of fishing villages which are as picturesque as anything in

Britain. Such places as Crail, Anstruther, Pittenweem, St Monance and Elie have retained much of their old-world charm, with their tiny harbours and stone quays crouching below steep, cobbled streets, crow-stepped gables and a confusion of chimneys on pantiled roofs (Plate 14). In this area building-stone was plentiful, so brick was rarely used. St Monance's Auld Kirk has a stone octagonal tower set low to avoid the North Sea gales. Although these coastal villages are perched on low rock cliffs or tucked into hidden bays with yellow sands, they are backed by wide expanses of both late-glacial and post-glacial raised beaches. The light sandy soils thus produced are even more fertile than those at higher levels, so that the coast road winds through unfenced fields of wheat and barley. Woodlands are uncommon in this well-farmed landscape, generally surviving only in the narrow rocky ravines, or 'dens', where the streams have cut down through the ubiquitous drifts.

An important exception to the coastal treelessness of East Fife is the extensive Tentsmuir Forest on the dunes to the north of St Andrews Bay, where the land is of no use for farming. The same conditions might have existed on the sandy peninsula of Out Head which juts north from St Andrews itself, but here the dunes have been developed for a different use, to make the most famous golf course in the world. The simple Norman church of local sandstone in neighbouring Leuchars vies with the Romanesque tower of St Rule's Priory Church, St Andrews, as the oldest building in East Fife. But St Andrews has the added advantages of its ancient castle, university and cathedral, grouped round a stone-built harbour, which combine to produce a greater aesthetic appeal than that of the extensive concrete runways of the Leuchars military airfield, carefully located on the flat raised-beach terrace of the Eden estuary.

Our excursion around the basin of the river Forth is now virtually at an end, for beyond the wooded volcanic hills of the northern horizon the Firth of Tay lies hidden from view. The andesitic lavas of Old Red Sandstone age which built this line of northern hills are really an eastward extension of those of the Ochils, although nowhere do they achieve such high relief. Their steep northern slopes drop abruptly to the Tay, making a coast road impossible and forcing the major routes far to the south of the crestline. Where these volcanic hills reach the eastern coast an interesting archaeological excavation has recently revealed a detailed picture of the technology and economy of some of Scotland's earliest known inhabitants. On a low andesite volcanic plug, covered by blown sands from the neighbouring Tentsmuir dunes, an early prehistoric site has been investigated near Morton. Dated by radiocarbon methods to an age of 7,400–6,400 BP, these simple Mesolithic hut circles include evidence of stone-tool manufacture. Half the microlithic industry utilized flint pebbles (from an unknown source), while the remainder used tough semi-precious stones such as agate, carnelian and opal, brought from the Tay gravels or the local lavas.

7. The Midland Valley – North

After the urbanized and industrialized landscapes of the Clyde and Forth, the scenery of the Lower Tay valley comes as something of a relief. To travel through Strathmore in springtime is a revelation, for the rich red soils and the vernal green of this sandstone vale create a picture more reminiscent of Herefordshire than of a Highland border county. Even Burns, whose fondness for scenery has more of the farmer than of the romantic poet about it, could not fail to be stirred by the Tay countryside as he wrote of '. . . the fine, fruitful, hilly, woody country round Perth' and '. . . the rich harvests and fine hedgerows of the Carse of Gowrie'.

The Lower Tay region is relatively simple to define in terms of its physiography, being bounded on most sides by major physical features. In the north the prominent scarp of the Highland Front parallels the Highland Boundary Fault; in the east the coastline of the North Sea is a no less distinctive boundary; while to the south the bulky line of the Ochils divides Tay from Forth. Only in the west is the boundary less certain, for south of Strathearn the watershed between the Allan Water and the Earn is low and ill-defined.

It has already been demonstrated that the Midland Valley, in its simplest form, is an ancient rift valley, bounded by two parallel fault-systems between which the surface rocks have subsided. In addition we have seen that, broadly speaking, as we move away from the central coal basins the rocks get progressively older as they are traced towards the bounding faults. Thus, in the case of the Lower Tay, the lowland which stretches north-eastwards from Dunblane to the North Sea at Montrose is devoid of Carboniferous rocks and corresponds very largely with the Strathmore syncline. This asymmetric downfold, some 15,000 feet (over 4,500 metres) in depth, affects a great thickness of Lower Old Red Sandstone which was folded and faulted along a north-easterly axis in mid-Devonian times. The north-western limb of the syncline is almost vertical, or even slightly overturned in places, while between the downfold and the Highland Boundary Fault a steep anticline brings a narrow strip of Lower Palaeozoic rocks up to the surface in Angus and Kincardinshire. Such mid-Devonian earth-movements also resulted in the metamorphic rocks of the Highland massif being driven south-eastwards towards the downfaulted Old Red Sandstone of the Midland Valley. Thus the Highland Boundary Fault can be regarded for much of its length as a steeply

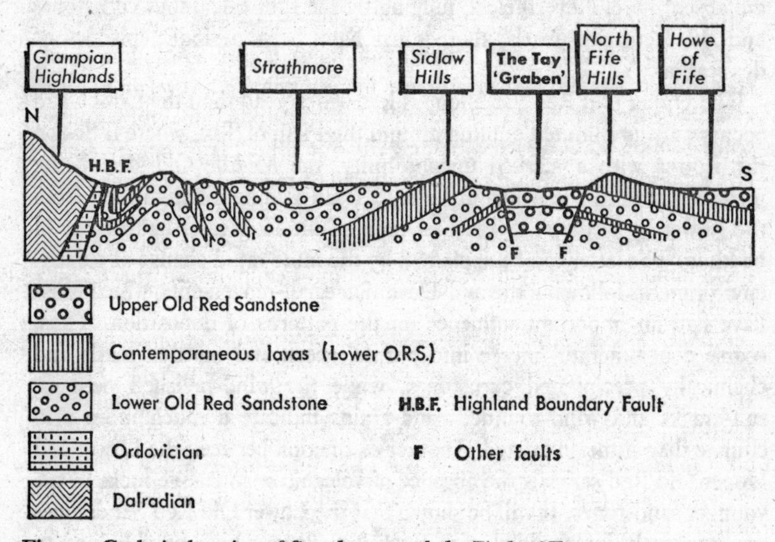

Fig. 23. *Geological section of Strathmore and the Firth of Tay.*

inclined reversed or thrust fault (Figure 23), although the fault appears to be normal near the Firth of Clyde.

There seems little doubt that the Highland Boundary Fault was not initiated by the mid-Devonian earth-movements, for it seems to have been in existence earlier in Palaeozoic times when the Caledonian mountain-building episodes first created the Grampian massif in central Scotland. From these early highlands powerful rivers must have flowed southwards, carrying detritus into the lower basins of the incipient rift. As they crossed the forerunner of the modern Highland Line their flow would have been checked by the sudden easing of the gradient. Thus, enormous fans and deltas of coarse material from the Highland metamorphics mark the geological boundary, and today these form the conglomerates and coarse grits of the Lower Old Red Sandstone. The finer sediments of sand and mud were light enough to be carried farther into the basin, so that the conglomerates give way southwards to the ubiquitous thick brown sandstones which floor the Strathmore lowland. The rocks show very little evidence of great aridity during Lower Old Red times, since the salt deposits which characterize the English New Red Sandstone are absent here. Nevertheless, judging by the thin calcareous cornstones and the lacustrine marls, there must have been periodic episodes of dessication.

The Upper Old Red Sandstone is less widespread than the Lower and occurs only as a limited outcrop around the Firth of Tay, where it lies on the Lower with a marked unconformity, the Middle Old Red being unrepresented in the Midland Valley. It can also be distinguished from the Lower by virtue of its brighter redness and its greater irregularity of bedding. The latter can be explained by the more rapid change of palaeoenvironments following the mid-Devonian earth-movements, which must have had an important influence on the patterns of deposition. Thus, coarse conglomerates merge into purple, green and mottled marls and chemically precipitated cornstones, while the wind-polished pebbles, sun-cracks and wind-rounded sand-grains indicate a much more arid climate than hitherto existed. Another distinction between the Lower and Upper Old Red series is the absence of volcanic or intrusive rocks in the younger sandstones. It will be shown that the Lower Old Red Sandstones are frequently interbedded with thick lava flows which have helped to form both the Ochils and the Sidlaw Hills because of their general resistance to denudation. They are mainly basaltic lavas in which the formerly gas-filled cavities subsequently became lined with secondary minerals such as calcite and quartz. Erosion has now carried many of the chalcedonic stones into the local rivers, where they are treasured as 'Scotch Pebbles', famed for their distinctive concentric agate banding.

The Lower Tay region can be divided fairly readily into three distinctive districts, each of which is delimited very largely on geological grounds: first, the synclinal lowland of Strathmore, with its red soils and rich farming lands; second, the eroded and faulted anticline now occupied by the Firth of Tay, which is overlooked by the craggy escarpments of the Ochils and Sidlaws (Figure 23); and third, the western lowlands of Strathearn. These will be looked at in reverse order, so it is to the last-mentioned that we must now turn.

Strathearn and Perth

One of the traditional routes from the heavily populated Scottish heartland to the Highlands lay north-eastwards via Stirling Bridge and Perth. A motorway now leads the traveller rapidly past the splendours of Stirling and Dunblane Cathedral into the prosperous farmlands of Strathallan, which skirt the north-western flanks of the Ochils. Here one is so impressed by the deep glacial trough of Glen Eagles which dominates the southern skyline that in no time the Forth–Tay watershed has been crossed. Before us is a vision of all that is best in Scottish lowland scenery, with far-ranging views across regular field patterns, neatly manicured hedgerows, shelter-belts and dark clumps of conifers, all dotted with neat farms and rolling woodlands and backed by the often snow-clad peaks of the Highlands. Little wonder that this stretch of the country abounds with extensive estates, country houses and baronial castles. Near to Crieff, for example, the formal gardens of Drummond Castle are a miniature Versailles, approached through verdant avenues of beech and lime. Not far distant stands the luxurious Gleneagles Hotel, located in the heart of the vale not only for the extensive views which its position provides but also because of the magnificent golf courses which surround it. It is no accident that these were located here, for the kettle moraine of the local area provides a suitable terrain for the golfer (Plate 15). These sandy and gravelly hollows and hummocks, covered in a profusion of gorse and heather, were useless for ordinary farming, the slopes being too steep and the soils too acid for anything but rough grazing. But these same features provided ready-made hazards for a golf course, and it is interesting to speculate how many other zones of undulating morainic relief have been converted into golf courses at minimum expense.

This hummocky drift around Auchterarder can be traced eastwards into the widespread Strathearn glacio-fluvial outwash plain, both having been formed during the decay of the ice-sheet known as the Perth Readvance. But before turning to examine the limits of this important glaciation it would be instructive to look briefly at the efficiency of glacial erosion in the vale, since this was in part the source of the glacial drifts. Professor D. L. Linton has noted the contrast in form between the hard-rock landforms of the volcanic Ochils and the relatively soft-rock features of the Strathallan–Strathearn lowland. An ice-sheet moving generally eastwards towards the Firth of Tay not only moulded and truncated the spurs of the Ochil northern slopes to an elevation of some 900 feet (274 metres), but also succeeded in pushing a lobe of ice forcibly into the valley of Glen Eagles. Ice erosion was sufficiently powerful to lower the pre-glacial watershed by some 750 feet (230 metres) near St Mungo's, so that today the summit col of the Glen Eagles through-valley is half a mile (0·8 kilometres) farther south and at a considerably lower elevation than before. Such breaching of a pre-glacial watershed by ice is known as 'glacial diffluence'. The spectacular steep-sided form of Glen Eagles (Plate 15) is largely a result of the resistance shown by the Ochil igneous rocks; by comparison the sandstones of Strathallan and Strathearn were more easily eroded. The former divide between these two valleys has been substantially lowered in this way – the depth of rock removed by glacial erosion has been estimated at between 300 and 400 feet (90–120 metres).

The ice-sheet which sent a lobe through the Forth gap at Stirling (see p. 119) is the same one that swung round the western end of the Ochils before entering the basin of the Tay via Strathearn (Figure 21). A well-known drift section in the river bank of the Almond, two miles (3 kilometres) north-west of Perth, has suggested to some authors that this was a major ice-readvance (it was termed the Perth Readvance), although there is now a suggestion that the kame-and-kettle zone near Perth may represent only a still-stand in the general retreat from the maximum extent of the last major glaciation in Scotland. Nevertheless, the extensive sheet of outwash can be traced downstream to Perth from Almondbank, and also down Strathearn, where it runs into the Main Perth Raised Beach at an elevation of about 115 feet (35 metres) above sea-level.

Overlooking the junction of the Earn and Tay, two ancient boroughs are located on the foot-slope of the Ochils: Newburgh, nestling beneath its vitrified fort, which stands atop the 500-foot (150-metre) Clatchard cliffs of blue igneous rock; and Abernethy, which boasts one of the two Irish-style round towers extant on the mainland of Scotland, built with local grey and cream sandstone as a refuge against ninth-century Danish attacks. But our attention is constantly attracted by the narrow gap to the north, cut by the Tay as it breaks through the line of the hard volcanics of the Sidlaws, and it is no surprise to discover that such a strategic site was selected, in much the same way as at Stirling, for the location of a major town. Yet Perth, the Fair City of literary fame, fails to match the stature of Stirling either in the quality of its townscape or in its situation on the valley floor. The city had a great architectural heritage from the past – it once contained the greatest concentration of religious establishments in Scotland – but many of its finest buildings have now been destroyed, although the Fair Maid's House, Hal o' the Wynd's House and the ancient water-mills survive. Nevertheless, the medieval street-plan and the original street names remain to remind us of the town's antiquity and of its commercial importance when it commanded the lowest bridging-point of the Tay. The river-water was not only used for milling: its softness (due to the dearth of calcareous minerals in its catchment) has made Perth the site of important bleaching, dyeing and whisky-distilling industries.

Upstream from Perth the river banks were once the scene of major industrial activity, for water power became the most important factor in the location of Perthshire's textile industry during the Industrial Revolution. At the outset cotton-spinning spread here from the Clyde, but more important was the later establishment of the linen industry based on local flax. The location of the mills on the Tay, Ericht and Almond was determined by the narrow zone of rapids and cataracts where the rivers, having already crossed the Highland Line, had commenced downcutting into the lowland sandstones of Strathmore. The cataracts themselves, such as that at the Linn of Campsie, near Stanley, occur where the Tay is busily engaged in lowering one of the tough igneous dykes which slice through the Old Red Sandstone hereabouts. In much the same way as the knick-points of the Clyde Falls were once utilized at New Lanark, the physical advantages of the Tay and its tributaries were adapted by Perthshire's earliest industrialists. The presence of meanders helped to increase the efficiency of the water-power, the most spectacular example being at Stanley. Here the Tay, deeply incised into a 150-foot (46-metre) gorge, executes a sharp eastward loop. Some of the river's water is led away from a weir by a tunnel cut through the core of the incised meander, whence it emerges to power the mills farther downstream. At a much earlier date an artificial diversion was also made at Almondbank, where some of the Almond water was diverted along the Town's Lade to power the corn-mills of medieval Perth. But the foundation of this settlement goes a great deal farther back in time: the Romans founded their fort of Bertha (hence Perth) at the junction of the Almond and Tay. Of particular interest in this context is the line of Roman signal stations sited along the conspicuous Gask Ridge, itself formed in part from a particularly thick igneous dyke of quartz-dolerite, running westwards from Perth to the river Earn near Crieff.

The Firth of Tay

A scramble to the summit of either of Perth's flanking bluffs, Kinnoull Hill (728 feet, 222 metres) and Moncrieffe Hill (725 feet, 221 metres), will reveal to the traveller not only the narrowness of the Perth Gap between these volcanic hills, but also the way in which the Tay channel widens eastwards as it enters the Firth. The valley too widens considerably below the confluence of the Earn, as the steep hill slopes of the Ochils and Sidlaws stand back from the river estuary and its flanking carselands.

The structural history of this tract is noteworthy because it is the best example of a true rift valley in Scotland, if not in the whole of Britain. We have already seen how the massive lava flows and associated volcanics of the Old Red Sandstone volcanic episodes helped to build the Ochil Hills which lie to the south of the Firth of Tay. Research has shown that rocks of the same type also contribute to the stature of the Sidlaw Hills to the north of the estuary. The dip of the Ochil lavas is to the south-east and that of the Sidlaw lavas to the north-west, which demonstrates that they are in fact the opposing limbs of an anticline, known as the Tay Anticline (Figure 23). It would be possible to conclude that the lavas of the intervening crest of the arch have been removed by denudation, which is very often the case in this type of upfold. But the steep, river-facing slopes of the Ochils and the equally steep Braes of the Carse across the valley have been shown to be parallel fault-line scarps at the borders of a rift valley. Thus, the missing volcanics from the highest point of the upfold have not been destroyed but merely downfaulted, being now buried by a layer of Upper Old Red Sandstone, to say nothing of the Pleistocene and post-glacial infill (Figure 23).

A great deal of the superficial material within the Firth of Tay is of post-glacial age, forming the well-known Carse of Gowrie (Plate 16). This flat stretch of marine clay, lying between the Sidlaws and the reed-fringed estuary, is one of the richest and most fertile tracts in Scotland, famed for its abundant corn harvests. Near to the river banks themselves, in both the Tay and the Lower Earn valleys, the soils get heavier and cattle-pastures are therefore more prevalent. But it is for its wheat, barley, sugar-beet, peas, beans and soft fruits that the Carse is best known, so it is little wonder that near-by Dundee has an important preserving and canning industry.

There has been a tendency to compare the carselands of the Tay and Forth with the English Fenlands because of their similar land-use patterns. The analogy is inappropriate, however, for, as Dr Sissons has pointed out, the Fenlands are at sea-level whilst these Scottish carselands have been raised a long way above sea-level by post-glacial isostatic warping. Thus, they are rarely flooded by their rivers, even during the highest floods, and now stand high and dry above the incised river channels. Nevertheless, the Carse has not always been dry, for there is evidence to show that until the draining and agricultural improvements of the eighteenth century the land surface was dotted with pools and marshes because of the impeded drainage of the uplifted marine clays of this post-glacial raised beach. The numerous place names with the prefix 'Inch' (island), such as Inchmartin and Inchmichael, show where there were dry sites on the Carse prior to the drainage schemes.

Curiously, unlike the carse clays of the Forth which they resemble in every other way, the clays of the Carse of Gowrie seem never to have carried a major cover of peat on their surface, so there is no history of peat-cutting around the Firth of Tay. Above the main post-glacial raised beach of the Earn and Tay, which rises in elevation as it is traced up-stream from 28 feet (8·5 metres) at Dundee to 36 feet (11 metres) near Bridge of Earn, several conspicuous river terraces can be seen. Foremost of these is the one on which Newburgh is located and which declines in elevation eastwards until it disappears beneath the post-glacial carse clay as a buried raised beach. Like their counterparts in the Forth, these represent late-glacial features created from the outwash of the decaying Perth Readvance ice-sheet, their staircase of terraces having been carved by the frequently rejuvenated river during the pulsatory uplift of the land as the weight of the ice-cap was gradually reduced (Figure 21).

There is no close agreement concerning the dominant locating factor of the original settlement at Dundee. Some authorities believe that the earliest inhabitants merely moved downhill from the vitrified Iron Age fort which crowns the prominent volcanic plug of Dundee Law (571 feet, 174 metres), another example of a crag-and-tail; others point to the significance of the Stannergate Mesolithic kitchen-midden deposits nearer the shore, for these 'Larnian-type' artifacts imply a much older fishing community. Of one thing we can be sure, however, and that is the importance of a dolerite exposure in the siting of the medieval town, for it provided a dry platform above the marshes of the post-glacial raised beach and the valley of the Scouring Burn. This and the neighbouring Dens Burn now flow underground in culverts, but these two provided the important advantages of power and bleaching ability in the early years of Dundee's textile industry. Of similar importance was the occurrence of local wells, aligned along the junction of the Old Red Sandstone and the intrusive dolerite.

Some 5 miles (8 kilometres) to the north of the city the Sidlaws rise abruptly above the low plateaux of Old Red Sandstone, which hereabouts are dotted with prominent basic igneous hills carved from a variety of petrographic types. The Sidlaw Hills themselves reach only modest heights (1,492 feet, 455 metres) in comparison with the Ochils, partly due to the fact that the Ochil–Sidlaw lava group becomes less thick as it is traced north-eastwards away from the major volcanic centre near Stirling. Thus, a thickness of 6,500 feet (nearly 2,000 metres) in the western Ochils declines to 3,000 feet (900 metres) in the Sidlaws and a mere 700 feet (200 metres) on the Angus coast. Nevertheless, the sudden thickening of the basaltic lavas near Montrose suggests that a separate volcanic centre must have functioned here in Lower Old Red times, although no vents have been found. Because of their base-rich rocks the soils of the Sidlaws, like those of the Ochils, contain important nutrients such as calcium, phosphorus and potassium, so that montane grasslands are more widespread than on the more acid soils of the granitic Highlands farther north. But heather moor is still very extensive on the Sidlaws, and hardy Blackface sheep compete with forestry plantations for the remaining acreage of rough grazing. One hilltop which has survived the extensive afforestation is that of Dunsinane in the western Sidlaws, from which it is possible to look across the expanse of Strathmore to the equally famous Birnam Wood, immortalized by Shakespeare in *Macbeth*.

Strathmore and the Eastern Coastlands

Where the Tay enters Strathmore at the Pass of Birnam this great sandstone vale is some 8 miles (13 kilometres) wide, a width it retains virtually intact north-eastwards through Angus into Kincardineshire for a distance of nearly 50 miles (80 kilometres). We have seen that, generally speaking, this lowland tract corresponds to the outcrop of the Lower Old Red Sandstone. Closer examination of the rocks themselves demonstrates that the dark brown sandstones, grits and conglomerates are virtually identical with those of similar age in South Wales and the Welsh Borderland. It is paradoxical, therefore, that in the latter region the Old Red Sandstone of, for example, the Brecon Beacons is one of the most resistant formations in the area, whilst in Strathmore the sandstone coincides with a zone of lowland. The explanation lies not only in downfaulting along the Highland Boundary Fault but also in the juxtaposition of the even more resistant Dalradian rocks to the north and the tough Sidlaw volcanics to the south.

This is an excellent example of 'differential erosion', where denuding agencies (including ice-sheets) have been able to lower the relatively less resistant sandstones more effectively than the tougher metamorphics and volcanics of the neighbouring regions. It must be remembered, however, that within the Old Red Sandstone certain formations are extremely hard, so that some hill masses have survived even to the south of the Highland Boundary Fault. The most prominent of these is Uamh Beag (2,173 feet, 662 metres), overlooking Strathallan from the north-west and owing its bulk to the resistance of the Old Red basal conglomerates, known as the Dunnottar Group, which in places achieve thicknesses of 6,900 feet (2,104 metres). The Dunnottar conglomerates also form the distinctive line of low foothills which stretch north-eastwards from Blairgowrie through the Hill of Alyth (968 feet, 295 metres) and Tullo Hill (1,031 feet, 314 metres) to Edzell on the North Esk. Between these foothills and the Grampians themselves a narrow, discontinuous linear valley separates the basal Old Red Sandstone conglomerates from the Highland Boundary

Fault. Once more we can see how differential erosion has picked out a narrow outcrop of less resistant Ordovician faulted wedges and Downtonian (Lower Old Red/Upper Silurian) rocks which occur at intervals along the Highland Boundary Fault (Figure 23).

The soils of Strathmore are normally fertile red loams renowned for their high yields of hay, corn and root crops. In addition, the sheltered south-facing slopes of the vale, especially around Kirriemuir and Blairgowrie, support intensive soft-fruit growing, mainly raspberries and peas. It is wrong, however, to look for a direct relationship between the solid geology and the land use, for, except on the highest hills, the land surface carries a thick layer of glacial drift, derived partly from the Highlands and partly from Strathmore itself. Thus, if we are to interpret the present land use it is important to know something of the glacial history of this region.

Three sources of glacial drift have been recognized, although the very fertile shelly clays of the eastern coastlands around Montrose, produced by means of a Scandinavian ice-sheet from the bed of the North Sea, are of only limited extent. More important are the widespread bright red drifts of Strathmore, derived from a major ice-sheet which moved north-eastwards down the vale. Along the Highland border there are signs of a third source of superficial materials, for here a locally restricted south-easterly advance of Highland ice brought grey ground-moraine from the Grampians, together with a good deal of glacio-fluvial outwash, during the melting phase of the Perth Readvance.

Dr Sissons has drawn attention to the extensive 'sandur', or plains of outwash, which occur in Strathmore at the mouths of most of the Highland glens. To the south of Blairgowrie, for example, the moors, woods and golf course between Muirtown and Rosemount indicate the gravelly soils of a sandur, interrupted by the 'myres' and lochs of many kettle-holes. Similarly, around Edzell the forests and abandoned airfield are located on the flat surface of an outwash plain created by the Pleistocene forerunner of the North Esk river system. Here the modern river now follows a channel through a magnificent suite of five terraces, eroded by former meltwater streams in the glacial outwash sands and gravels. Old channels and linear depressions can clearly be seen along the feet of the erosional bluffs which delimit the terrace fronts.

If we were to follow this outwash downstream we would find that its surface declines gently southwards away from the mountain front, whilst at the same time the coarseness of the deposits gets progressively less. An analogy is at once apparent between this glacial outwash of Pleistocene age and the lithology of the Strathmore Old Red Sandstone itself, where the conglomerates are gradually replaced by finer sediments as they are traced away from the Highland border (see p. 132). Although the climatic conditions of the two palaeoenvironments were vastly different, the principles of sedimentation are very similar. If, on the other hand, we trace the glacio-fluvial outwash upstream into the glens themselves they can be seen to emanate from a tumbled terrain of kames and dead-ice hollows which mark the terminal zone of the Perth Readvance valley glaciers. The glaciers emerging from the valleys of the Bran, Tay and Ericht appear to have combined to form a piedmont (see Glossary) ice-lobe in western Strathmore which left its grey Highland till above the red sandstone till of the vale. Farther east, however, the glaciers failed to emerge from Glen Prosen, Glen Clova and Glen Esk during this particular glacial episode.

Wherever the ice-sheets left extensive sandur plains or tumultuous kame-and-kettle topography the land use of Strathmore changes in no uncertain fashion. From the cornfields and cattle-crowded pastures of the damper claylands the landscape alters abruptly to gorse- and heather-clad moorlands and conifer plantations on the well-drained sands and gravels. One of the best examples can be seen to the north of Glamis Castle, where a tract of pine-studded 'muirs and myres' interrupts the rich farmlands between Kirriemuir and Forfar. One interesting point relating to these two sandstone-built market towns is that two of the world's most eminent geological pioneers were born there – John Playfair at Forfar and Charles Lyell near Kirriemuir, although the latter town is better known as the 'Thrums' of the works of J. M. Barrie. The last of the line of market towns in Strathmore is Brechin, on the banks of the South Esk, although flax and jute spinning have turned it into something of an industrial town. Its red sandstone 'Irish' round tower, of similar age to that of Abernethy, now forms part of the cathedral, where its slender form refuses to match the more robust fourteenth-century tower and spire.

Beyond Brechin, Strathmore tapers into a hill-girt basin known as the Howe of the Mearns. The reason this extensive lowland trough fails to reach the sea at Stonehaven is the increasing importance of the volcanic rocks in the succession as the Strathmore syncline is traced north-eastwards. Thus the line of hills along the southern flank near Laurencekirk and the vale's culminating Bruxie Hill (710 feet, 216 metres) at the eastern end are built from lavas. Nevertheless, the larger Strathfinella Hill (1,357 feet, 414 metres) which overlooks the Howe from the north, has been carved from an outcrop of the Dunnottar Group conglomerate (see p. 139). We are now not far from the type-site of this resistant rock, for on the near-by coast of Kincardineshire the ruined Dunnottar Castle, '... swilled with the wild and wasteful ocean', stands atop its 160-foot (49-metre) cliff of conglomerate.

The coasts of Angus and Kincardineshire exhibit an interesting variety of forms which are related to the alternating igneous and sedimentary rocks within the Old Red Sandstone succession. As we might expect, the harder basaltic lavas, intrusive dykes, Old Red conglomerates and coarse grits are the chief cliff-formers, whilst the relatively soft sandstones, shales and superficial deposits correspond with the lower and less spectacular coastlines. To the north of the Tay estuary lies the large triangular raised-beach foreland of Buddon Ness, whose dunes and shingle ridges help to build the renowned Carnoustie golf links. Northwards from here the coast is generally low and featureless to beyond Arbroath, where the appearance of the igneous rocks brings an irregularity to the cliff scenery which we associate with differential erosion by marine agencies. Around the sea stack of the Deil's Heid are outstanding examples of how the jointing of the Upper Old Red Sandstone has influenced wave erosion. The fine blow-hole of Graylet Pot, which terminates a narrow 100-yard (90-metre) sea-cave, is one of the more noteworthy features as we traverse the high conglomeratic cliffs on which Lud Castle is precariously perched. Between here and Montrose the main exposures of the Ochil/Sidlaw lava group reach the North Sea and, since the coastline has been cut obliquely across their various outcrops, the result is a series of alternating bays and headlands. Red Head, with its 250-foot (75-metre) basaltic lava cliffs, is succeeded by the beautiful curving sandy beach at Lunan Bay. The dunes and raised beaches of this attractive haven have been derived from the unconsolidated glacial and glaciofluvial deposits infilling the broad Lunan valley, which itself corresponds with the outcrop of the less resistant red, grey and blue shales and flagstones of the so-called Carmyllie Group of the Lower Old Red succession.

Northwards, however, the axis of the major Tay anticline (see p. 136) brings andesites and basalts back into the picture to create rugged but unspectacular sea cliffs around Fishtown of Usan. The Kincardineshire coast in general repeats the pattern, although its cliffs and raised beaches have the added attraction of the St Cyrus salt-marsh and dune complex, a National Nature Reserve.

Before we leave the Midland Valley of Scotland and at last cross the important boundary fault into the Highlands, mention should be made of the drainage pattern of this important transition zone. In this context few regions have excited so much interest from geologists and geomorphologists as that consisting of the basin of the Tay and its neighbouring streams. This may well be the result of the general discordance exhibited between the rivers and the underlying structural trends. While the general structures run from south-west to north-east (as the Caledonian 'grain'), the drainage is largely from the west or north-west, showing scant regard for the structural grain as it makes its way east and south-eastwards to the North Sea.

Many of the earliest writers (including Sir Archibald Geikie, Sir Halford Mackinder and Messrs Peach and Horne) believed that all the major consequent streams were extremely ancient and originally followed south-easterly courses across an uplifted and tilted peneplain from the North-west Highlands to the ancient Rhine delta. At this period the Midland Valley was apparently unexcavated, whilst the easterly-flowing anomalous rivers were virtually ignored. Later writers, such as Professor Linton and Dr Bremner, formulated a completely new hypothesis in which they asserted that the former drainage was an easterly-flowing one, initiated on a tilted cover of Chalk strata which was subsequently eroded (Figure 24a). In this theory the south-easterly flowing drainage components were regarded merely as secondary developments (generally known as 'subsequent streams') related to the gradual excavation of Strathmore and the Tay 'graben'.

In both hypotheses a great deal of stress was laid on the concept of superimposition from former 'cover' rocks which have now disappeared, leaving the streams partly adjusted to the underlying structures and partly discordant and therefore 'superimposed'. Although it is tempting to regard the anomalous Perth gap and the discordant nature of the North and South Esk as products of superimposition onto the resistant

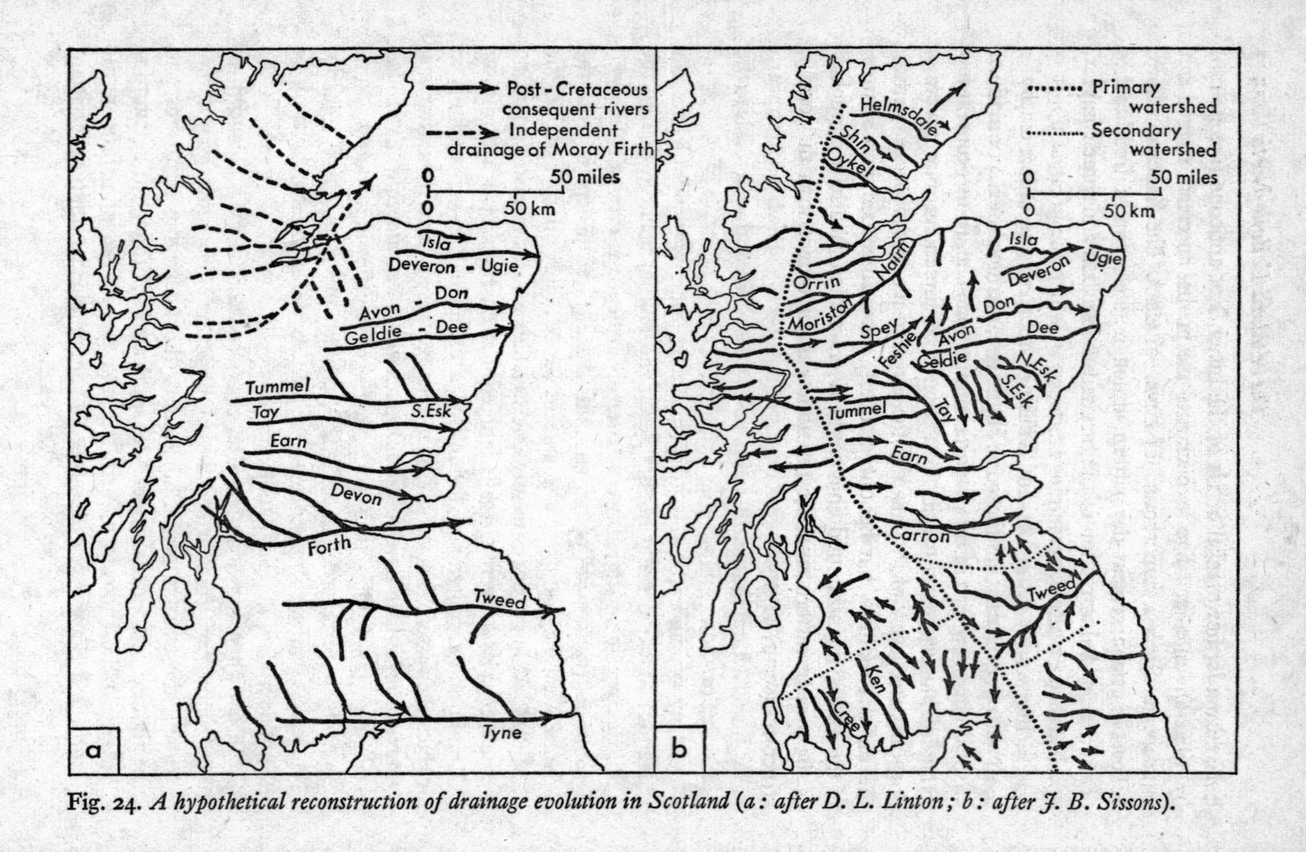

Fig. 24. *A hypothetical reconstruction of drainage evolution in Scotland (a: after D. L. Linton; b: after J. B. Sissons).*

lavas from a former blanket of Upper Old Red Sandstone (as suggested by Geikie), an alternative explanation might be regarded as more satisfactory. Such a hypothesis has been offered by Dr Sissons, who believes that the present-day pattern is not so far removed from that of the initial drainage after all (Figure 24b). The latter may well have developed on a gradually emerging landmass in which the greatest uplift was in the west. This gentle but pulsatory uplift was accompanied by local warping, so that as the consequent streams developed upon the successively emerging coastal platforms during Late-Cainozoic times they extended themselves seawards in a variety of directions, but always down the steepest slopes. Thus the former drainage was *accordant* with the earlier coastlines, but became gradually more *discordant* with some of the underlying structural lines as the rivers incised themselves, much as the present eastern coastline hereabouts is discordant with the local structures (see p. 142).

8. The North-east Coastlands

Kincardineshire is an unusual county in that its boundaries rarely coincide with the lines of physical features. Indeed, it is little more than a transition zone at the junction of three distinct topographical regions, Strathmore, the Grampians and the North-east Coastlands. Nothing reflects this transitional character better than the agricultural landscape, for amongst its mixed farming-patterns the barley and potato crops remind us of Strathmore, the sheep-rearing of the Grampian Mountains and the beef-cattle of the Buchan lowlands. But the observant traveller will also have noticed that as one journeys northwards from the Howe of the Mearns the colour of the ploughed fields changes from bright red to greyish-brown. Were he sufficiently interested to ask the local farmers he would discover that this indicates where the drier, less stony and more fertile red soils of Strathmore give way northwards to heavier, wetter, stonier and more acid soils. As if this were not enough to indicate that we are in fact crossing an important geological boundary, the gradual disappearance of Old Red Sandstone as a building material beyond the town of Stonehaven is further evidence to demonstrate that we are now finally leaving the Lowlands of Scotland as we cross the Highland Boundary Fault.

Since the Fault brings the tougher Dalradian grits, schists and gneisses suddenly into the scene, it is not only the cultural landscape which reflects the transition. The terrain also changes as the long shingly bay of Stonehaven is left behind and the roads climb two or three hundred feet (60–90 metres) onto the windswept, almost treeless cliff-top of the Kincardineshire plateau. Here the foothills of the Grampians reach the North Sea coast, hindering coastal communications and settlement alike. The villages are either clustered along the rock-bound coast or hidden in the narrow and often tree-lined valleys which entrench the plateau. The trim whitewashed fishing village of Muchalls, with its smugglers' cove, vies with Portlethen as a tourist attraction, but the cliff-top village of Findon is more renowned because of its associations with smoked 'Finnan' haddock. The coastal cliffs between Stonehaven and Aberdeen have been described in detail by Professor K. Walton, who has explained the intricate detail of the rock structures and the ways in which these have influenced the formation of the stacks, cliffs, arches and 'yawns' (a type of geo – see Glossary). Not only have the textural differences within the metamorphosed Dalradian rocks been etched out by wave attack, but glacial drifts and linear igneous dykes have also played an important part in creating such landmarks as the Bridge of One Hair and Castle Rock of Muchalls and features with such distinctive names as Arnot Boo and Blowup Nose.

When looked at as a whole, the north-eastern corner of Aberdeenshire, right round to the Moray Firth, is seen to be delimited by remarkably rectilinear coastlines. When we consider how varied is the lithology of the coastal rocks between Stonehaven and Nairn it is difficult to understand the reasons behind this linearity. One explanation may lie in the possibility of gigantic submarine faults making the coastlines equivalent to large-scale fault-line scarps like that between Cromarty and Tarbat Ness (see p. 310). Another explanation could be sought in the relatively ineffective glacial erosion along these coasts, in comparison with that on the western shores of Scotland. It is noteworthy that some authors have even suggested that parts of Buchan remained unglaciated throughout certain episodes of the Ice Age. Herein also may lie part of the answer to a paradox which now faces us: although we have crossed the Highland Boundary Fault and found that here the geology is clearly that of Highland Scotland, the countryside to the north of Aberdeen belies this fact. We shall see that extensive coastal lowlands, with some of the richest beef-cattle farming in Britain, characterize this region where, Professor Walton has concluded, apart from local inequalities of relief, '. . . the complex basement has behaved under erosion almost as if it were homogeneous.' In the succeeding pages there will be an attempt to explain how rocks which make no major contribution to positive relief in places like Buchan can, as they are traced inland, create some of the highest mountains in Britain.

Aberdeen and its Rivers

On our northward journey, no sooner has the quarried granite shoulder of Craighill (344 feet, 105 metres) been passed than the ancient Bridge of Dee can be seen below, backed by the shimmering silvery-grey townscape of Aberdeen, with its pinnacled skyline of spires, towers and multi-storey blocks. As most Scots people know, the city grew from the gradual coalescence of twin burghs, originally located on the neighbouring rivers of Dee and Don. Although the modern suburbs have now expanded westwards onto the granite-floored coastal platforms, the original settlements were greatly influenced by a north–south ridge of glacio-fluvial sands and gravels which separated the dunes and raised beaches of the coast from a series of waterlogged hollows which lay to the west of the present Gallowgate (Figure 25). This gravel ridge has been interpreted as a stage in the recession of the Dee glacier at a time when glacial meltwater was helping to carve the prominent gorge of the Den Burn.

The fishing village of New Aberdeen was to spring up at the southern end of the ridge, where the Den Burn creek of the Dee estuary provided a sheltered haven away from the currents and shifting sandbanks of the main river. After extensive reclamation and dredging, this site of Footdee now forms the nucleus of Aberdeen's important harbour complex. In contrast, the estuary of the Don is constricted and shallow and its outlet is variable, so its port facilities have never been significant. It was here, nevertheless, around the large meander at Seaton Park where the Don cuts through the northern end of the kame-like ridge, that the ancient burgh of Old Aberdeen grew up. On the better-drained gravel slopes of the ridge to the east of Old Aberdeen Loch St Machar's Cathedral was built, its earliest fourteenth-century pillars being constructed from Old Red Sandstone (as were the twin spires), but the remainder from local granite. The Chanonry (the ecclesiastical quarter) and the sixteenth- to nineteenth-century University 'village' also grew up near by, a grouping worthy enough to be scheduled as one of Aberdeen's urban conservation areas.

It is believed that St Machar's Cathedral provides the earliest example in Britain of granite being utilized for large-scale building enterprises, but so widespread did its use become that Aberdeen is now famous as the Granite City. It is interesting to note that the older buildings near the original twin burghs are constructed from roughly quarried blocks or water-worn boulders taken from the rivers. As we move away from these two nuclei, however, it can be seen that the quarried stone becomes more finely worked, culminating in the remarkable façade of the Marischal

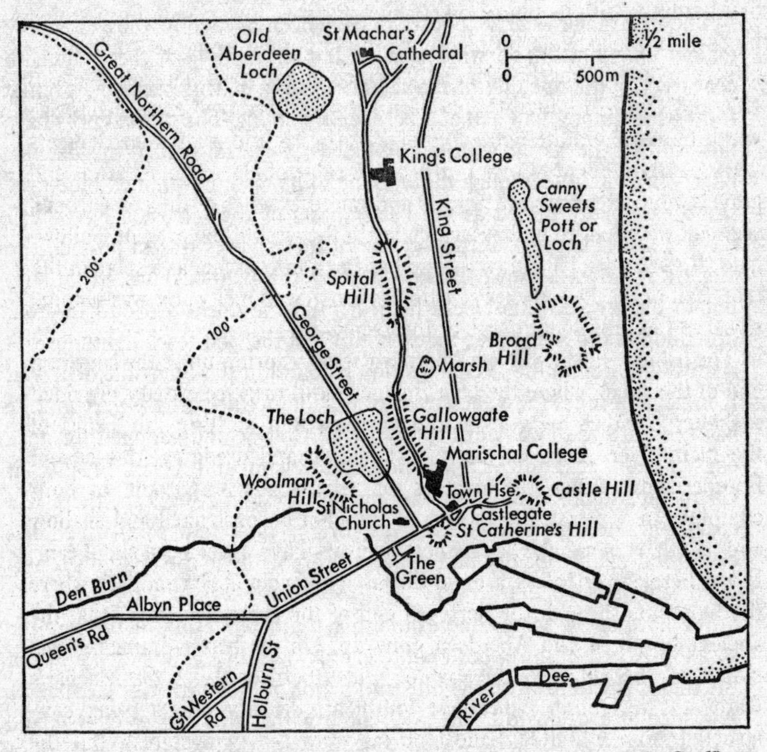

Fig. 25. *The site of Aberdeen in relation to physical features* (*after J. R. Coull*).

College. Up to and during Victorian times most buildings were constructed of solid granite, but because of rising costs many of the later houses of the inner suburbs are only granite-faced. Finally, as costs became prohibitive and granite quarries closed down, the outer suburbs, such as Craigiebuckler, were built of non-granitic materials, much as clay tiles have replaced slates as a roofing fabric. As if to anticipate this change, the spire of the nineteenth-century Triple Kirks was built of brick, in contrast with the glittering array of surrounding granite buildings. It is this 'glitter of mica at the windy corners' that has helped the other granitic minerals of blue-grey feldspar and quartz to create the 'shining mail' of this granite city, although the pink and red granites of Corrennie and Peterhead are sometimes seen.

The first major quarry appears to have been opened in 1604 to supply mainly door lintels and window sills, but the middle of the eighteenth century saw the opening of the famous quarry of Rubislaw, the granite from which more than half of Aberdeen was built. The city suburbs have now spilled out as far as this 465-foot (142-metre) crater (the deepest quarry in Britain), although the working of its medium-grained blue-grey stone has finally ceased as the better parts of the intrusion have been worked out (Plate 17). Fittingly, one of its last uses was to provide the stone for the well-known Bruce statue at the Bannockburn battle site, but in former years its blocks helped build the docks at Southampton, Portsmouth and Sheerness, to say nothing of the Bell Rock lighthouse. A slightly different mineral composition is found in the Don-side quarry at Kemnay, 13 miles (21 kilometres) away. The resulting light-grey granite has been utilized locally at the Marischal College and supplied for dock construction at Leith, Newcastle, Sunderland and Hull. Finally, at Corrennie, some 20 miles (30 kilometres) west of Aberdeen, there is a coarser-grained granitic-gneiss, with bright pink feldspars, which has been used mainly for decorative purposes, notably in the Glasgow municipal buildings.

We have spent some time examining the exploitation of the granitic rocks in Aberdeenshire, but granite is only one, albeit the best-known of the rocks which make up the hills and plains of the north-east coastlands. A glance at a geological map will confirm that this corner of Scotland possesses one of the most complex mixtures of lithologies and structures in the whole country (Figure 26). To understand it we must now look in some detail at the geological history of this tract during the Caledonian orogeny, for the widespread occurrence of igneous rocks (both intrusive and extrusive) is closely associated with this great period of folding and mountain-building.

From more than a century of field-mapping in the Scottish Highlands, and especially in the Grampians, it has become possible to reconstruct the geological events which, despite having taken place some 600 million

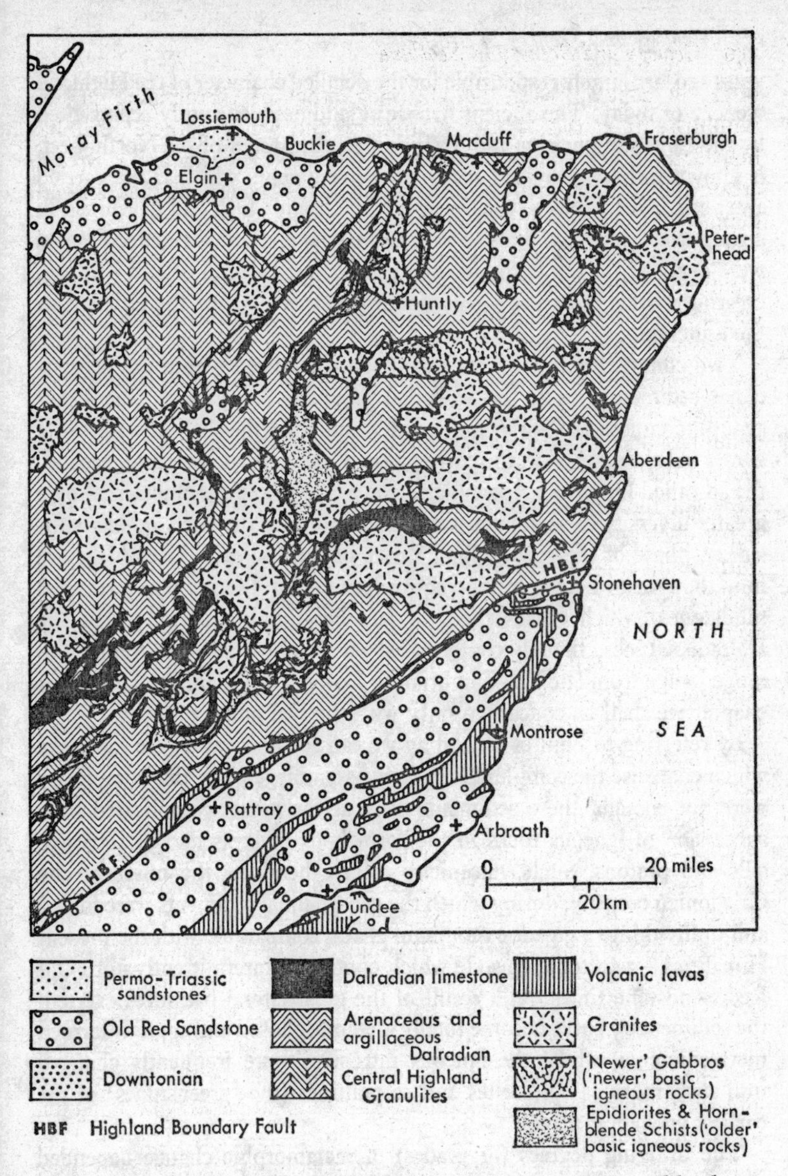

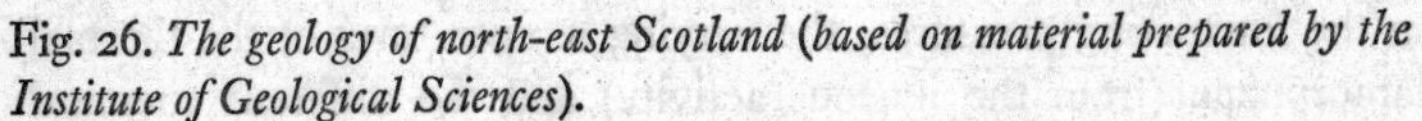

Fig. 26. *The geology of north-east Scotland (based on material prepared by the Institute of Geological Sciences).*

years ago, are largely responsible for the detailed character of the Highland scenery of today. The ancient Archaean landmass, currently represented by the Lewisian gneisses and Torridonian sandstones of the North-west, is known to have provided many of the sediments which rivers carried away south-eastwards to the linear oceanic basin which had developed in a zone stretching from Ireland to Scandinavia. As these various sediments were deposited in the basin, so their accumulated weight caused it to sag, creating what is known as a geosyncline – in this instance, the Caledonian Geosyncline.

Two contrasting rock facies have been recognized within the geosynclinal sediments: first, the so-called Moinian Assemblage (the lower group), composed of fairly uniform deposits from the shallow waters of a slowly subsiding zone in the early phase of the geosyncline; and, second, the so-called Dalradian Assemblage (the upper group), characterized by a greater diversity of sediments accumulated during the later, rapidly subsiding, phase of the geosyncline. Rocks of the Moinian Assemblage are now thought to be the marine equivalent of the Torridonian terrestrial sandstones, which makes them of late Pre-Cambrian age, whilst the Dalradian rocks, though conformable on the Moinian, are thought to range in age from late Pre-Cambrian to Lower Cambrian. In the present chapter we shall be concerned only with the Dalradian rocks.

By referring to Figures 26 and 27 we can see how varied are the rocks which comprise the complex Dalradian Assemblage. As if this complexity were not enough, these sediments were at the outset complicated by a succession of igneous rocks in the form both of lavas and of intrusive sills and plutons. Such vulcanicity merely heralded the onset of the Caledonian orogeny, during which the geosynclinal sediments were folded and uplifted into a gigantic mountain chain, comparable with the present Himalayas – a tectonic episode which continued intermittently until Old Red Sandstone times. As a result of the great lateral pressures exerted, the sedimentary rocks become totally metamorphosed and are now termed metasediments, whilst the igneous intrusives were frequently changed into such rocks as epidiorites and serpentines (the 'greenstones' of the earliest geological maps).

The differing degrees (or grades) of metamorphic change depended upon the fluctuating pressures, the varying temperatures of the thermal metamorphism (from the igneous activity) and the different chemical compositions of the original rocks. Where thermal metamorphism was limited or absent, for example, the rocks would have been affected by pressure alone, often resulting in the development of a slaty cleavage, as in much of Banffshire, where sporadic slate-quarrying was once carried on. Elsewhere, the combined effects of pressure and heat (dynamo-thermal metamorphism) caused recrystallization of the rocks, and platy metamorphic minerals, such as mica, were developed with a markedly linear preferred orientation. Where there was a preponderance of platy minerals the rocks sometimes took on a schistosity, and this was particularly true of the fine-grained argillaceous rocks whose beds became pressed into minutely puckered folds. These mica-schists range in colour from black or grey (often with graphitic properties), like those found between Huntly and the Lower Spey, to a lustrous silvery hue, such as those around Garron Point between Cullen and Portsoy. The less easily deformed siliceous ribs of the coarser bands in the semi-argillaceous rocks, where the minerals were granular rather than platy, became stretched out into lenticles or 'eyes' within the schistose rocks. This was especially true in the so-called 'schistose grits' or 'quartzose schists' of the Upper Dalradian metasediments, which probably represent the former greywackes of the deep sea-floor. Such rocks make up much of the Kincardineshire plateau and the undulating coastal platforms of Formartine to the north of Aberdeen.

The effect of the metamorphism on the quartz-rich sandstones of the geosynclinal sediments was to cause recrystallization into strongly lineated quartzites, whilst the pebbles of the coarser sandstones were elongated and squashed. In north-east Scotland the quartzites are both flaggy and massive and are particularly important in the coastal cliff scenery between Buckie and Cullen and in the low hills to the south of Fraserburgh. Since quartzite is a particularly resistant rock, we shall find that in most cases it forms a conspicuous eminence wherever it occurs in the succession.

The last of the more important metasediments which we can distinguish in the Dalradian Assemblage are the metamorphosed limestones. Generally these occur only as thin ribs amongst the schists, but they can occasionally be found as thick beds of crystalline limestone or marble, although these are rarely of building or ornamental quality because of their dull-grey colour and irregular texture. The best ornamental marbles in Scotland are found only in some of the Western Isles (see pp. 250 and 265).

CLASSIFICATION	BALLAPPEL AREA	SOUTHERN GRAMPIANS NAPPE COMPLEX			AGE
	BALLACHULISH AND LOCHABER	LOCH AWE AND ISLAY	CENTRAL PERTHSHIRE	BANFF AND N.E. GRAMPIANS	
Upper Psammitic Group		Loch Avich Slates and Grits	Leny Grits and Ben Ledi Grits	Macduff Slates	UPPER DALRADIAN
Pelitic Group		Tayvallich Slates and Lavas	Aberfoyle Slates Pitlochry Schists Dunkeld Slates	Whitehill Group	
Upper Calcareous Group		Tayvallich Limestone	Loch Tay Limestone	Boyne Limestone	
Lower Psammitic Group		Crinan Grits and Quartzites	Ben Lui Schists	? Gneisses of Ellon, Downside, etc.	
Pelitic and Calcareous Group		Ardrishaig Phyllites and Shira Limestone	Ben Lawers Schists		
Carbonaceous Group	Cuil Bay Slates	Easdale Slates	Ben Eagach Schists	Portsoy Group	

Quartzitic Group	Appin Phyllites and Quartzites	Islay Quartzite Series	Central Highland Quartzite Series	Durn Hill Quartzite	
Lower Calcareous Group	Ballachulish Slates and Limestone	Portaskaig Conglomerate Islay Limestone	Schichallion Boulder Bed Blair Atholl Series	Sandend Black Schists and Limestone Group	LOWER DALRADIAN
Pelitic and Quartzitic (Transition) Group	Leven Schists Glencoe Quartzite Binnein Schists and Quartzite Eilde Schists and Quartzite	Mull of Oa Phyllites Moal an Fithich Quartzite SLIDE	? Schists and Quartzites of Rannoch Outlier	Garron Point Group Crathie Point Group Findlater Flags West Sands Group Cullen Quartzite	
	Eilde Flags (Moinian)	Lewisian and Torridonian	Struan Flags (Moinian)	Moinian	

Fig. 27. *The Dalradian succession of the Grampians (adapted from J. G. C. Anderson). (Note: There is no direct connection between the Ballappel and the Southern Grampians Nappe Complex.)*

Despite the fact that thick metamorphic limestones crop out on Deeside, near Aboyne, they are of such poor quality that they have little value except as a roadstone. In contrast, the county of Banffshire is well endowed with limestones of high quality for agricultural purposes, so they have been quarried on a considerable scale. It will be realized that in a mountainous environment where heavy rainfall, peat bogs and silica-rich rocks combine to produce extremely acid soils, the addition of lime in some form is essential for satisfactory crop growth. The best-known Dalradian limestones in this region are the so-called Sandend Group, named from their coastal outcrop near Portsoy. From here they can be traced inland as a discontinuous band past Keith and Dufftown for a distance of 40 miles (65 kilometres).

Because many basic intrusions and lavas had become interbedded with the sediments of the Caledonian geosyncline, they too became severely metamorphosed, resulting in the formation of hornblende schists and epidiorites. The earliest granitic intrusions, such as those near Keith and Portsoy, were also subjected to great regional pressures. Some of these so-called 'Older Granites' were converted into gneisses by dynamothermal metamorphism, whilst a second phase of invading granitic material (either as magma or as hydrothermal solutions) diffused through all the existing metamorphic rocks to create what are known as 'migmatites', similar in character to the banded gneisses. The third and final phase of Caledonian igneous activity in this area coincided with the end of the main folding and metamorphic episodes, and the resulting rocks can be grouped into two main types – the gabbro intrusions (see Glossary) and the 'Newer Granites'. Since we have already looked in some detail at the latter, we shall conclude this survey of the Caledonian orogeny by looking briefly at the gabbros of this area (Figure 26).

The newer gabbros occur as thick, sill-like masses which have been intruded into the Dalradian rocks of Buchan and Strathbogie. Their discontinuous outcrop runs in a semicircle from Portsoy, through Huntly and Insch, to Maud in central Buchan (Figure 26). Unlike many of the sills that have been described in earlier chapters, these gabbroid sills have been denuded more easily than the rocks into which they were injected, with the result that they almost invariably correspond with tracts of lower land. More important, since these gabbroid rocks are richer in such minerals as calcium, phosphorous and iron than the surrounding granites and slates, their overlying tills have weathered into extremely fertile soils. Studies have shown that in the Insch valley, for example, the inhabitants as far back as Neolithic times made good use of this fertility, which was added to by the growth in post-glacial times of a broad-leaved deciduous woodland. It is no surprise to discover, therefore, that in the district known as the Garioch a well-farmed lowland basin penetrates deeply past Insch into the Grampian foothills. Framed to the south by the granite of Bennachie (1,733 feet, 528 metres) and the schists of the Correen Hills (1,599 feet, 487 metres), and to the north by the slaty ridges of Foudland (1,530 feet, 466 metres) and Tap o' Noth (1,849 feet, 564 metres), this amphitheatre is now overlooked by the Clashindarroch Forest. Such forestry plantations are found extensively on the foothills above 1,000 feet (300 metres), where the argillaceous slates and schists have weathered into free-draining loams, but where the heavier rainfall has also caused widespread podzolization (see Glossary). Were it not for the fact that there are no large boulders to interfere with 'forest ploughing', these tree-clothed slopes would have been left as open grouse moor and sheep runs.

Before we leave this interesting embayment of Insch, it is important to note that its westernmost end is formed from a narrow basin of Old Red Sandstone corresponding to a narrow lowland which runs southwards past Rhynie to the renowned Kildrummy Castle. This is one of several such depressions (for example, around Glen Livet and Cabrach) that have been created where structural basins of Old Red Sandstone have survived in patches above the Moinian and Dalradian rocks. The disposition of the larger Old Red basins around the Moray Firth leads one to ponder on the former extent of these cover rocks prior to their almost complete removal from the Grampian region. Their relics in this north-eastern corner of Scotland suggest that a blanket of Old Red Sandstone may once have covered all the coastal plateaux. If this were true it would go a long way towards explaining the form and low elevation of the coastlands. We may remember that earlier in this chapter (see p. 147) we noted the problem of why the same granites, slates and schists of the Grampian Mountains fail to produce any positive relief when traced north-eastwards into Buchan. A possible answer may be that these plateaulands of Buchan are little more than the exhumed surface of the pre-Devonian floor after the unconformable Old Red Sandstone 'veneer' has been stripped away.

To gain an impression of the Aberdeenshire countryside there is no better viewpoint than Brimmond Hill (869 feet, 265 metres), to the west of Aberdeen itself. From its summit it is possible to look eastwards over the city, northwards across the Don to the rolling moors and farmlands of Formartine, westwards to the forested granites of Bennachie and Hill of Fare (1,545 feet, 471 metres) and southwards to the famous braes of Deeside. Immediately below us are the so-called 'freedom lands', which were once 'a dreary waste . . . a wilderness of marshy bog and stony crag, a rude primeval, undeveloped heath'. Today, however, these granite hill-slopes support a patchwork of flourishing plantations and well-cultivated fields, all demarcated by gigantic walls of boulders cleared from the fields. The walls are known as the 'consumption dykes', and were constructed from the fourteenth century onwards as these farmlands were painfully reclaimed from the waste. Down on the riverbank is Peterculter, the only industrial community on Deeside, built on terraces of glacio-fluvial outwash beyond the line of the Dinnet ice-limit (a possible local equivalent of the Perth Readvance). The mills are actually on a tributary, the Leuchar Burn, which has cut a rocky gorge as it enters the Dee, having lost its pre-glacial valley, which is now drift-plugged. In the Dee gorge near by, the cataracts of Corbie Linn and the Thunder Hole add character to the sylvan scene.

Many visitors travelling the Deeside roads have compared this royal river with the French Loire, because of its many imposing castles, such as Crathes, Aboyne, Balmoral and Braemar. But our excursion will demonstrate that the scenery of the two valleys is vastly different, given the contrasting geology, soils and vegetation. As we journey upstream through the thick pine forests and granite hills the Dee is fairly restricted in its valley near to Banchory. As we approach Aboyne, however, the hill-sides retreat as the influence of a limestone outcrop makes its presence felt (see p. 156). This is not, however, markedly reflected in changes of vegetation and land use, for not far upstream is the Dinnet kame-and-kettle terrain and thus the haughlands of the valley hereabouts are under-lain by very coarse boulder-strewn outwash from these acid highland drifts. On the north bank the tiny Dess Burn has cut a gorge into the softer limestone where its acid waters leave the Cromar granite exposure, while the Falls of Dess at the head of the attractive gorge are caused by the tough granite margin.

Before venturing into the beautiful valley of Upper Deeside, the visitor would be well advised to make a short detour northwards at Dinnet to see the spectacular Burn o' Vat. Here a small tributary stream, known as the Vat Burn, follows a deep ravine down to a gigantic circular pothole cut 50 feet (15 metres) into the solid granite. This is the Vat itself, and at first sight it seems difficult to believe that so tiny a stream could have produced such a phenomenal pothole and gorge. To understand its formation one must look below the end of the channel to where kames and kame terraces can be traced out along the hillsides above the kettle-holes now occupied by Loch Kinord and Loch Davan. It is possible to conclude that the kames and outwash terraces were created by glacial meltwaters which were also responsible for the excavation of the Burn o' Vat channel.

Not far to the north, where the Howe of Cromar lies cradled beneath the towering hills, the so-called Queen's View is another worthwhile vantage point. Far away to the south the sharp, isolated cone of Mount Keen (3,077 feet, 938 metres) rivals the 'steep frowning glories of dark Lochnagar' (3,786 feet, 1,154 metres), both owing their eminence to the resistance of the Newer Granites. The poet Byron, who lived for a time at Ballater, was equally impressed with 'Morven of the Snow', although he was no doubt blissfully unaware that its 2,862-foot (873-metre) summit and its 'crag-covered wild [above] the billows of Dee's rushing tide' were carved from epidiorites and hornblende schists, rather than from the more famous granite. Such a lithological contrast means that the base-rich soils of Morven give it a grassy rather than a heathery vegetation.

On returning to the Dee valley, it becomes noticeable that the valley floor has now narrowed to some 400 yards (around 370 metres) width as it is traced upstream, for we are now in the Highlands proper, where mountains loom on every side and the river gradient steepens. Thus our excursion must terminate here, where the tremendous waterfalls of the upland glens and some of the finest remnants of the ancient Caledonian pine forests serve to give us a foretaste of the splendours of the high Grampians. Suffice it to say that Balmoral Castle is a final tribute to the building materials of Aberdeenshire, for its light-grey granite walls are roofed by slates from the Hill of Foudland (see p. 157).

The contrasts in land use between the Dee and Don valleys have been highlighted in a local jingle: '... The River Dee for fish and tree, the River Don for horn and corn.' The difference can be accounted for by comparing the acid, bouldery outwash of Deeside with the more fertile and less acid argillaceous soils of Donside. But the well-fenced fields along the Don, with their sleek herds of Aberdeen Angus beef cattle, give way downstream to a ribbon of industrial development which has now linked up Dyce with Aberdeen itself. Where the Don has entrenched itself into the coastal platforms, the numerous rapids between Cothall and Woodside have long been harnessed to provide power for linen, cotton and woollen mills. The abundance of water power and local linen were also responsible for the location of the well-known paper-making industry of Donside.

Once across the Don it soon becomes evident that the soils of the coastal plateaux are more fertile than those of south-west Aberdeenshire. The acid, bouldery granitic soils of Deeside gradually give way northwards to more freely draining loams associated with the schists and basic intrusions of the north-eastern corner. As a direct reflection of this transition, the moorlands and forests of the thinner, poorer soils are gradually replaced by the well-ordered farmlands of Formartine and Buchan. In this prosperous triangle some 20 per cent of Scotland's arable farming is carried on, with root-crops and oats being produced as winter feed for the pedigree beef cattle. Barley is also widely grown because of the distilleries' enormous demands, but we are now too far north for wheat to be successful.

Buchan

Several geomorphologists have studied the coastlands of this north-eastern corner and have concluded that in general it is possible to identify a staircase of coastal platforms between sea-level and 1,000 feet (300 metres). Their flatness and regularity have suggested, furthermore, that they may be of marine origin, for here the drifts are thin enough to reveal that agents of denudation have succeeded in cutting extensive, gently sloping platforms right across the complex lithology and structure of the region.

Although the highest of the platforms, at around 1,000 feet (300 metres) above sea-level, is very extensive and easily recognized as a sort of plinth from which the high Grampians rise, it has been regarded by some as an uplifted and dissected peneplain and by others as being of marine origin, possibly created in Pliocene times. The lower rock platforms, such as the so-called Grampian Valley Benches (between 850 and 1,000 feet, 260–305

1. *Siccar Point, Berwickshire. The unconformity of the nearly horizontal Upper Old Red Sandstone on the underlying steeply inclined Silurian strata was seen by James Hutton in 1788, and helped to establish some of the fundamental principles of geology. The hammer indicates the plane of unconformity.*

2. *The Eildon Hills and the Tweed valley. The rough moorland and steep slopes of the igneous rocks which make up the Eildon laccolith contrast with the well-farmed fields on the gentler slopes of the sedimentary rocks in the Tweed valley, whose meandering river can be seen in the foreground. This was the countryside in which Sir Walter Scott spent his later years.* (*See also Figure 5.*)

3. *The Grey Mare's Tail, Dumfriesshire. The waterfall is part of the stream known as the Tail Burn, which drains a hanging valley before plunging to the glacially overdeepened valley floor of Moffatdale.*

4. *The Devil's Beef Tub, Dumfriesshire, and the Southern Uplands. The deep embayment in the left foreground has resulted from the partial removal of less resistant Permian sandstones from the floor of a Pre-Permian trough carved in the Silurian rocks of the Southern Uplands. Note the grass-covered, treeless hills and the distant eminence of Tinto Hill.*

5. Sweetheart Abbey, Kirkcudbrightshire. This ruined thirteenth-century building is principally constructed from New Red Sandstone. The silvery-grey Criffell Granite, from a neighbouring quarry, has been used as a rubble infilling of the walls and to repair damage. Dressed granite blocks were later used to construct the local cottages, one of which can be seen to the left.

6. *Ailsa Craig, Firth of Clyde. Note the columnar structure of the igneous rock, and the dolerite dykes which cut through these southern cliffs and are now picked out by gullies. The buildings to the right are located on the post-glacial raised beach.*

7. Drumadoon Point, Isle of Arran. The columnar cliffs have been carved from a thick intrusive sill of quartz-porphyry which was intruded in Tertiary times into the horizontally bedded Mesozoic sedimentary rocks. Triassic sandstones may be seen below the sill to the left, although the overlying sedimentary strata have been denuded.

8. *Glen Rosa, Isle of Arran. The glacially deepened U-shaped valley is ringed by Arran's highest granite peaks. The distant ridges, the sharp, cloud-dappled summit of Cir Mhór and the A'Chir ridge of Beinn Tarsuinn* (left) *are notched and gullied where invading dykes have been picked out by erosion. Note the granitic peak of Goat Fell* (right), *and the ice-lowered col of The Saddle at the head of the glen.*

9. *Dumbarton Rock, Dunbartonshire. This prominent volcanic rock, rising sharply from the mudflats of the Clyde estuary, was a defensive settlement long before Glasgow was founded. The industrial settlement along the banks of the Leven River can be seen extending as far north as Loch Lomond, where the bulk of Ben Lomond* (right) *can be distinguished beyond the line of the Highland Boundary Fault.*

10. *Glasgow and the Clyde. This view over Scotstoun, looking down-river towards Clydebank, shows the way the dredged channel has facilitated ship-building. The wooded drumlin of Victoria Park* (right centre) *matches the neighbouring wooded drumlin beyond. Note the far-reaching suburbs on the raised terraces of the Howe, and the distant Kilpatrick Hills.* (*See also Figure 16.*)

11. *Edinburgh. The Castle on its igneous crag can easily be distinguished in the centre foreground, while the sill of Salisbury Crags and the ancient volcano of Arthur's Seat rise beyond the city. Note the location of the railway in the Waverley depression, the geometrical layout of the New Town* (left) *and the Firth of Forth and East Lothian in the distance.* (*See also Figure 20.*)

12. *Shale 'bings' in West Lothian. The tips of red shale are conspicuous reminders of the once flourishing oil-shale industry of West Lothian, which ceased to function in 1962.*

13. *The Ochil Hills and the Forth valley at Stirling. The contrasting topography has resulted from the line of the Ochil Fault, which divides the resistant Ochil lavas, of Old Red Sandstone age, from the Carboniferous sedimentary rocks of the Clackmannan plain* (right). *Note the Wallace Monument on its doleritic crag, and the meanders of the Forth in the foreground.*

14. *Crail, Fifeshire. This attractive fishing village displays the typical crow-stepped gables and pantiles of the eastern coastal settlements.*

15. *The Gleneagles golf course, Auchterarder, Perthshire. In the left foreground a lake-filled kettle-hole contrasts with the tree-dotted, hummocky drift terrain of the golf course. Note the ice-eroded breach of Glen Eagles in the distant Ochil Hills.*

16. *The Carse of Gowrie and the Firth of Tay. The fertile farmlands are located on the carse clays of the uplifted post-glacial shoreline, demonstrating how eighteenth-century drainage schemes have allowed agricultural improvement (contrast Plate 35). Dundee can be seen in the distance.*

17. *Rubislaw Quarry, Aberdeen. The enormous granite quarry is now surrounded by the suburbs of Aberdeen. Its blue-grained rock supplied much of the city's urban fabric.*

18. *The Bow Fiddle, Portknockie, Banffshire. This unusual coastal arch has been carved from beds of steeply dipping quartzite.*

19. *The Culbin Sands, Morayshire. The shifting coastal dunes at Culbin are now largely hidden by conifer plantations. This old photograph demonstrates their character at an early stage of the afforestation. Note the extensive offshore sand-banks and the deflection of the mouth of the River Findhorn.*

20. *The Cairngorms. This view, looking southwards, shows the deep glacial breach of the Lairig Ghru across the main pre-glacial watershed. The summit of Ben Macdhui can be seen in the centre of the photograph and that of Braeriach to the right. Note the snow-patch in Coire an Lochain* (left), *the incision of the stream into the glacial deposits, and the upper limits of tree-growth in the Glenmore basin.*

21. *Rannoch Moor, Argyllshire. The wilderness of blanket bog in the foreground overlies the Rannoch Moor Granite, whose outcrop has been lowered by erosion into a gigantic amphitheatre surrounded by harder metamorphic rocks of Moinian and Dalradian age. These form the distant Grampian peaks south of Loch Tulla (shining in the left background). Note the main road right of centre.*

22. *The Highland Boundary Fault at Loch Lomond. The remarkable alignment of the landforms along the Highland Boundary Fault is demonstrated by the linear islands and the fault-line scarp of Conic Hill* (centre). *The tougher Dalradian rocks of the Grampians* (left) *have here been brought into contact with the less resistant Upper Palaeozoic sedimentary rocks which form the lowland to the right.*

23. The Easdale Slate Quarries, Island of Seil, Argyllshire. Slate-quarrying has ceased in this formerly important village of Ellanbeich, and the quarries are now flooded by the sea. Note the Isle of Kerrera and the Firth of Lorn in the distance.

24. Raised beaches on the Isle of Jura. The late-glacial beach ridges of bare shingle in the foreground have been uplifted to over 60 feet (18 metres), and have in turn been cliffed by waves of the post-glacial raised beach (25 feet, 8 metres) which now stands above the modern high-water mark. Note the quartzite peaks of the Paps of Jura beyond the inlet of Loch Tarbert.

25. *Loch Etive, Argyllshire. This view, looking north-eastwards, shows the headward reaches of this lengthy fjord. The granitic peak of Ben Starav dominates the skyline on the right, while the high summits of the Glen Coe volcanic complex create the distant highlands.*

26. *The Parallel Roads of Glen Roy, Inverness-shire. The three strand-lines of a former glacially impounded lake can be distinguished on the left-hand mountainside. Note the near-by deeply gullied delta-terrace, which formed during the closing stages of the pro-glacial lake's existence.*

27. *Loch Ness, Inverness-shire. The renowned fault-guided valley of Glen Mor is occupied by a number of linear lakes, the largest of which is Loch Ness. This view, looking north-eastwards, shows the glacially oversteepened valley sides.*

28. *Loch na Keal from Ulva, Isle of Mull. In the foreground the columnar basalt of Ulva is strikingly displayed, while the Gribun basaltic precipices of the southern shore, aproned with screes, can be seen in the middle distance. Mesozoic rocks occur beneath the Gribun precipices. Beyond lies Ben More, built from a gigantic pile of Tertiary lavas.*

29. *The Isle of Staffa. The renowned sea-caves of Staffa, including Fingal's Cave* (right) *have been carved in the lower, columnar horizons of a basaltic lava flow. The slaggy crust of the lava now forms the cave roofs and the main plateau surface of the tiny island.*

30. *The Sgurr of Eigg. The long ridge of volcanic pitchstone is thought to have been isolated by denudation of the surrounding basalts, the horizontal beds of which may be seen in the right foreground below the cliffs of An Sgurr. Note the curving base of the pitchstone where it overlies the basalt, suggesting that it may have been extruded into a former valley.*

31. *The Cuillins of Skye. Seen from the south across the waters of Loch Scavaig, the jagged gabbro ridge of the Cuillins may be ranked among Britain's finest views. Mesozoic sedimentary rocks can be distinguished to the right of the cottage, which is part of Elgol village.*

32. *The Red Hills of Skye. The smooth conical peaks of the granitic Red Hills contrast with the jagged outlines of the gabbro in the Black Cuillins.* (*See Plate 31.*)

33. *The Storr, Isle of Skye. The basaltic plateau of north-east Skye is truncated by a series of eastern cliffs where the lavas have been eroded into buttresses and pinnacles, such as the Old Man of Storr depicted here. Below this remarkable pinnacle the terrain is characterized by a jumbled mass of landslips.* (*See also Figure 45a.*)

34. *The eastern cliffs of the Isle of Raasay. These high cliffs are composed of Jurassic sedimentary rocks underlying a cap of basalt which forms the peak of Dun Caan* (left). *Incompetent beds of Liassic shales and clays beneath the calcareous sandstones of the main scarp have assisted the collapse of the overlying rocks to form the major landslips around the lake.*

35. *A* machair *landscape at Balmartin, North Uist. Coastal dunes in the background are largely stabilized by a cover of closely grazed turf. Beyond the road a few striped fields can be distinguished on the 'white land' of the* machair, *but the settlements are located amidst narrow unfenced fields at the edge of the 'black land', where the shell-sand has blown onto the peat. The stone walls* (bottom left) *mark the edge of the rough grazing on the moorland.* (*See also Figure 47.*)

36. *The lake-dotted blanket bog of central Lewis. Most of the lakes in this trackless wilderness occupy hollows in the glacially scoured rocks, where ice-sheets have picked out structural elements in the underlying gneisses. This view, looking southwards, shows Loch Langavat in the distance.*

37. *Beinn Eighe and Liathach, Torridon Highlands, Wester Ross. The almost horizontal layers of Torridon sandstone form the summit ridge of Liathach* (foreground) *and the slopes of Beinn Eighe beyond. The latter peak is capped with a layer of white Cambrian quartzite which creates the conspicuous veil of screes along its summit ridge.*

38. *Suilven, Sutherland. This most spectacular of British peaks has been carved from an outlier of Torridon sandstone which lies unconformably upon a basement of Lewisian gneiss (the actual unconformity can be seen to the left of the photograph). Note the ice-scoured gneissic platform and the neighbouring peak of Canisp.*

39. *Inchnadamph and Loch Assynt, Sutherland. The conspicuous light-coloured outcrop of the Durness Limestone can be traced southwards from the left foreground, behind the settlement and into the prominent scarp. The western limit of thrusting is marked approximately by the line of the road. A complex of Lewisian gneiss and Cambro-Ordovician sedimentary rocks forms Ben More Assynt in the background* (left).

40. Smoo Cave, Durness, Sutherland. This wide cave in the Durness Limestone opens onto the beach but was partly fashioned by an underground river which enters the back of the cave as a waterfall.

41. *Nigg Bay oil-rig construction, Cromarty Firth. The presence of deep water and a flat shoreline has encouraged the location here of oil-rig construction yards. Note the contrast between this massive industrial undertaking and the traditional occupation of fishing* (see foreground) *which has survived in the unspoiled village of Cromarty.*

42. A flagstone fence, Dounreay, Caithness. Few of the Caithness flagstone fences survive today in a landscape which exhibits the impact of twentieth-century technology in the form of the nuclear power station of Dounreay, seen in the background.

43. *The Old Man of Hoy, Orkney. This world-famous sea-stack of Old Red Sandstone rises from a layer of contemporaneous lavas which may be seen in the wave-cut bench just above sea-level.*

44. *Scapa Flow, Orkney. The intermingling of land and water which is so typical of the Orkney archipelago is well illustrated by the deep natural harbour of Scapa Flow. Note the higher hills of Hoy on the western skyline.*

45. *Herma Ness sea cliffs, Isle of Unst, Shetland. The spectacular cliffs near Scotland's northernmost point are made up of steeply dipping beds of Pre-Cambrian gneisses and schists with intervening sills of epidiorite. Note the white staining from the guano of the countless seabirds.*

metres), are also a source of controversy, because as they are traced up the valleys of the Dee and Don they rise in elevation, with every appearance of having been formed sub-aerially by river action. On the other hand, the widespread 'Buchan platform' at around 350 to 400 feet (105–120 metres) O.D. and the 'coastal platform' of some 200 feet (60 metres) elevation have generally been regarded as uplifted wave-cut benches of Plio-Pleistocene age. There is a continuing reluctance on the part of many geologists to accept the implications of such high sea-levels just prior to the Ice Age, although the alternative, of stepped sub-aerial peneplains, with the concomitant process of rapid scarp retreat, is also a difficult hypothesis to conceive. It is on the 'Buchan platform' that some puzzling superficial deposits of quartzite pebbles and Cretaceous flints have been discovered. These gravels, which are overlain by the oldest till of the district, are thought to be of Pliocene age and to represent a possible pre-glacial marine submergence in this area of at least 400 feet (120 metres).

Whatever their derivation, we have seen that the platforms were carved by erosional agents indiscriminately across the complex lithology, but there is one rock type which seems to have resisted the general levelling. We shall see the important effect which quartzite has had on the scenery of the coastal cliffs, and its influence on the inland scene is equally manifest. Wherever the rock occurs a bold hill has resulted, as at the Bin of Cullen (1,050 feet, 320 metres) and the Hill of Mormond (768 feet, 234 metres) behind Fraserburgh. Not only is quartzite resistant to weathering but it is also deficient in soil-forming nutrients. Thus, its sporadic occurrence is marked by strongly leached infertile pockets of podzolic soils which give patches of moorland amongst the rolling farmlands of the region.

It would be wrong to suppose that the landscape of Buchan has always been one of prosperous husbandry, for some two centuries ago these rolling hills were nothing but a barren moorland. The visitor has only to travel the byroads between Mormond Hill and Hill of Fishrie (748 feet, 228 metres), for example, to realize that here is a newly created farming landscape, painstakingly reclaimed from a heathery waste of bogland. The presence of so many planned villages gives us a hint of the agricultural history, for they represent an attempt to resettle many of the Highland crofters who were evicted during the infamous 'Clearances' in the aftermath of Culloden. We can see the degree of success achieved by visiting such places as New Byth, Strichen, Mintlaw, New Deer, Cuminestown or the ambitiously named New Leeds, all created during the so-called 'Miracle of Buchan' between 1763 and 1803. Above all it is perhaps New Pitsligo, perched on its granitic hill, which best epitomizes the stature of the enterprise, for its tree-lined approach-roads have replaced a trackless bog where villagers once dug peats out of their back gardens in an attempt to cultivate the land. Today some of its inhabitants are still employed to 'harvest' the neighbouring bog of Middlemuir, but it is now exploited commercially by machinery, and the peats sent to distilleries to be used in whisky manufacture.

If we were to look closely at the light-grey porphyritic granite outcrops near New Pitsligo, we would see that the granite surface has been broken down to a fine crystalline gravel. It has been claimed that this is an example of deep-weathering, in which certain less resistant mineral components of the granite were destroyed by chemical rotting during an earlier climatic period when the temperatures were considerably higher than at present. The suggestion that the coalescing piedmont ice-sheets (see Glossary) from the Moray Firth, Strathmore and the Dee valley failed to override central Buchan during the last glaciation has led to its title of 'moraineless Buchan'. If this is true, deeply rotted rocks may have survived from an interglacial, or even a pre-glacial, episode, although most modern authorities believe that the disintegrated granite represents nothing more than severe frost-shattering during a period of periglacial activity as Buchan became uncovered by the melting of the ice-sheet. Nevertheless, the long smooth slopes of the hills of central Buchan carry few traces of ice-sheet erosion or glacial drift, whilst the survival of the ?Pliocene gravels suggests that glacial erosion must have been very selective during the Upper Pleistocene.

On turning to examine the coastlands of Buchan, it is not difficult to see a contrast between the high, irregular erosional cliffs of the north-facing coasts, which cut across the strike of the rocks, and the lower, smoother, mainly depositional shorelines facing the North Sea.

From the mouth of the Don to the mouth of the Ythan a 10-mile (16-kilometre) curve of sand dunes has grown up in front of the ancient cliffline of the post-glacial raised beach. So unbroken is the visibility along this low coastline, where neither tree nor hill breaks the monotony of the sandy shore, that it was used by the Ordnance Survey as the base line in the original 1817 triangulation for the mapping of Scotland. The greatest development of dunes on this coastline is found at the well-known Sands of Forvie, where the river Ythan breaks through this miniature Sahara to reach the sea. The Sands of Forvie have been investigated in great detail and a complex picture of ice advances, pro-glacial lakes, fluctuating sea-levels and vegetation changes has been reconstructed by Professor W. Kirk and others. The 4-mile (6-kilometre) stretch of dunes, some of which achieve heights of up to 187 feet (57 metres), overlies a layer of glacial drift which in turn buries the solid rocks. The redness of the glacial tills and sands suggests that they were brought from the Old Red Sandstone tract by glaciers moving along the coast from Strathmore. As the ice-sheet waned, so late-glacial seas invaded the area to a height of 75 feet (23 metres), leaving a series of raised beaches in the Ythan valley between this elevation and 30 feet (9 metres) O.D. during the subsequent phase of falling sea-level. A further rise of sea-level saw the formation of the post-glacial raised beach on which some of the earliest traces of human habitation in Scotland have been found. This final marine transgression brought with it vast quantities of silt and sand which, together with the earlier glacio-fluvial outwash, was then fashioned by southerly winds into seven large east-west 'waves' of dune sand. Although the three southern 'waves' are immobile, the four northern sand masses are still moving northwards as parabolic dunes. Such movements have been fairly well documented throughout historical time as the Errol estate has been gradually overwhelmed. Old ridge-and-furrow patterns are occasionally uncovered from beneath the sand, as are the ruins of Forvie church, which have been exposed for some years. Although legend would have us believe otherwise, the inundations have probably been gradual and spasmodic rather than catastrophic.

Northwards again, past the old Castle of Slains, the schists recur as the main cliff-formers until the pink Peterhead granite begins to dominate. But, before we examine the interesting cliff scenery of the granite, a word is necessary about the tiny Bay of Cruden, where the easily excavated dune sands encouraged the North Sea oil companies to construct their first historic landfall for the Forties Field pipe-line. Excavation and restoration costs would have been extremely high in the neighbouring granite, since environmental concerns necessitated the pipe-line being underground as it crossed the coast. The master jointing of the local porphyritic, flesh-coloured granite (which is highly prized for its ornamental qualities) can be seen to run in a north–south direction, with the secondary joints at right angles. As we have come to expect, marine erosion has carved these joint patterns into an intricate maze of stacks, caves and 'yawns', in addition to the natural arches at The Bow and Dunloss. A good deal of quarrying has disfigured the cliff-top near Boddam, although the spectacular chasm of the Bullers of Buchan is not man-made. Here, where the sea has cut an arch and a cauldron-like blowhole, and where Dr Samuel Johnson was moved to comment on 'the perpendicularly tubulated' rocks, the powers of marine erosion are awesome, especially during an easterly gale.

Although the granitic outcrop extends farther north, its last major influence can be seen at Peterhead itself, where gleaming red granite houses cluster around its harbour and rocky peninsula. Beyond this important oil service port the coastline returns to a series of long, curving dune-fringed strands all the way to Fraserburgh. Professor Walton has made a detailed study of the coast near Rattray Head and has been able to show how longshore drift succeeded in sealing off a former arm of the sea between the rocky outcrops at Rattray Head and Inzie Head. There is evidence that during late-glacial times a bay existed here, the old cliff-lines of which can be traced to the west of the Loch of Strathbeg. Later on, the erosion of an exposure of boulder clay at Inzie Head nourished the southward growth of a long shingle spit, fashioned by waves whose maximum 'fetch' was from the north-east. In time the spit reached almost to Rattray Head, so that during medieval times there was here a harbour of some importance, trading regularly with the Low Countries. By the seventeenth century, however, the narrow outlet had become badly silted, and the entrance was finally sealed about 1720 by a major storm which is reputed to have trapped a shipload of slates in the now isolated lagoon. Thus the freshwater Loch of Strathbeg was formed and the significance of Rattray declined.

The Coastlands of the Moray Firth

Once round Kinnairds Head, where the famous fishing port of Fraserburgh stands four-square against the uncomprising North Sea storms, the coast road quickly takes the traveller past the largely deserted fishing harbours of Sandhaven and Rosehearty. From this point westwards the landscape assumes an altogether different character. The granites and sweeping dune coasts of Aberdeenshire are now behind us and ahead lie the glories of the so-called 'Banffshire Riviera'. This north-facing coastline is almost certainly one of the least known and hence one of the most underrated of the Scottish coasts. Its combination of magnificent cliff scenery, picturesque villages and a riot of wild flowers in the roadside verges is reminiscent of the more famous northern coasts of Devon and Cornwall. The presence of the Old Red Sandstone between New Aberdour and Gardenstown extends this comparison, for the rich red soils which flank the moorlands of Windyheads Hill and the deeply incised valley of the Tore of Troup could quite easily be mistaken for Exmoor. Crouching on the tiny ledges beneath the towering 300-foot (90-metre) cliffs of Troup Head, where metamorphic grits and phyllites crop out, the tiny fishing harbours of Pennan, Crovie and Gardenstown are claimed by many to be the prettiest in Scotland. Each has to be approached by hairpin bends down fearsome gradients where the tough Old Red conglomerates have weathered into slabbed and pinnacled precipices. The long, narrow line of sandstone cottages at Pennan circles a picturesque harbour; at Crovie the dwellings turn their gable ends to the northern seascape; at Gardenstown the niches and ledges of the bright red cliff itself are crammed with a tumbling succession of brightly painted houses perched in improbable positions above the harbour.

To the west of Gardenstown the wall-like form of the coastal cliffs changes as the Old Red Sandstone gives way to the metamorphic rock once more. Now they display more intricate detail, reflecting the less uniform lithology of the metamorphic succession. Banffshire has been described by C. Graham as 'a rugged bridge between the austerities of Buchan and the lovely Laich or Lowlands of Moray'. Because its scenic attraction is found largely along the shoreline it is important to examine the Banffshire Riviera in some detail.

The irregular character of the next 25 miles (40 kilometres) of coastline is due to the rapidly changing lithology and the tight folding of the Dalradian rocks. Structurally, this whole complex succession is part of the recumbent fold of the so-called Iltay Nappe (see p. 171), the upper limb of which has itself been torn off from its 'roots' and transported bodily by large-scale Caledonian thrusting along a basal plane of sliding known as the Boyne Lag. Considering that the whole area has also been affected by faulting and igneous intrusions, it is little wonder that the stratigraphy is so confusing. But, thanks largely to the detailed work of Professor H. H. Read, it is now possible to identify the Dalradian rock groups with some confidence, since the coastline cuts discordantly across the strike of a virtually complete succession of the Upper and Lower Dalradian (Figure 26), making this a classic area for their study.

The complex folding of the Macduff group of metasediments clearly demonstrates the principles of differential erosion. Between Gamrie Bay and the mouth of the river Deveron the precipitous coastal cliffs have coves excavated in the less resistant blue slates and promontories made of the steeply dipping flagstones. Most of the larger headlands, however, are carved from the almost vertical beds of pebbly grit, especially at Knock Head near to Banff itself. Along the foreshore of Boyndie Bay a prominent scatter of large glacial erratic blocks is called the Tumblers, those of black diorite being known locally as the Boyndie Heathens. The town of Banff itself has sometimes been compared with Bath, more because of its Regency town houses than for its actual stonework. This lovely old Scottish town, with its many fine churches and public buildings, is built on three levels which correspond to river terraces of the Deveron. In the Deveron's wooded valley stands the well-known Adam-designed Duff House, 'the most ornate of all Georgian baroque mansions'.

The pebbly grits, limestones and flags to the west of Whitehills produce a coastline of lower elevation, the site of a new nuclear power station, whilst Boyne Bay is carved in an even less resistant limestone which marks the base of the Upper Dalradian succession. Here we cross the line of discontinuity known as the Boyne Lag (see above), for the headland of Cowhythe Hill has been carved from Lower Dalradian gneiss. This same rock also forms the craggy cliffs of East Head, overlooking the picturesque haven of Portsoy, whose trimly painted houses surround the old 'marble' factory on the shore. In reality this 'marble' is a narrow intrusion of serpentine, now worked only for small-scale ornamental purposes. Westwards again, the rapid alternation of the Durn Hill quartzite and the Sandend dark schists results in the crenellated coastline between Redhythe Point and Garron Point. The coast road now sweeps down to Cullen, the colourful 'gem' of the Riviera, where another fine Adam mansion also sports the carvings of Grinling Gibbons. Near by are the 'singing' white sands, whose spherical quartz grains emit a muffled squeak when trodden, and the prominent quartzite sea stacks of the Three Kings. We are now in the realm of the tough Cullen quartzite, which has been sharply folded into steep anticlines and synclines. The quartzite creates not only the isolated wooded hill of the Bin of Cullen but also the spectacular Bow Fiddle rock at Portknockie (Plate 18), whose harbour has been carved from a breached anticline. But ahead lies an almost continuously built-up settlement from Findochty to Portgordon, and after that a very different coastal scene prevails beyond the waters of the Spey.

As we cross into Morayshire, the Dalradian rocks are replaced by those of the Old Red Sandstone, and once again this change is reflected in the topography. Except for the New Red Sandstone cliffs between Lossiemouth and Burghead, solid rock is not seen again at the coast until Inverness is reached. The low coastal plateaux of Banffshire are replaced by the tree-girt, sandy bays of Spey, Burghead and Culbin, one of the finest areas in Britain for studying coastlines of deposition.

The ever-shifting mouth of the Spey, Scotland's fastest-flowing river, displays good examples of channel-braiding. The river has sufficient energy to carry enormous amounts of coarse material from the uplands but where the gradient slackens near sea-level this load is dumped, to be variously worked by stream currents into numerous islands or by marine waves into shingle spits along the coast. Such beach ridges are common along the coasts of Moray and Nairn, frequently deflecting the river mouths to the west (for instance, at Lossiemouth and Findhorn). Superb examples of spit formation can be examined in Burghead Bay, although at Culbin the finest shingle ridges have been obliterated by the thick mantle of conifers which has been planted to stabilize the shifting dunes (Plate 19). Nevertheless, beyond the Culbin Forest there is an insular ridge complex, known as The Bar, which is well worth a visit. A journey along the forest footpaths to the straw-coloured sands and lonely salt-marshes will appeal to all naturalists and to those who wish to escape from the bustle of urban life. The marshes are swarming with bird-life, whilst sea-pinks (*Armeria maritima*), sea-asters (*Aster tripolium*) and sea-spurreys (*Spergularia salina*) give patches of brilliant colour amongst the grey-green salt pans. It appears that dominant waves are causing The Bar to migrate slowly westwards, at the rate of about one mile (1·6 kilometres) per century, leaving the ends of eighteen re-curved spits terminating in the salt-marshes. Such migration once deflected the river Findhorn westwards in a course now marked by the silted Buckie Loch; a major storm then breached the spit in 1702, forming a new river mouth and causing Findhorn to be rebuilt in its present location.

Today's traveller will carry away an impression of endless forests, for between the Spey and the Nairn three major plantations have changed these sandy shores beyond all recognition. It is something of a surprise, perhaps, to see forests growing so close to the sea. This is only possible because the prevailing winds are off-shore, thus mitigating the inhibiting effects of salt spray on tree-growth. The same south-westerlies helped to create the formidable coastal dunes of this tract, as the numerous glacio-fluvial deposits and the post-glacial uplift of the shorelines combined here to create exposed sand flats of great magnitude (Plate 19). Before planting by the Forestry Commission in the 1920s, the Culbin dunes were the most extensive in Britain, but, as with the Forvie dunes (see p. 163), there is a record of constant inundation of surrounding farmland during the Middle Ages, culminating in the gradual burial of Culbin village in the seventeenth century.

On the rock-bound coasts farther east, raised-beach shorelines have played little part in the coastal scene. West of the Spey, however, both late- and post-glacial beaches are very extensive, possibly because of the greater supply of glacial deposits and the greater efficiency of the rivers as agents of transport. The attractive town of Elgin, for example, is located on a delta fan built out by the Lossie into the late-glacial sea amidst a decaying ice-sheet. This sandy flat was subsequently uplifted to 100 feet (30 metres) above sea-level. Raised beaches of similar origin are found near the mouths of all the major rivers along this coast, all up-tilted to the west in the manner of the Forth and Tay beaches.

The post-glacial marine transgression is thought to have inundated the low-lying tract to the north of Elgin known as the Laich of Moray. Here Old Duffus castle, the finest Norman motte and bailey in Scotland, now stands on its mound of boulder clay. The rocky islands created at Burghead and Branderburgh have now become tied to the mainland by reclamation of the intervening raised-beach marsh. The cave-dotted northern cliffs of these former islands are interesting, being formed from Permian and Triassic sandstones, some of which have been quarried for building-stone at the well-known Hopeman quarries. It is noticeable how the land use reflects the changing soils and geology in this tract, the dark arable land of the Spynie flats (where Loch Spynie is the last remnant of the coastal marsh) contrasting with the rocky coastal ridge of sandy, heath-covered soils.

A few miles west of Forres the Findhorn valley runs close to the borders of Nairnshire. It was at this point that Johnson and Boswell noted that 'fertility and culture' were left behind, whilst they reckoned that at Nairn '... we may fix the verge of the Highlands', for here they first heard Gaelic spoken and encountered their first peat fires. The transition is not so marked today, although this is a convenient point to turn our attention to the *real* Highlands. We have skirted the outer ramparts and even ventured through the portals into Upper Deeside, but the time has now come to look at the Grampians themselves.

9. The Grampian Highlands

The mountains and glens of the Scottish Highlands are amongst the greatest glories of British scenery and have been highlighted by poet, writer and artist so frequently that few are unaware of their varied landscapes. The Grampians form the heartland of these well-known uplands, and whether one views them romantically, like Scott, as '... Caledonia! stern and wild ... land of brown heath and shaggy wood, land of the mountain and the flood' or rather more prosaically, like Sir Archibald Geikie, the geologist, their scenery cannot fail to leave a lasting impression. Geikie captures the scenic detail so precisely that his 1865 description is worthy of repetition:

> The mingling of mouldering knolls with rough angular rocks, the vertical rifts [sic] that gape on the face of crag and cliff... the strange twisted crumpled lines of the stratification, the blending of white bands of quartz with dark streaks of hornblende that vary the prevailing grey or brown or pink hue of the stone, the silvery sheen of the mica and the glance of the felspar or the garnets, the crusts of grey and yellow lichen or of green velvet-like moss ... these are features which we recognize at once as distinctively and characteristically Highland.

All that remains is to explain some of the ways in which these important constituents of the Grampian landscape were created.

In the first place, the two main lithological divisions, the Moinian Assemblage and the Dalradian Assemblage (see p. 152), can be contrasted. The former, occurring in the northern and north-western Grampians, comprises well-stratified thicknesses of fairly uniform argillaceous and arenaceous metasediments. The Dalradian, on the other hand, is much more diverse in both lithology and thickness of strata, so that differential erosion has caused a more rapid alternation of topographic detail in the Highland Border tract which stretches south-westwards from Banffshire to the Clyde. Generally speaking, therefore, it is the southern Grampians which exhibit the most diverse relief, whilst the Moinian areas farther north have yielded to denudation in a more uniform manner, producing somewhat featureless plateau lands such as the Monadhliath. It must be remembered, however, that wherever igneous rocks appear in the Grampians the apparently simple relationship between lithology and relief outlined above is certain to be affected. It will be shown, for example, that the granites of Cairngorm and Rannoch make a significant impact on the Grampian scene, as do the volcanics at Glencoe and Ben Nevis (see Chapter 11).

Because of the lack of fossils in the Moinian and Dalradian rocks, their structures have long defied analysis. Although it was soon realized that these rocks had been severely folded and faulted by the Caledonian mountain-building episode, for many years it was impossible to determine whether the succession was the right way up (Figure 28a), was inverted by overfolding (Figure 28b) or was re-duplicated by numerous isoclinal folds (Figure 28c). However, by the use of such criteria as the orientation of current-bedding and graded-bedding (see Glossary) it ultimately became possible to decipher the complex structures and stratigraphy, although there is still controversy among geologists concerning some of the tectonics. Thanks largely to the work of Dr E. B. Bailey, it can now be demonstrated that the Grampians have been carved from a number of enormous recumbent folds, similar to Alpine nappes (see Glossary), often separated from each other by low-angle thrusts, known as 'slides', along which overlying folded strata have been transported bodily across the underlying rocks for considerable distances towards the north-west (Figure 28d).

Initially it was believed that two major structural units could be identified: in the south, the so-called 'Iltay Nappe' ('Iltay' meaning that it belongs to the geographical tract between *Islay* and Loch *Tay*); and, in the Lochaber district of the north, the 'Ballappel Foundation' (from *Ball*achulish, *App*in and Loch *Eil*de), the two units being separated by the 'Iltay Boundary Slide'. It is now thought more appropriate to call the two structural units the 'Southern Grampians Nappe Complex' and the 'Northern Grampians Nappe Complex' respectively (Figure 29). In view of these structural differences between north and south, it has been found convenient to describe the Grampian scenery in three parts: the northern

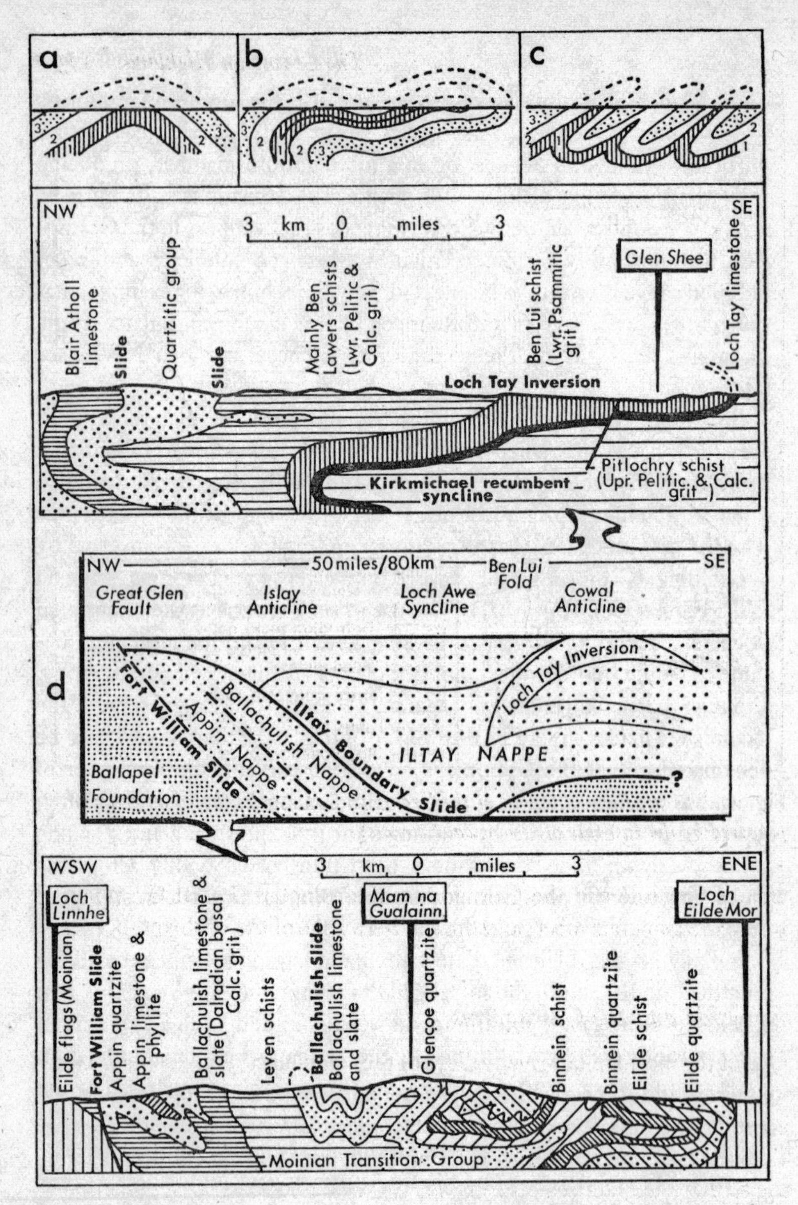

Fig. 28. *The geological structure of the Grampians (after J. G. C. Anderson and T. R. Owen): (a) The succession right way up. (b) The succession inverted by overfolding. (c) The succession re-duplicated by isoclinal folds. (d) Diagrammatic reconstruction of the Iltay Nappe (top) and the Ballapel Foundation (bottom).*

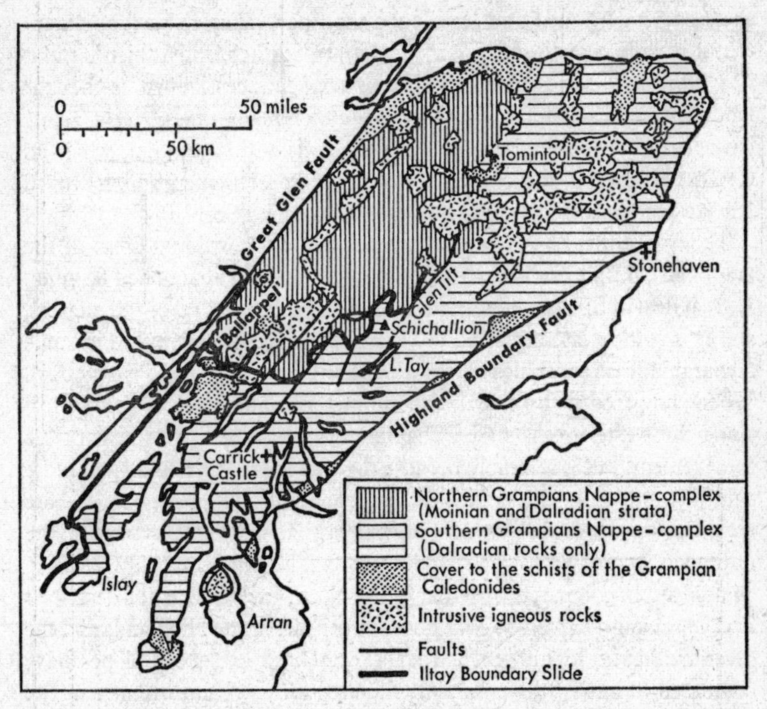

Fig. 29. *The structural divisions of the Grampian Caledonides (based on material prepared by the Institute of Geological Sciences).*

area of Speyside and the Cairngorms; the complex Central Grampians where the two units meet; and the southern zone of the Highland Border.

Speyside and the Cairngorms

One of the most famous river valleys of the Grampians, long renowned for its salmon and its pearls, Strathspey not only serves as a broad avenue of access to the interior but also exhibits a great variety of striking scenery. Like the neighbouring Nairn and Findhorn, the Spey has cut deeply into the schists and granites of the massif, assisted in part by the fact that their courses are closely adapted to the regional strike, so that they appear to follow the SW to NE Caledonian 'grain' for much of their length. Ice-sheets have deepened and broadened the Spey valley to such an extent that it is wide enough to be termed a 'strath', rather than a 'glen', and it now exhibits a series of broad glacio-fluvial terraces known as 'haughlands', which can be traced up the valley for many miles. Such factors have had an important influence on the soils and vegetation, which in turn have helped to encourage a narrow tongue of settlement and improved land to penetrate deeply into the mountainous interior.

Today Strathspey is one of the most densely populated areas of the Highlands, despite its long history of emigration and abandoned farming. A great deal of Speyside lies between 700 and 1,000 feet (210–300 metres), and it is somewhat surprising to visitors to find that cultivation in some favourable areas continues to elevations as high as 1,200 feet (370 metres). When compared with the cultivation limits of some 300 feet (90 metres) along the western coasts (see Chapters 10 and 11) this is a remarkable achievement, but it is largely because the east has lower rainfall and more sunshine and lacks the degree of dissection and preponderance of bare rock due to ice-erosion which prevail farther west. Nevertheless, the improved land of Strathspey, significant as it is, comprises only a trifling amount when compared with that of the North-east Coastlands described in the previous chapter. We are now in the heart of the Highlands and the severer climate, highly leached soils, peat bogs and steep slopes have combined to inhibit the possibility of extensive land improvements. The mere scatter of plots of oats and potatoes dotting the infrequent hayfields has been won from the rough grazings and forests of the valley floor only with difficulty.

Where the Spey leaves the Highlands at Rothes the land use of the valley floor, with its large stretches of arable, is reminiscent of the neighbouring coastlands, while the occasional field of barley serves to remind us that this is one of Scotland's most famous whisky-distilling areas. The east-bank tributaries of Glenlivet and Glenfiddich, for example, are renowned for their malt whiskies, the quality of which is reputed to be greatly influenced by the local geology. Experts claim that water draining off granite through peat gives a superior whisky to that produced either near Elgin, where the water has drained through Old Red Sandstone, or near Keith, where a band of Dalradian limestone affects the quality.

As one journeys upstream towards the district of Badenoch, the stone walls and scattered farms give way to a zone of thick forests, where conifer trees flank the meandering river and carpet the neighbouring valleys and hill-slopes. Grantown stands amid this sea of pines, its long street broadening to green walks between classical stone houses, but interspersed among the darker conifers are large patches of silver birch which help to relieve the sombreness of the pine woods. The birches not only clothe the cleared forest land but also survive on the thinner, stonier soils of the bouldery hillsides. In this zone the valley floor is no longer flat, for a still-stand of the Spey glacier has left a confusion of glacio-fluvial hummocks on which heather vies with whortleberry as a ground cover. Between the kame-like hummocks lie kettle-holes and ill-drained flats, many of which contain lochs, especially in the tract from Boat of Garten to Kincraig. Loch an Eilein with its insular castle, Loch Pityoulish with its climbing woodlands and the attractively embowered Loch Alvie are all tourist haunts, but none is so famous as the kettle-hole of Loch Garten in Abernethy Forest, which has a much-frequented osprey nesting-site.

Above Loch Insh the valley floor changes once more into a broad, flat and almost treeless strath. Here the gradient of the Spey is extremely low, so that the river meanders for several miles across a marshy plain of permanently rush-choked pasture. This river flood-plain lives up to its name so frequently that the Spey has been canalized between artificial levees in an attempt to control the constant inundations. The flatness of the valley floor hereabouts has been explained in terms of a former valley lake of which Loch Insh is the shrunken remnant. Originally delimited downstream by a rocky barrier which crossed the valley near to Alvie, this narrow, ice-overdeepened basin of upper Strathspey became an 8-mile (13-kilometre) ribbon lake similar to the modern Windermere in the Lake District of England. Infilling by sedimentation from upstream, however, has obliterated all but the easternmost end of the former lake basin. Above Newtownmore the headwaters of the Spey abandon the Caledonian trend of Strathspey, and we shall return later to look at the upper Spey in the context of river capture and drainage evolution (see p. 181). It is now time to turn our backs on the valleys and look upwards to the surrounding hills, for here are some of the most renowned mountainlands of Britain.

Before examining the Cairngorms we must glance briefly northwards to the rolling peat-covered uplands of the Monadhliath (Gaelic – the Grey Mountains). Carved largely from uniform layers of mica schists, these flat-topped mountains are almost devoid of corries and are virtually featureless and trackless, a widespread wilderness of peat and moorland rising gently from around 2,500 feet (750 metres) to heights of some 3,000 feet (900 metres), as part of the so-called Grampian Main Surface (see p. 179). The Monadhliath have always been a great barrier to communications – both road and rail routes to Inverness utilize the fortuitous glacial meltwater channel of the Slochd to climb out of Strathspey beyond Carrbridge – and their bulk still forms a physical and cultural boundary between Badenoch and the lands of the Great Glen (see Chapter 11).

In contrast to the Monadhliath, the Cairngorms are granite-cored and corrie-scarred, and form the most extensive area of land above 3,500 feet (1,000 metres) anywhere in Britain (Plate 20). Despite their bulk, they have become one of Scotland's best-known tourist attractions, largely because of their magnificent scenery. In recent years tourists have flocked to their ski-slopes, adding to the climbers and naturalists who in earlier years had seen these mountains as their special preserve. Today, however, the Cairngorms form a National Nature Reserve which, at over 64,000 acres (more than 25,900 hectares), is not only the largest in Britain but also one of the most extensive in Europe. Here a great variety of wildlife flourishes in the numerous habitats, but none is more exciting than the reindeer recently reintroduced on the Cairngorm summits, having become extinct in Scotland some 800 years ago. The 4,000-foot (1,200-metre) plateaux, so similar in character to those of the Scandinavian mountains, produce sufficient growth of the lichen *Cladonia rangiferina* (reindeer moss) to support an extensive herd, but the summits are also the home of an interesting Arctic–Alpine flora, relics of a vegetation which flourished elsewhere in Britain during the Ice Age. Anyone who has seen the beautiful starry saxifrage (*Saxifraga stellaris*) or the bright pink flowers of the moss campion (*Silene acaulis*) growing strongly on the bare, inhospitable granite will treasure the memory.

The rose-coloured Cairngorm granite is part of a gigantic pluton whose outcrop, covering 160 square miles (410 square kilometres), is intruded into the surrounding Moinian metasediments. The red feldspars give the overall colouring to the bare rock exposures, but large quartz crystals and flecks of mica add to the glistening appearance of the coarse-grained granite. During the cooling phase of the granitic magma, cavities became filled with 'smoky' quartz inclusions, varying in colour from yellow to black, and these crystals now constitute the famous 'cairngorm' gem-stones.

Two types of jointing have developed in the granite exposures: a set of vertical joints, which have often been picked out by glacial erosion to form the precipices of the corrie headwalls; and a type of laminar pseudo-bedding which occurs as horizontal sheeting in the surface layers of the rock. It has been shown that in all instances the pseudo-bedding lies parallel with the rolling hilltops of the massif, but that it is clearly truncated by the cliffs of glacial erosion.

Leading on from this observation, Dr D. E. Sugden has demonstrated that a clear distinction may be made between the *pre-glacial* elements and the *glacial* elements of the Cairngorm scenery. He sees the subdued rolling summits, with their tors and their deep layers of regolith or rotted rock, as survivals from a pre-glacial landsurface in which the granite was exposed to deep chemical weathering under a warmer climate than that of today. Thus, such features as the smooth summit of Cairn Gorm (4,084 feet, 1,245 metres), the bouldery plateau of Ben Macdhui (4,296 feet, 1,310 metres) and the tors of Beinn Mheadhoin (3,883 feet, 1,184 metres) were created before the Pleistocene and were little modified by an ice-sheet which subsequently moved across them in a north-easterly direction.

In contrast, the landforms of glacial erosion were selectively impressed upon this landsurface when the ice-sheet's power to erode was greatly influenced by the configuration of the existing topography. Glacial troughs, such as Glen Einich, Glen Geusachan and that which cradles Loch Avon, were excavated along those of the pre-glacial drainage lines that were suitably orientated to take the north-easterly discharge of the Grampian ice-sheet. Where pre-glacial valleys stood at right angles to the ice-movement, however, as in Glen Quoich, Glen Derry and even the north-south element of Glen Geusachan, the valleys managed to retain their original fluvial form. That is not to say that they remained unfilled by ice, for the great gash of the well-known Lairig Ghru (Plate 20), across the centre of the pre-glacial watershed, was bulldozed by glacial diffluence when ice filling Glen Dee could find no suitable outlet to the north. Another good example of glacial breaching can be seen at The Saddle, to the east of Cairn Gorm. Here an ice-sheet flowing down Glen Avon lowered the former divide at the head of Strath Nethy, thus sending an ice-stream northwards as well as north-eastwards.

Making an analogy with existing ice-sheets, such as those in East Greenland, Dr Sugden suggests that watershed-breaching on this scale is unlikely to have resulted merely from the local Cairngorm ice-cap, which created only the corries and modified the major glacial features at a later date. He also claims that the Cairngorm corries show evidence of having been formed at different stages of the Ice Age, the older, larger corries having been overrun by the regional ice-sheet subsequent to their excavation. These larger corries are usually grouped around north-east-facing pre-glacial valley heads, such as those which isolate Braeriach (4,248 feet, 1,295 metres) from Cairn Toul (4,241 feet, 1,293 metres). The younger corries, however, exhibit no such preferred orientation, being found on the flanks of the troughs and even within the larger corries themselves, on west-, north-, and east-facing slopes. Two examples of such composite corries are those which carry the well-known lingering snow patches below the Cairn Gorm–Cairn Lochan summit ridge (Plate 20). Deeper and more precipitous than the ski-slopes of Coire Cas, these spectacular hollows of Coire an-t-Sneachda and Coire an Lochain carry tiny moraine-dammed lochans (small lochs) which are backed by great amphitheatres of bare granite slabs. Less accessible but more remarkable is the corrie lake of Lochan Uaine, which is cradled in a scoured rock basin beneath the precipices of Cairn Toul, at over 3,000 feet (900 metres).

There has been a great deal of controversy about the late-glacial history of the Cairngorms during the phases of their deglaciation. Some authors believe that the Cairngorm ice-cap disappeared progressively in a single phase of down-wasting, interrupted only by a slight re-activation during the late cold period associated with the Loch Lomond Readvance. Others believe that during the last interstadial, or break in the glacial conditions, the Cairngorms, together with the rest of Scotland, became totally ice-free before the rapid re-birth of a new ice-cap (equivalent to the Loch Lomond Readvance) created widespread hummocky moraines, meltwater channels and outwash terraces. It seems probable that on the ice-free ground of the Highlands much periglacial activity would have taken place during the late glacial, with frost-shattering freshening up the summit tors, cliffs and screes, whilst major solifluxion (see Glossary) transformed the gentler slopes. But as the climate was ameliorated the Cairngorms, like other Scottish massifs, saw a gradual change from periglacial to fluvial activity, with many of the former solifluxion phenomena becoming buried by an accumulation of post-glacial peat. Nevertheless, the climate of these high plateaux has remained the severest in Britain, and it is no surprise to discover that large stone 'polygons' and 'stripes' have continued to form at intervals, especially during the so-called 'Little Ice Age' of the eighteenth-nineteenth centuries, and that this 'patterned ground' (see Glossary) may be forming on a small scale even today.

It is now possible for the tourist to reach the summit of Cairn Gorm quite easily by ski-lift, and there is ample opportunity to view many of the phenomena described above if adequate precautions are taken in this Arctic–Alpine environment. There is no better viewpoint than these high plateau-tops from which to ponder the concept of uplifted peneplains, because from here the roof of Scotland appears as a vast dissected table-land, with all the highest Grampian peaks rising as isolated residuals above an accordant 2,400–3,000-foot (730–915-metre) plateau of the 'Grampian Main Surface' (the 'High Plateau' of Peach and Horne). Although there is no agreement so far as their genesis is concerned, most authors are agreed that the complex Caledonian fold structures have been planed right across, to create such a remarkable accordance of summits and spurs that it has been said of the eastern Grampians that they exhibit more flat ground on the hilltops than on the valley floors.

From the high tops it is also possible to look down on the far-spreading forests which envelop the north-western flanks of the Cairngorms like a blue-green ocean. In the basin of Glenmore, remnants of the old natural Scots Pine forest have survived around Loch Morlich and in the Pass of Ryvoan, although newly planted stands include many acres of spruce as well as pine. By a study of the ancient pine roots beneath the mountain peat it has been possible to show that the natural tree-line of the Cairn-gorms reached an elevation of some 2,600 feet (790 metres) during the post-glacial climatic optimum, four to five thousand years ago. Today, however, because the climate is cooler the forests extend only up to the 1,500-foot (460-metre) contour (Plate 20). Like the neighbouring Aber-nethy Forest, those of Glenmore and Rothiemurchus were once part of the great Wood of Caledon, which formerly extended 'from Glen Lyon and Rannoch to Strathspey and Strathglass and from Glencoe eastwards to the Braes of Mar' (D. Nairn). But the axes, fires and domestic livestock of mankind have depleted the Grampian pine forests as surely as the climatic deterioration.

The earliest burnings were connected with Norse raiding about a thousand years ago, although in this area many of the woods were destroyed by the notorious fourteenth-century rebel, the Wolf of Badenoch. Nevertheless, their systematic exploitation took place in succeeding centuries when trees were felled for boat-building on the Spey as well as for charcoal in the local iron-smelting farther west. But the greatest single detrimental factor appears to have been the 'Coming of the Sheep' during the Highland Clearances after Culloden. Thus by 1790 flockmasters from the Southern Uplands had completed their occupation of the Grampians, forcing the unfortunate inhabitants to emigrate or move to the crofting townships of the coastal margins. Hundreds of thousands of acres of moorland were burnt regularly, with the sheep ultimately destroying the tree seedlings which survived. This was especially true of the native birch woods which grew on the margins of the pine forests.

The Central Grampians

To the south of Strathspey the road and railway climb steadily across the wastes of Badenoch to Drumochter Pass. Here, in the very centre of the Grampians, away from the forests and the spectacular Cairngorm cliffs, there is a chance to concentrate on the three elements which constitute so great a part of the Highland scene – the ubiquitous moorland, the mountain stream and the trackless bogland.

Dull and monotonous most of the year, until the heather transforms them into a blaze of colour, the Scottish moorlands have a character of their own: lonely and desolate to some, silent and romantic to others. It will be remembered that in the Southern Uplands heather moor is uncommon, for, as in the western Highlands, the grasses and sedges have ousted it from its dominant position. But in the eastern Highlands heather dominates the acid soils on the sweeping slopes of moorland and peat bog alike up to heights of some 2,700 feet (820 metres), above which only blaeberry and crowberry survive as darker patches in the tawny colour of the *Nardus* grasslands. Three species of heather abound – *Calluna vulgaris* and *Erica cinerea* on the drier sites, and *Erica tetralix* in the damper situations – and these have been perpetuated in the eastern Highlands by regular moor-burning. In the wetter western Highlands, however, continued burning is tending to eliminate heather in favour of its grass and sedge competitors.

Today's systematic moor-burning to improve grazing-value merely perpetuates the wanton forest and scrub destruction of earlier centuries which, more than anything, is responsible for the emergence of the Highlands as a landscape of moorlands rather than forests. Modern rotational moor-burning is practised as much for the benefit of the grouse as for the sheep, since the red grouse is very dependent on heather as its major food supply. Deforestation has extended the territory of the grouse partly at the expense of the red deer, which prefers the forested or scrubby habitats at certain seasons of the year. Since burning for grouse in the wetter west has merely reduced the amount of heather moor, the 'Glorious Twelfth' means more in the Grampians than it does in the Western Highlands.

Turning now to look at the streams of the central Grampians, we cannot fail to be struck by the fact that once away from Speyside the drainage pattern exhibits very little regard for the underlying Caledonian structures. Apart from the valley of Loch Laggan, which is virtually a south-west extension of the line of Strathspey, or the fault-guided trench of Loch Ericht, the majority of the river valleys run discordantly across the structure. Certain writers, notably Professor D. L. Linton, have concluded that the original drainage of the Grampians, and indeed of most of Scotland, was dominated by east-flowing rivers. This writer has also shown how in the Central Highlands, the headwaters of the east-flowing Don have been captured by the Avon, and similarly those of the Geldie–Dee by the Feshie–Spey (Figure 24a and b). Is it also possible that the discordant west–east section of the Spey headwaters may once have flowed eastwards before being captured by the piratical lower Spey working rapidly headwards along the strike? Endless speculation is possible, since farther north the Isla could be linked with the Ugie by means of the middle Deveron, whilst farther south the Tummel aligns with the South Esk (Figure 24a).

Professor Linton believed that the initial streams once flowed eastwards on an unconformable cover of Chalk deposited on a marine-cut surface which trimmed the underlying Caledonian structures. He further believed that the highest Scottish peaks are remnants of such a surface that was tilted gently towards the east, thereby initiating the Scottish river system. After denudation had removed the Cretaceous cover, the drainage would have become superimposed onto the underlying rocks (thus explaining the present discordance), but would subsequently have been dismembered by streams such as the Spey which had already become adjusted to structure. Unfortunately for this interesting hypothesis, it is possible to show that while many discordant streams do flow eastwards, an equal number of equally discordant rivers flow in other directions, notably those flowing south-eastwards to Strathmore (Figure 24b).

Other writers, such as Professor T. N. George, have attempted to explain the drainage anomalies by recourse to the idea of a major marine submergence of the Scottish massif in mid-Tertiary times. According to this hypothesis the various erosion surfaces, such as the Grampian Main Surface and the Valley Benches, were formed during pulsatory uplift. At the same time the rivers were initiated, flowing in a variety of directions across the newly emerged shore platforms towards the slowly evolving coastlines and being constantly rejuvenated into their sometimes discordant courses as the landmass was tectonically uplifted (see also p. 143).

The third element in a typical Grampian landscape is the peat bog. Distinguishable from the raised bog and valley bog of lower altitudes, the extensive blanket bogs of the Scottish Highlands add to the desolation of the lonely upland plateaux and basins. We have already encountered numerous examples of raised bog in the Scottish Lowlands (see Chapters 6 and 7), where the peaty carselands are known as 'mosses'. It is now necessary to explain how the type of peat bog which develops in any area will depend on the plant association which has created it and how this in turn depends on the mineral nutrients available in both the ground-water and the rainfall.

The raised bogs of the lowlands and the blanket bogs of the Grampians and the Northern Highlands differ not only in form but also in genesis and age, for they were created in different ways. Let us look first at an example of raised-bog formation.

After the ice-sheets had retreated from the Midland Valley, many badly drained basins of boulder clay were left behind. In addition, the former coastal mud-flats of calcareous clays in the Forth estuary were soon to be uplifted by post-glacial isostatic warping. Together, these waterlogged landscapes provided an ideal environment for the growth of fen-peat. The large tussocks of certain fen plants appear ultimately to have built up the surface of the fen above the level of the alkaline ground-water, so that the natural acidity of the decaying vegetation was no longer neutralized.

This allowed the bog moss known as *Sphagnum* and other plants, such as cotton grass (*Eriophorum vaginatum*), which thrive only in more acid water habitats, to gain a hold on the fen surface. In this way the peat grew upwards as a raised bog, the surface of which became typically convex and which now required sufficient atmospheric humidity if its upward growth was to be maintained. It must be emphasized, however, that raised bogs grew by vegetational increments as described above and that their title 'raised' is not related to the fact that some of them occupy raised shorelines, as on the carselands; many raised bogs are to be found in areas which have not been tectonically upwarped.

On turning to examine the blanket bogs, we find that, unlike raised bogs, they have remained largely independent of ground-water supplies, while depending more on high rainfall and atmospheric humidity. Thus the Central Highlands and the western seaboard of Scotland exhibit ideal conditions for the growth of blanket bog, which has become a typical vegetation type, or 'climatic formation', in these heavy-rainfall areas. As it did not need to arise from local fen basins, the ubiquitous acid peat gradually blanketed much of the terrain (except on steeper slopes) during the cool, wet phases of the post-glacial: hence its name – blanket bog.

The featurelessness of many of the landscapes of Badenoch, Atholl and the Monadhliath is partly explained by their extensive blanket bogs; these areas also stand in an important transition zone in relation to the bog-vegetation character. To the west of this zone the main constituents of the blanket bogs are ling, bog myrtle, cotton grass, moor grass (*Molinia caerulea*) and deer's hair grass (*Scirpus caespitosus*), in addition to the *Sphagnum* mosses. As we move into the lower-rainfall areas farther east, however, the blanket bogs will be found to possess much more ling, crowberry, blaeberry and cloudberry – especially on the plateau summits of the Cairngorms and Lochnagar – whilst bog myrtle is usually absent.

One of the best areas to view blanket bog is the Moor of Rannoch, acclaimed by many as one of the most desolate yet awesome landscapes to be found anywhere in Scotland. Today a journey by rail or road across Rannoch Moor takes the traveller swiftly across the dreary wastes of dun-coloured grasses, flecked with grey granite exposures, blotched with black peat-hags and dotted with pools of dark water (Plate 21). But travellers who know their Robert Louis Stevenson will vividly recall its description in *Kidnapped*: it was virtually trackless, and 'a wearier looking desert man never saw.' Although Rannoch Moor is forbidding to most visitors, it has been designated a National Nature Reserve, since its inner recesses are nearer to being a true wilderness than most other areas in Britain.

On examining the geology of the area, it is somewhat surprising to discover that the great amphitheatre occupied by the Moor is floored by a medium-grained grey granite, whereas the surrounding mountains are composed mainly of schists and quartzites. In earlier chapters we have noted that granite has been responsible for many of the highest uplands, whilst to a layman it is generally regarded as the least yielding of rocks. For what reason, therefore, does the Rannoch granite correspond with a basin-like topography, albeit at an average elevation of 1,000 feet (300 metres)? It is not merely a matter of age, for other Caledonian granites help to build the Cairngorms and Lochnagar. The answer seems to lie in the lithology of the rocks which surround the perimeter of its pluton: to the east and north are the Moinian quartzites, to the south the Dalradian quartzites and quartzose mica schists, to the west the volcanic rocks of Glencoe (see Chapter 11). All of these peripheral rocks exhibit a resistance to denudation greater than that of the particular granite in question, so that once a shallow upland basin had been fashioned here, the ice-sheets of Pleistocene times would rapidly have removed the thick accumulations of rotted rock as they overdeepened the – relatively – 'weak' granitic terrain.

It is with some relief that we turn our backs on this spongy, streaming scene and move eastwards from Rannoch railway station, retracing the so-called 'Road to the Isles' along the valley of Tummel and Loch Rannoch. The lochside circuit is one of the finest in the Highlands, with the peaks and wooded braes reflected in the sun-dappled waters of Loch Rannoch. On the steep north-facing slopes the well-known Scots Pines of the Black Wood of Rannoch represent the largest remnant of the great Wood of Caledon to have survived south of Rothiemurchus, although they barely escaped wholesale felling at the beginning of the last century. Resin and turpentine from their red trunks were once used for primitive lighting in the Central Highlands, before paraffin was imported from the West Lothian oil shale refineries (see p. 116). The south-facing slopes of the valley are not as steep as those opposing them across the loch; indeed, they are called in Gaelic *An Slios Min* – The Side of Gentle Slopes. The contrast in gradient is partly reflected in the land use, for patches of improved land are common on the gentler (and also sunnier) south-facing slopes.

The asymmetry of this west–east valley is largely a result of the geological structure here, for, whilst the slopes to the north of Loch Rannoch are virtually dipslopes, lying almost parallel with the forty-degree south-easterly dip of the strata, those to the south of the loch oppose the angle of dip and are therefore modified escarpments. Here the shapely peak of Schichallion (3,547 feet, 1081 metres) dominates the southern skyline, its conical form resulting from the steeply dipping, tough Dalradian quartzite (Figure 29). Its summit is easily accessible from the minor road which cuts obliquely across its northern flanks, which may explain why in 1777 the Astronomer Royal chose this peak for experiments to discover the weight of the Earth. Some 600 feet (183 metres) above the road, careful examination of the rock exposures of these northern slopes will reveal a linear outcrop of the Schichallion Boulder Bed, a conglomerate which occurs low down in the Dalradian succession (Figure 27). Embedded within a layer of metamorphic grit, numerous cobbles and boulders of 'erratic' lithology can be seen, all exotic to the rocks of the Central Highlands. In its general disposition the Boulder Bed is exactly like an indurated boulder clay, and it is now usually regarded as a metamorphosed till of the same age as that at Portaskaig in Islay (see p. 208). This Pre-Cambrian 'boulder clay', or tillite, is thought because of its wide distribution at a constant horizon to have been deposited from floating icebergs.

We are now in the realms of the Iltay Boundary Slide, where the Northern and Southern Nappe Complexes meet (see p. 171). Not only is the succession complicated by overfolding and thrusting, but it has also been severely sliced through by major north-north-east trending wrench or tear-faults (Figure 29). Lateral movements of up to 5 miles (8 kilometres) took place – mainly in Hercynian times, for igneous dykes of Lower Old Red Sandstone age have been off-set by the faulting. The latter has often resulted in the formation of belts of shattered rock, which in turn have been picked out by agents of denudation as zones of weakness. It is therefore no surprise to find that the alignment of Glen Tilt with the kink in Loch Tay, as well as the linear shapes of Loch Ericht and Loch Laidon (across Rannoch Moor), is due to the presence of major faulting (Figure 29).

Loch Tummel and Loch Rannoch are not fault-guided and have often been used as an example of the discordant drainage pattern of the Central Highlands. When viewed in its entirety, this 30-mile (48-kilometre) valley can be seen to change its form from wide lake-filled alluvial basins to stretches which are rocky, narrow and constricted. Examples of the latter can be seen both upstream and downstream of Loch Tummel where the hard Schichallion quartzites cross the valley; they made it difficult for the valley glaciers to cut a uniform U-shaped trough from west to east. Nevertheless, the Rannoch valley was seen by Professor Linton as one of fifteen major glacial troughs in the south-west Grampians which illustrate the concept of radiating valleys in glaciated lands (Figure 30). He was able to reconstruct the former limits of the most powerful centre of ice-dispersal in Scotland by plotting the stream-lines of a former ice-movement which '... accommodated itself in several instances to the existence of ready-made depressions following the northeast–southwest Caledonian strike of the Dalradian rocks'. He noted that, if the existing valley pattern would not allow radial ice-dispersal, glacial breaching would create new valleys by overriding watersheds. Examples of this type of glacial breaching can be seen in the main east–west Grampian watershed itself, now occupied by the through valleys of Loch Ericht and Loch Treig (Figure 30).

The Highland Border Country

To the south of the main Grampian watershed, now crossed at Drumochter Pass (1,484 feet, 452 metres) by the highest railway in Britain, the river Garry flows discordantly south-eastwards towards Strathmore, joining first the Tummel and then the Tay near Pitlochry. In so doing it crosses from the Moinian schists onto the Dalradian rocks near Blair Atholl, where thick layers of grey and white limestones are interbedded with the slates and schists. The effect of these less resistant rocks on the scenery is quite striking, for, below the desolate moorlands around Drumochter, Glen Garry descends into the wide, fertile basin of Atholl. Although green-veined marble occurs sporadically in near-by Glen Tilt, the main Blair Atholl Limestone is quarried only for agricultural purposes and road construction. The effect of the calcareous rocks on the soils and vegetation, however, is manifest: the grassy slopes surrounding the beautiful park-

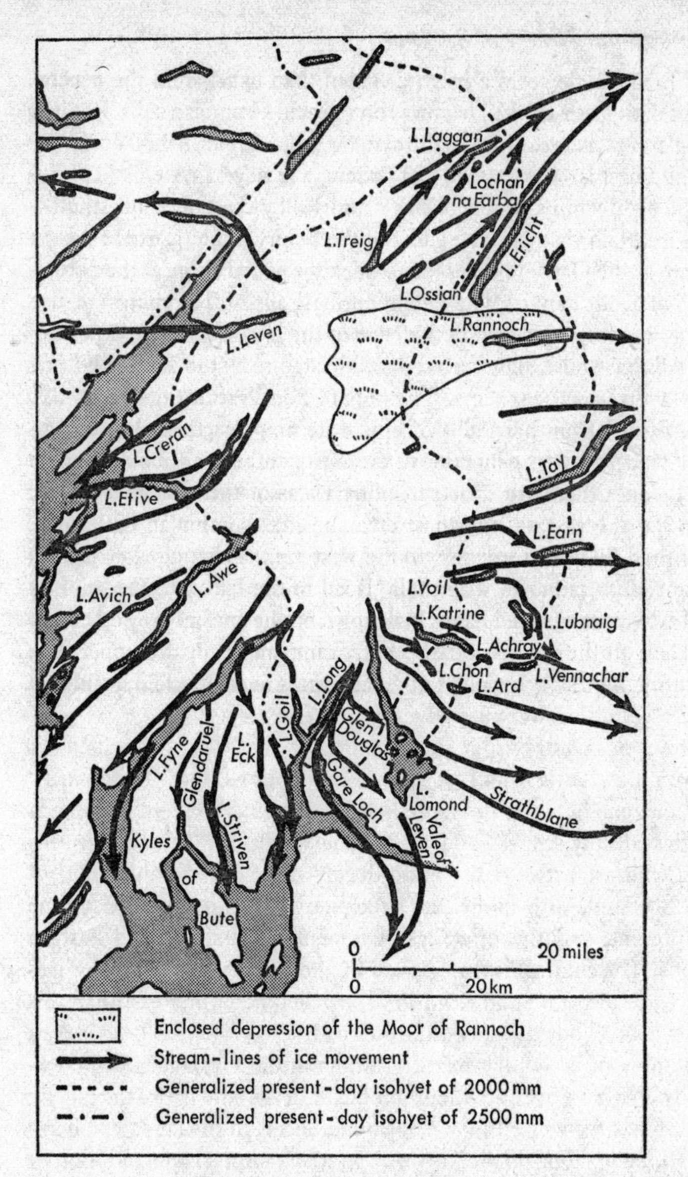

Fig. 30. *The influence of ice-movements on the pattern of glens and water bodies in the Western Grampians (after D. L. Linton).*

lands of Blair Castle seem a brighter green than usual amid the general picture of dark, rush-choked pastures on the acid Grampian soils. But it is the woodland which really takes the eye, for around Blair Atholl occur the oldest and finest larch plantations in Britain, a legacy of systematic planting in the eighteenth and nineteenth centuries by the Dukes of Atholl.

A few miles downstream the effect of the Schichallion quartzite is seen once more as the Garry plunges into a thickly wooded gorge at the notorious Pass of Killiecrankie. The neighbouring Falls of Tummel were also created as a knick-point held at the edge of the tough quartzite exposure, but the falls have now, alas, been reduced in stature by the Tummel–Loch Faskally hydro-electric scheme. The peak of Ben Vrackie (2,757 feet, 841 metres), carved from hard sills of epidiorite amongst the schists, is an excellent viewpoint at the junction of the Garry and the Tummel; from its heathery slopes there are wide-spreading vistas of the Highland Border country. To the east and north stretch the high Grampian tablelands, deeply scored by glacial troughs; to the west are the shining ribbon lakes of Tummel and Rannoch, with their 'Road to the Isles'; to the south is Strath Tay, whose forested flanks look down on the ancient grey cathedral of Dunkeld; to the south-west the burrowing waters of the upper Tay thread through its sinuous loch to the farmlands near Aberfeldy, and it is in this direction that we must now turn.

The tract of country from Breadalbane to Aberfeldy, dominated by Glen Lyon, Ben Lawers and Loch Tay, is thought by Dr Joy Tivy to mark a transition zone between the more heavily glaciated Western Highlands and the less deeply scoured and eroded landscapes farther east. She sees this as a contrast between the more deeply dissected 'weather side' of western Scotland, with its heavier precipitation both in the Pleistocene and at present, and the drier, less fragmented massifs of the Eastern Highlands. The contrast is emphasized by the fact that the last major ice-advance (the Loch Lomond Readvance) reached no farther east than this transition zone. Thus the outwash from the last ice-front has been carved into a number of broad, flat terraces, which in the Fortingall–Kenmore–Aberfeldy area have been an important factor in the land use of the region.

Dr Tivy has recognized four major relief facets in the landscape hereabouts which epitomize the broader straths of this Highland Border country, where fingers of lowland farming and settlement infiltrate into the mountainland. The first of these is the 'haughland' of the alluvial floodplain and its associated river terraces, where the relatively fertile soils support crops of oats, turnips and hay – fodder for the cattle and sheep of these virtually 'lowland' farms. The second facet, known as the 'braes', comprises the rocky benches and patches of hummocky glacial drift which flank the valley floor, where variable soils and slopes give generally poorer grazing land on which smaller marginal stock farms wage a battle against the encroaching bracken and scrub. Such rough grazing is even more characteristic of the glacially oversteepened slopes which constitute the third facet, but this land is now generally blanketed with Forestry Commission plantations which, in the early unenlightened days, frequently cut off the valley stock farms from their mountain grazing above. It must not be forgotten, however, that before the Commission's conifers began to dominate the hills, patches of semi-natural oak and birch wood (including Burns's 'Birks of Aberfeldy') clothed these valley sides, while the massive deciduous trees of Weem and the splendid yews of Fortingall are renowned for their longevity. The final facet of these Highland Border landscapes is made up of the open plateaux and summits above the tree-line. These peaty uplands with their acid soils and heather moor have been described earlier in the chapter, but here the narrow outcrops of the Blair Atholl and Loch Tay Limestones bring sweeter soils and bands of greener vegetation to the Highland scene. The limestone exposures can often be picked out not only by the vegetation contrasts but also by abandoned limekilns and the sites of former 'shielings', which were the temporary summer dwellings of the crofters and their livestock during the once important but now defunct practice of transhumance.

Nowhere is the influence of calcareous soils on the mountain flora better seen than at Ben Lawers (3,984 feet, 1,215 metres), whose lofty summit towers above Loch Tay. Now owned by the National Trust for Scotland, this Grampian giant is known to thousands of visitors because of the easily accessible information centre high up on its western slopes. But before its recent popularity Ben Lawers had long been famous amongst botanists because of the richness of its Arctic–Alpine flora. In this case the vital factor is not so much the limestones themselves as the calcareous Ben Lawers Schists and their associated epidiorite sills, which have weathered into the basic minerals conducive to the growth of a calcicole (calcium-loving) flora. Furthermore, the softness of the rocks and the steepness of the cliffs have ensured a constant replacement of mineral salts as compensation for the heavy leaching of these mountain soils. The Arctic–Alpine flora which was common in much of southern Britain during the later phases of the Pleistocene has now been ousted by forests and 'warmer' species elsewhere; its last outpost is here on Ben Lawers, where it survives only by virtue of the favourable ecological factors. Today its major enemies are man and his attendant flocks.

From Ben Lawers it is only a short step to an even more famous tourist haunt in the Highland Border country, The Trossachs. Visited by countless thousands of people because of their easily accessible scenic charm, The Trossachs are often referred to as the 'Scottish Lake District'. The rugged mountains, the tree-clad slopes and the sparkling waters of Lochs Katrine, Achray and Vennacher are grouped in that pleasing combination of rock, wood and water that has appealed to all types of visitors from the Wordsworths onwards, although the latter compared Loch Katrine unfavourably with Ullswater. But, as the Lake District has its supporting poets, so The Trossachs have no less a bard than Sir Walter Scott, who immortalized the area in *Rob Roy* and *The Lady of the Lake*. The literal meaning of the name 'Trossachs' is 'The Bristly Place'. When Scott writes, 'The rocky summits, split and rent, formed turret, dome or battlement', he is making a perceptive appraisal of the geological character hereabouts some fifty years before Archibald Geikie attempted to explain such scenery. Geikie was the first to see that actual lithology was more important than absolute height in the creation of high-quality scenery. He says of The Trossachs: '... where the rocks are harder and more quartzose they give rise to a gnarled craggy outline, and though the mountains may not be lofty ... their ruggedness gives them a wildness which almost compensates for their want of height.'

The major relief has been carved from steeply dipping Dalradian schistose grits which here form part of the inverted limb of the so-called Iltay Nappe. The two most important members are the Ben Ledi Grits and the Leny Grits, which together help to build the twin sentinels of The Trossachs – Ben Ledi (2,873 feet, 876 metres) and Ben Venue (2,386 feet, 727 metres). Much of the lower, wooded ground between the Loch Ard Forest and Loch Achray has resulted from the appearance of thick beds of red, purple and green slates within the Dalradian succession, since these are generally less resistant than the intervening grits. The slates have long been worked as a roofing material in the quarries near Aberfoyle, where their sixty- to seventy-degree dip on the skyline illustrates how weathering and erosion can produce a 'bristly' detail even in rocks which are reputedly of poor resistance to denudation. It is the steepness of the dip, allied with the rapid alternation of lithology, therefore, which helps to give the region its rugged character, but Pleistocene ice has also played a significant part.

The water bodies of The Trossachs are part of the radiating pattern of glacial troughs recognized by Linton and referred to above (Figure 30). Loch Chon and Loch Ard follow one radiating line, whilst Lochs Katrine, Achray and Venachar follow another. Glacial scouring has over-deepened the valleys at all these localities, but it is noteworthy that wherever the grits occur the valleys narrow (as at the Pass of Leny and at The Trossachs Hotel) where hard rock bars cross the valley floors. Like Lochs Ericht and Treig farther north, however, Loch Lubnaig, near Callander, occupies a depression cut right across a pre-glacial watershed by diffluent ice unable to find a suitable exit elsewhere.

A glance at Figure 30 will reveal that Scotland's best-known stretch of water, Loch Lomond, is also part of the radiating pattern of glacial troughs created by the most powerful centre of ice-dispersal in Britain. Its shape reflects this genesis in part, for the narrow, linear northern stretch of the loch is a repetition of all the water-filled troughs of the south-west Grampians, whether fresh-water lochs or sea-lochs like neighbouring Loch Long. But as Loch Lomond approaches the Highland Boundary Fault at Balmaha it changes from a glacial trough 600 feet (180 metres) deep to a broader, shallower lowland lake a mere 75 feet (23 metres) in depth. The main change occurs at Luss, where a zone of slates crops out, and although the Leny Grits cross the loch at Strathcashell Point it is not long before the Old Red Sandstone fashions the scenery at the southern end of the loch. Here the wooded islands with their backdrop of uplands are reminiscent of Derwentwater, although the geology is very different. The alignment of the landforms along the Boundary Fault is particularly noteworthy, the sharp ridge of Conic Hill (1,175 feet, 358 metres) continuing south-eastwards through Balmaha and the islands of Inchailloch, Torrinch, Creinch and Inchmurrin to the western shores (Plate 22).

Away to the north, the Grampian hump of Ben Lomond (3,192 feet, 973 metres) – the most southerly mountain over 3,000 feet (900 metres) in Scotland – blocks the northern skyline, but the low southern shores, with their prosperous farmlands and neat oak woods, mark the sudden transi-

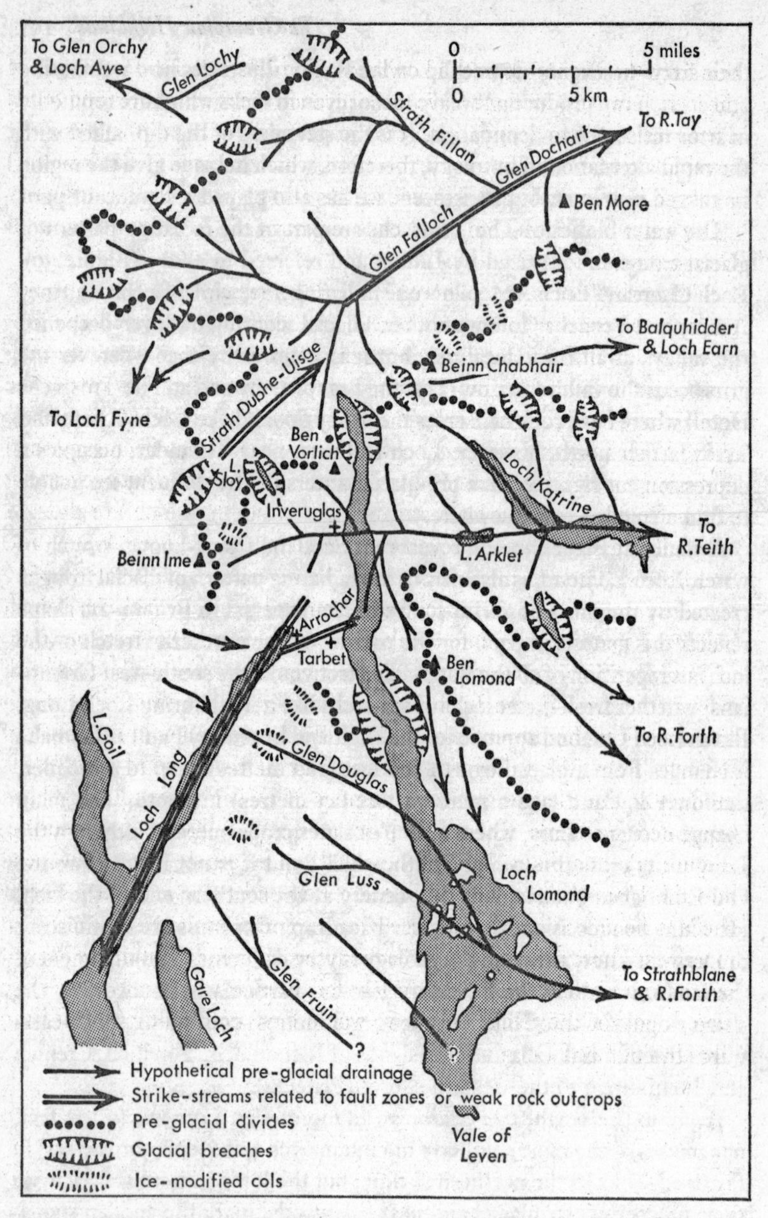

Fig. 31. *A hypothetical reconstruction of the evolution of Loch Lomond (after D. L. Linton and H. A. Moisley).*

tion from the Highlands to the Lowlands. To understand the formation of the loch, however, we must leave these pleasant southern shores and travel northwards into the craggy, pine-clad steeps near to the northern end. At Inveruglas, where the waters narrow to less than one mile (1·6 kilometres) across, a power station makes use of the falling water from a tributary valley (in which Loch Sloy has been impounded) to generate some of Glasgow's electricity.

The work of ice is everywhere manifest. Its greatest feat was to trench through the former watersheds hereabouts, thereby cutting off the pre-glacial headwaters of Inveruglas from their former eastern outlet through Inversnaid to Loch Katrine. The original west–east valley now carries the artificial reservoir of Loch Arklet in a 'hanging valley' 500 feet (150 metres) above Loch Lomond's present trench (Figure 31). It has been suggested that similar glacial interference also disrupted the former drainage pattern of the Douglas and the Luss streams farther south, which are thought originally to have flowed south-eastwards to the Endrick Water and Strathblane. Their former courses have, however, been obliterated by the erosional capability of the major Loch Lomond Readvance glacier, which bulldozed southwards before fanning out as a piedmont lobe to the south of Luss about 10,300 years ago. An exact contemporary of the Menteith glacier (see p. 120), the Loch Lomond glacier left its mass of terminal moraines near the present lake outlet in the Leven valley to the south of Balloch. Thus, the valley now occupied by Loch Lomond is largely a product of the Pleistocene ice, a number of earlier valleys having been integrated by some 2,000 feet (600 metres) of glacial down-cutting (Figure 31). There is evidence to show that just before the last ice advanced into the basin the sea flooded into the Lomond hollow, leaving marine shells to be picked up by the succeeding ice and deposited in the terminal moraines. These morainic ridges have succeeded in holding out the present ocean from Loch Lomond, but the lake height is only 27 feet (8 metres), so were it not for the glacial debris we would have here another linear fjord like that of Loch Long.

10. The Lands of South Argyll

Once we are past Loch Lomond we can sense the presence of the sea. To the west the Grampians become fragmented as the oceanic environment permeates the scene. Sinuous channels of sea-water infiltrate the soaring peaks, whilst bony fingers of mountainland poke resolutely south-westwards into the Atlantic waves. Here is the ancient kingdom of Dalriada, which has given its name to the Dalradian rocks of the geological succession. Peopled by an ancient folk whose affinities lay, not surprisingly, with Ireland rather than Scotland, these peninsulas and islands of south Argyll were once a focal point in the seaways of the Celtic 'fringe', at a time during the Dark Ages when civilization had virtually disappeared elsewhere in Europe. The area now possesses such a vast number of archaeological treasures that it has been referred to as the historic cradle of Scotland, remaining the seat of Scottish power until Kenneth Macalpin conquered the Highland Picts and moved his capital to Perthshire in the ninth century A.D. Today the islands, hills and glens of ancient Dalriada are scattered with old churches, Celtic crosses, megaliths, duns, vitrified forts and carved stones as a testimony to its former importance. But these same hills and glens are now virtually deserted; only the tiny coastal settlements survive amidst the ubiquitous moorlands and scattered forests of this mountainous terrain. It is a land where the geology is as complicated as the remarkable coastline, a land of rocky knolls, hidden lochs, forested braes and tiny cultivated fields, where every turn in the road brings a fresh vista of seaweed-draped shorelines, rugged islands and lingering sunsets reflected in the restless ocean waters.

We have seen in the previous chapter (see p. 171) that the Southern Grampian structures are dominated by a gigantic recumbent overfold known as the Iltay Nappe. This same major overfold can be traced south-westwards into south Argyll, where the upside-down succession forms the

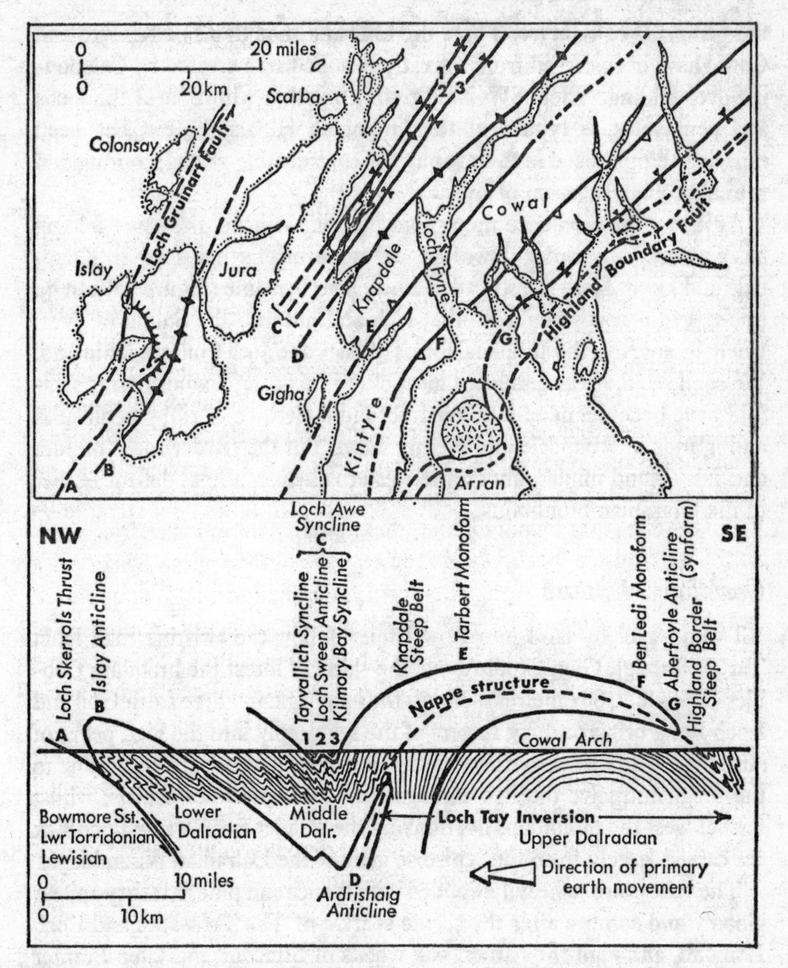

Fig. 32. *The structure of south Argyllshire (after J. L. Roberts). The notation of letters and numbers on the map refers to the geological section below.*

so-called Cowal Arch (Figure 32). The axis of this arch or antiform can be traced from the northern end of Loch Lomond, through Ben Vorlich (3,088 feet, 941 metres) and Ben Ime (3,319 feet, 1,012 metres) in Arrochar, south-westwards through the hills of Cowal and across Loch Fyne into the peninsula of Kintyre. Thus, the hills and mountains of Kintyre and Cowal have been carved from a greatly denuded arch created by Caledonian over-folding. Their SW–NE 'grain', together with that of the lochs and peninsulas, is typical of the Grampian Highlands, but has been especially emphasized in this area by the considerable glacial scouring and the subsequent invasion of the sea.

As if to counterbalance the upfold described above, the same folding has produced a complex downfold, or 'synform', farther north in Knapdale and Lorn. Here the so-called Loch Awe Syncline (Figure 32) can be traced south-westwards from that loch and out into the Sound of Jura, where it separates the anticlinal forms of Islay and Jura from the mainland. We shall see that the less resistant rocks of this complex northern downfold have been denuded into low hills and a plethora of ribby peninsulas and islands in Knapdale and Nether Lorn. But the harder rocks of Jura and Islay stand implacably on the western skyline, a final defiant outlier of the Grampian mountains.

Cowal and Kintyre

All who travel by land into Cowal must follow the twisting road from Tarbet through Glen Croe. Above the skirts of forest the brooding Cobbler (2,891 feet, 881 metres) surveys the lonely glens where Loch Goil and Loch Long bring probing fingers of the sea deeply into the high peaks of Arrochar. The distinctive shapes of the mountains here are partly due to the tough intrusive diorites amongst the quartzose mica schists, whilst farther west the mountains overlooking the slender fresh-water Loch Eck are carved largely from the schistose grits of the Dalradian Assemblage.

The valleys are so heavily wooded with spruce and pine that they appear gloomy and sombre after the scenic sparkle of The Trossachs and Loch Lomond, although the natural oak woods of Strachur and Glen Branter bring some relief. The tiny farms are few and far between in this region of steep slopes, acid soils and heavy rainfall; except on the narrow raised beaches, improved land is rarely found in the southern peninsulas of Cowal. Two exceptions occur, however, on the eastern and western perimeters. In the east, around the smart tourist resort of Dunoon, a belt of more friable slaty soils has encouraged the growth of market gardening and the presence of dairy herds on the pasturelands in response to the demands of tourism. Along the western shores, facing Loch Fyne, the Loch Tay Limestone reappears and this, together with the iron- and magnesium-rich rocks of the so-called Green Beds (whose colour is due largely to the mineral chlorite), helps to create a narrow belt of flourishing agriculture. This is best seen between Otter Ferry and Ardlamont Point, where the normally sour pastures of rush and sedge give way to bright green grasslands and ploughed fields. On the whole, however, Cowal will be remembered for its forests – the Argyll National Forest Park was the first of its kind to be created in Britain, some forty years ago.

When the broader stretches of Loch Fyne are reached, colour returns to the landscape. In addition to the multi-coloured fodder crops and white-washed farms of the wider raised beaches, the shoreline contributes its red, brown and green covering of sea-weeds on the grey, knobbly rocks. The tiny fishing villages which dot the coastline remind us of the former importance of Loch Fyne in the history of the Scottish herring industry. But after a nineteenth-century boom the herring has now virtually disappeared, leaving the former fishing cottages as modern holiday homes for nearby Clydesiders.

Other important industries, too, have gone from the shores of Loch Fyne. The first of these gave its name to the village of Furnace, on the north shore: during the eighteenth century iron-smelting was commonplace on this forested south-western seaboard. Following the establishment of ironworks at Bonawe in 1711 and at Invergarry in 1730, that on Loch Fyne was set up in 1754, depending largely, like its forerunners, on imported Cumberland ore. Although some local haematite was used, from the valley of the Leacainn river, the location of the industry was decided by the extensive natural woodlands of oak, ash, beech, birch and holly, all excellent for producing charcoal. In addition, although the deep sea-lochs provided a cheap means of transport, the short-term exploitation meant that thousands of acres of Scottish woodland had been lost before the Carron works at Falkirk began to utilize coal instead of charcoal and steam instead of the ordinary water-power of the burns (see p. 118). The neighbouring village of Crarae is also noteworthy, since its quarry in the intrusive porphyrite of Beinn Ghlas (1,377 feet, 420 metres) provided many of the setts which now pave the streets of Glasgow.

Like its smaller eastern counterparts in Arrochar and Cowal, the great sea-loch of Loch Fyne is an example of a fjord. In contrast with the broad estuaries of Scotland's eastern coast, the mountainous western seaboard is deeply funnelled and scored by these sinuous water bodies (Figure 30). It has long been realized that they represent the drowned seaward ends of a pre-glacial drainage system that has experienced severe overdeepening by ice-sheets during the Pleistocene, the ice-sheets having been centred much closer to the western than to the eastern coasts. Thus, the major fjords of Loch Fyne and Loch Long have been deeply eroded along the 'grain' of the Caledonian folding by ice from the important dispersion centre of the south-west Grampians, noted in Chapter 9. Loch Awe follows a similar orientation, although the post-glacial marine submergence has failed to inundate its deeply ice-scoured trough. Farther south the change in direction of Loch Fyne, together with the Kyles of Bute, Loch Striven and Loch Goil, has been taken as evidence of a former south-easterly-draining river network now drowned by the post-glacial transgression. It seems extremely speculative, however, to link these south-easterly-trending fjords with the evolution of the Upper Clyde and Tweed, as some earlier authors have done. It is surely better to regard them merely as a glacially modified branching network of a normal west-flowing drainage system.

Where Loch Fyne reaches the open sea to the north of Arran, the long, narrow peninsula of Kintyre protrudes some 40 miles (65 kilometres) out into the Atlantic to the south of the fjord of West Loch Tarbert, which almost cuts through the neck of the peninsula. Composed essentially of the same lithology and structures as Cowal, the backbone of the peninsula nevertheless fails to achieve the high relief of its Grampian neighbour across Loch Fyne. For this reason many would regard Kintyre as being outside the Highlands, despite its location to the north of the Highland Boundary Fault. Indeed, as if to demonstrate this transitional character, its southern limits include exposures of Old Red Sandstone and Carboniferous rocks. Furthermore, from a historical and sociological standpoint Kintyre has few affinities with the Highlands, as befits a region in the same latitude as Ayrshire. Since the Campbells of Argyll chose to settle the peninsula with successive plantations of Lowlanders, it is not surprising that the traditional Highland way of life is missing. Thus, crofting villages will not be found, since they have long been replaced by medium-sized farms and large fields. The low-lying western coasts with their reduced rainfall and better soils contrast with the hilly eastern margins, many of which have been given over to forests. Agriculturally, these western and southern coastlands of Kintyre, together with the fertile island of Gigha, are the most favoured parts of Argyll, with arable land more extensive than elsewhere. But the main concern is with dairying, so rotation grasses and fodder crops dominate the rural scene on the loamy soils of the extensive raised-beach terraces. The metamorphic limestones and mineral-rich Green Beds also crop out along the western margins, bringing more fertile soils wherever they occur. Finally, in the Laggan, the lowland around Campbeltown, Carboniferous Limestone adds to the richness of the soils, whilst Old Red Sandstone has a similar effect in the south-eastern corner around Southend.

Only at the Mull of Kintyre itself do the wild moorlands and steep 1,400-foot (430-metre) hills remind us that, geographically, we are still in the mountainous west, although glimpses of Arran away to the east help to foster a feeling of the Highlands. But our doubts are raised again when the Macrihanish colliery looms into view, for here the Limestone Coal Group provided a major source of employment in south Kintyre until the pit was forced to close on economic grounds in 1967. This anomolous outlier of Carboniferous rocks, which can be matched with the Ballycastle coalfield of Ulster on the other side of the North Channel, also includes calciferous sandstones, contemporary basalts and great thicknesses of Coal Measures which have yet to be exploited. Its preservation within the Highland province, surrounded on all sides by Dalradian rocks, appears to be the fortuitous result of downfaulting. Any analogy with the Lowland scene is soon dispelled, however, for the Atlantic coasts of Kintyre are windswept and treeless, whilst the ceaseless ocean swell on the sweeping dune-backed beaches of Macrihanish Bay gives a foretaste of the Hebridean landscapes which await us.

Knapdale and Nether Lorn

We have seen that Cowal and Kintyre are carved very largely from the inverted succession of the Iltay Nappe, where the younger Dalradian rocks lie beneath the older Dalradian rocks because of the overfolding (Figure 32). To the north of Loch Fyne and Loch Tarbert, however, are the regions of Knapdale and Nether Lorn. These are north-west of the Iltay Nappe, so their lithological succession is the right way up. This is the zone known as the Loch Awe Syncline (Figure 32), although the minor folding within this structure is enough to warrant the term synclinorium (see Glossary).

In Knapdale the detailed folding is so tight that a series of steep anticlines and synclines dominate the Caledonian structures. Denudation has picked out the latter to create a perfect example of the phenomenon known as 'inversion of relief' (Figure 33). Briefly, this means that all the anticlinal arches, being most vulnerable to erosion because their joints are under tension, have ultimately been worn down to form lowlands or valleys. Conversely, the more resistant synclinal structures, their joints

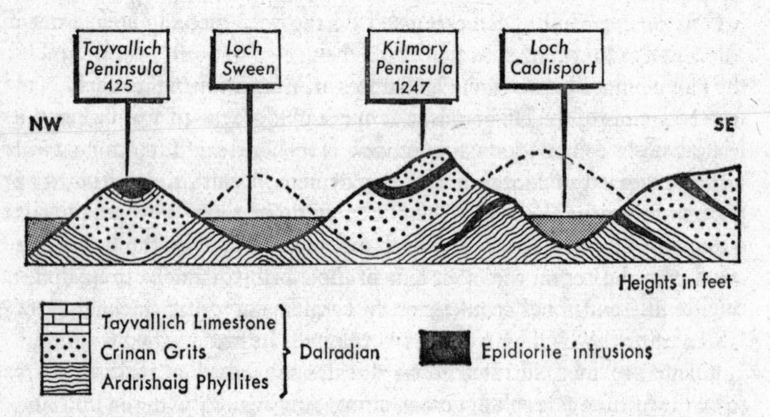

Fig. 33. *Diagram to illustrate the inversion of relief in Knapdale, Argyllshire.*

closed by compression, are frequently left upstanding to form tracts of higher land. So it is in Knapdale, where the invasion of the sea into the glacially overdeepened valleys has merely served to emphasize the influence of structure upon relief. Thus, Lochs Sween and Caolisport have been eroded along the anticlinal axes, whilst the rugged peninsulas of Tayvallich and Kilmory correspond with the synclinal arrangement of the tough Crinan Grits.

Although narrow exposures of metamorphic limestone in the Tayvallich peninsula and near Loch Crinan help to alleviate the general acidity of the soils, the widespread outcrop of the Crinan Grits in Knapdale militates against major land improvement. The scene is therefore one of derelict or poor-quality farming, and it is fortunate that the Forestry Commission has made great efforts to increase the land-potential by planting the extensive forests of Knapdale and Kilmichael. These have augmented the natural oak woods of the ribby peninsulas around the head of Loch Sween to create a sylvan scene rarely encountered on the western coastlands. It must be remembered, however, that Knapdale is sheltered from the searing Atlantic winds by the high mountains of Jura. The Caledonian trend is clearly reflected in the land use around Loch Awe, where the rough grazing and birch scrub often mark the linear outcrops of the quartzites and the countless epidiorite sills that protrude as rocky ridges above the forests and scattered strips of improved land, which in turn are restricted to the narrow parallel depressions that have been glacially etched out in the less resistant limestones.

The contrast between the landscapes of Kintyre and Knapdale could not be more marked. The virtually treeless southern peninsula, with its coastal cliffs and far-ranging seascapes, is replaced by the thickly wooded crenulate coastline farther north, characterized by its hidden bays, rocky ridges and restricted views. The better soils of the south, where the large prosperous farms have 50 to 60 per cent of their land under arable, contrast with the poorer acid soils and rush-choked pastures of the north, on whose tiny crofts and small farms the farmers succeed in ploughing only about one third of their holdings because of the rocky terrain.

Of all the land routes into Nether Lorn, none is more attractive than the traditional road through Inverary, since it combines the pleasures of that town with the beauty of Loch Awe and its overwhelming backdrop of Ben Cruachan (3,689 feet, 1,125 metres). Planned for the Duke of Argyll by the famous Adam family, the whitewashed buildings of Inverary are '... an entirety designed to delight the eye from every angle of approach, both by land and water'. This old drovers' route from the Hebrides to the Lowland markets has been travelled and described by Burns, Keats, Wordsworth, Johnson and Boswell, while Turner painted its bays, hills and woods '... with an artistic ecstasy that soared above all considerations of topographical detail'. The reasons for such popularity are not far to seek, for in Lorn we find high, steep Grampian peaks mirrored in the waters of Loch Linnhe, Loch Awe and Loch Etive. The cloud shadows on the tree-dotted slopes, the mottled crags, the sun-dappled waters and the ever-changing light of the rugged Atlantic coastline combine to create an unforgettable kaleidoscope of colour in this inspiring landscape – a landscape which prevails from here to Cape Wrath.

The tough granite of Ben Cruachan overlooks the ice-scoured and lake-dotted plateau of Nether Lorn, which lies south-east of Oban. This plateau is made up largely of successive sheets of basaltic lavas, lying upon and intercalated with beds of purple grits, grey shales and red sandstones. Archibald Geikie noted that these sediments were of Old Red Sandstone age and suggested that, since their basal conglomerates were composed of volcanic and metamorphic boulders of local origin, this Old Red outlier was formed as a shallow lake deposit, now surviving only by virtue of its thick blanket of contemporary Oban lavas. The conglomerates extend into Upper Lorn, where they form much of the coastline from Ganavan Bay to Dunstaffnage Castle, as well as Maiden Island at the mouth of Oban harbour. The ivy-smothered Dunollie Castle sits on a basalt-capped sandstone crag at the harbour entrance, whilst the town itself has grown in tiers like an amphitheatre around the bay. The handsome terraces of the Victorian esplanade were built on the post-glacial raised-beach platform, whose low rock cliff is pitted with sea-caves now uplifted to a height of about 40 feet (12 metres) above the present shore. Such caves once provided shelter for some of Scotland's earliest inhabitants: numerous Mesolithic artifacts have been discovered there, testifying to the hunting and gathering life-style of the so-called Azilian culture some 8,000 years ago.

The Islands

The terraced island of Kerrera, which acts as a breakwater for Oban Bay, is composed of a combination of Old Red Sandstone, igneous basalts and metamorphic graphitic schists. But there is also an outcrop of slates heralding the more extensive slate-belt that characterizes the inshore islands farther south. Here there are the islands of Seil, Luing, Easdale, Shuna and Torsa, known as the Slate Islands. Together with the Ballachulish quarries (see Chapter 11), these once supplied most of the roofing material for western Scotland. Approachable by road over Telford's hump-backed 'Bridge across the Atlantic' at Clachan Sound, the island of Seil exhibits little evidence of quarrying until its western shores are reached. Yellow flag irises line marshy hollows, although the friable slates break down to reasonable soils which support good grazing-land amid the low hillocks and ridges. On Easdale, the desolate quarries around the harbour are now flooded and closed, but the rows of industrial cottages are brightly painted as holiday homes, despite the unattractive litter of broken slates (Plate 23). When visited by Thomas Pennant in 1772, Easdale alone was exporting over 2 million slates to England and across the Atlantic each year, and less than a century later this annual output had passed the 7 million mark. All the quarries on the islands have now ceased working, the demand for slate having fallen because of competition from more easily and cheaply produced roofing materials.

The hilltop behind Ellanbeich village on Easdale Sound is a wonderful viewpoint for both Mull and the islands of south Argyll. Away to the south-west lies the angular outline of the Garvellach Isles, their cuesta-like form (see Glossary) presenting vertical cliffs north-westwards to the Firth of Lorn. These holy 'Isles of the Sea', composed of Pre-Cambrian quartzose tillites and easily weathered limestones, contain some of the oldest Christian ecclesiastical buildings surviving in the British Isles, equalled only by those of the Skellig Rocks in south-west Ireland. Linked with two early Celtic saints, St Columba and St Brendan, the chapels and beehive cells are remarkably constructed of mortarless, undressed stones, each overlapping the one below until a watertight corbelled roof is completed. It is a sobering thought that it was on such isolated oceanic rocks as the Garvellachs and Skelligs that Christianity managed to survive in the Celtic fringe, when all was turmoil elsewhere in Dark Age Europe. But the islands have been uninhabited for centuries, and only the ruins bear mute testimony to the Golden Age of the Celtic church. Their hoary stones are now surrounded by a carpet of beautiful flowers, including scarlet pimpernels, blue pansies, yellow flag irises and primroses, all of which flourish on the rich, dark, alkaline soils of the limestones.

To the south the dark pyramid of Scarba raises its head high above the low-lying slaty isles of Lunga and Luing, for it is bolstered by the tough Islay Quartzite which forms the backbone of Jura and Islay alike. The poor thin soils of the quartzitic rocks support only moorland on what Scott described as 'Scarba's Isle, whose tortured shore rings with Corrievreckan's roar'. Scott was refering to the notorious whirlpool between Scarba and Jura, the most dangerous of the many which occur in this cluttered archipelago of south Argyll, where tidal races bore

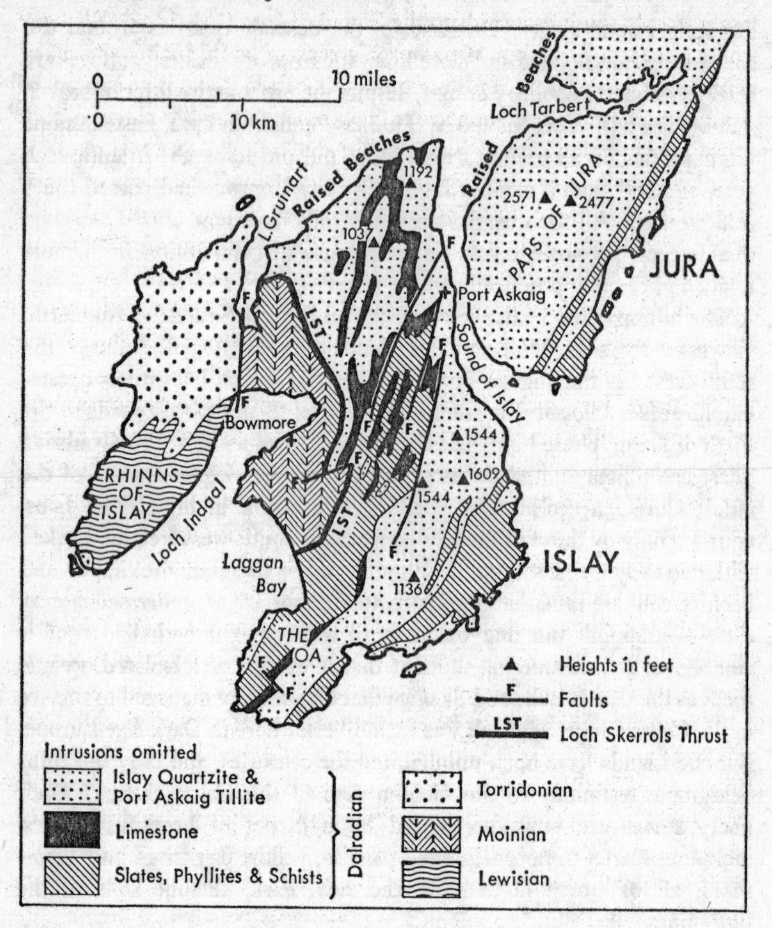

Fig. 34. *The geology of Islay and southern Jura (based on material prepared by the Institute of Geological Sciences).*

through glacially overdeepened channels carved in the relatively softer slates and phyllites of the Dalradian rock succession.

Except for a narrow band of slates, conglomerates and phyllites along its eastern margin, where the majority of the settlements are found, Jura is built entirely of an unyielding Middle Dalradian quartzite which dips fairly steeply south-eastwards to the Sound of Jura (Figure 34). We have come to expect this tough metamorphic rock to be invariably associated with exceptional upland landforms, and this barren island is no exception. Its mountainous spine culminates at the southern end in the three shapely cones known as the Paps of Jura – Beinn an Oir (2,571 feet, 784 metres); Beinn Siantaidh (2,477 feet, 755 metres); and Binn a'Chaolais (2,407 feet, 734 metres). In the opinion of the well-known Scottish writer, Alisdair Alpin MacGregor, their steep-sided elegance can be compared only with the famous Cuillins of Skye. They rise abruptly from the narrow coastal plain, and their conical summits are surrounded by shimmering aprons of angular screes which make any ascent to their massive cairns an arduous business.

The name 'Jura' means 'Deer Island', and even today it carries an overwhelming number of red deer (some twenty times greater than the human population), whose numbers, according to Sir Frank Fraser Darling, are too high for an ecological balance to be preserved. One result is that the scattered crofts of the east coast suffer severe depredations in an environment hardly conducive to anything but subsistence farming. The few bays and natural harbours along this eastern coast have been carved from the less resistant slates and phyllites; where, as at Loch na Mile and Lowlandman's Bay, offshore reefs of epidiorite sills occur they act as fortuitous natural breakwaters (Figure 34).

It is on Jura's west coast, however, that its greatest geological interest may be found, for here is a unique collection of raised-beach phenomena. First recognized by the Geological Survey officers, but mapped in detail by Dr S. B. McCann, this staircase of fossil erosional and depositional shore forms reaches its zenith in western Scotland (Plate 24). The oldest strandline stands at an elevation of about 100 feet (30 metres), and its broad rock platform is backed by high wave-cut cliffs. Although it is almost everywhere covered by raised-beach shingle, there is evidence to show that a glacial episode intervened between the cutting of the so-called 'pre-glacial' platform and the formation of the shingle. In places a glacial till occurs between the two phenomena, while glacial striations can be traced across the old platform, implying that the feature was overridden by mainland ice before the emplacement of the late-glacial beach deposits. The finest examples of the constructional beach ridges, at 80–100 feet (25–30 metres) O.D., are to be found north of the fjord of Loch Tarbert, which almost cuts the island into two halves. Here the shingle is unvegetated and as fresh as if it were formed only yesterday.

It has been suggested that while mainland ice still occupied the eastern margins of Jura and Islay (prior to the Loch Lomond Readvance), late-glacial seas were already fashioning the shingle ridges of the ice-free western shores which were later to be considerably uplifted by isostatic warping after the weight of the neighbouring ice-cap had been removed by melting. The exposed late-glacial coastline, open to the full fetch of Atlantic waves, appears to have been uplifted in stages, for a series of coastal lochans were later to be impounded by a lower shingle bank, now some 60 feet (18 metres) above present high-water mark. Below this elevation the whole late-glacial beach complex has been cliffed by the post-glacial raised beach, which itself now stands high and dry about 25 feet (8 metres) above the modern waves (Plate 24).

Islay has a much more complicated geological structure than Jura, and this is reflected in its unusual shape and topography (Figure 34). Nevertheless, certain lithological continuities between Jura and Islay can be recognized, since the Sound of Islay is very narrow, despite its glacial over-deepening. In the first place, the slates and phyllites of eastern Jura continue into south-east Islay, where they create a similar low-lying landscape of bare epidiorite ridges, slaty hollows of rough grazing and birch-oak scrub, all bounded by a broken coastline of linear bays and islands. A further scenic comparison is possible between the mountainous ridge of quartzite in Jura and its continuation south-westwards along the Caledonian 'grain' into the peaks of south-east Islay, although in this case the summits scarcely rise above the 1,600-foot (490-metre) contour (Beinn Cheigeir, 1,609 feet, 491 metres). Nevertheless, as with the Paps of Jura, their corries and drift-dammed lochans suggest that small ice-caps existed on both these islands during the Loch Lomond Readvance cold phase.

The steep south-easterly dip of the Islay Quartzite is similar to that in Jura, and together the structures of their highest hills may be taken as the south-eastern flank of the so-called Islay Anticline (Figure 32). The opposing north-westerly-dipping flank can be seen in northern Islay, where the quartzite reappears to form the sharp hills which terminate at the magnificent raised beaches on the lonely headland of Rubha a'Mhail. The intervening crest of the anticline has been denuded so that the quartzites have been destroyed and the underlying slates, limestones and phyllites uncovered (Figure 34). Since this complex succession of metamorphics (known collectively as the Mull of Oa Phyllites) has proved less resistant to erosion than the flanking quartzites, they now correspond to a lowland corridor between Port Askaig and Bowmore. As these relatively softer rocks are traced south-westwards past Duich and Laggan Bay into the southern peninsula of The Oa itself, they form a broad coastal lowland. Despite its low elevation, however, quite extensive peat bogs and coastal dunes have restricted the land use, so that there is mainly rough grazing in between the few scattered arable plots around the cottages. A distinctive feature of this lowland is the series of gigantic peat stacks which line the road from Bowmore to Port Ellen. Much of this peat is still used commercially in the production of the well-known Islay malt whisky, although Lowland coal is now imported for heating the distilleries, whilst island-grown barley is no longer used.

The other wide coastal lowlands which surround the sea inlets of Loch Gruinart and Loch Indaal are also thickly covered with superficial deposits of boulder clay, raised beach, peat bog and blown sand. It has been pointed out by Dr Margaret Storrie that, with this as a physical basis for agricultural improvement, Islay is much better endowed than some of the Outer Hebrides (see Chapter 14). Where the land is well drained, therefore, large stretches of arable and high-quality pastureland can be seen on the red loamy soils around Bowmore, Gruinart and Port Charlotte. In addition there are the broad bands of Islay Limestone, which bring swathes of brighter green to the pastures of the central corridor and have been quarried at Ballygrant for use both as a crop fertilizer and as a source of lime-wash for the houses. Because of this greater agricultural potential, and because of its proximity to the Lowlands of Scotland, Islay has received a greater influx of Lowlanders than the other Hebridean islands. Thus in Islay there are hardly any of the crofting townships which characterize the remainder of the Hebrides. Instead, the landscape, like that of Kintyre, has a Lowland aspect, with medium-sized and large farms dispersed among industrial and service villages: in the words of Dr Storrie Islay is 'a Hebridean exception'.

The redness of the soils around Bowmore is due to the fact that we have now passed across an important geological boundary. Although it cannot easily be seen beneath the superficial deposits, an important line of thrust-faulting occurs between Laggan Bay and Loch Gruinart. This is the so-called Loch Skerrols Thrust, which marks the line where Dalradian rocks were forcibly pushed westwards onto the older Bowmore Sandstone along a steeply angled plane of faulting during the Caledonian deformation (Figure 32 and Figure 34). It has been suggested that the Loch Skerrols Thrust is in fact the Islay equivalent of the Moine Thrust (see p. 224). Furthermore, the Bowmore Sandstone is now thought to be Moinian in age, although the western peninsula of the Rhinns of Islay is composed of Torridonian and Lewisian rocks, the oldest that we have so far encountered in our tour of Scotland and a further reminder that we are moving gradually towards the Archaean 'foreland' of the north-west. The Torridonian grits, slates and conglomerates occupy the northern end of the peninsula, where they have been worn down to a lake-studded, peat-covered coastal platform. Farther south, however, the true Rhinns of Islay are composed of acid and basic Lewisian gneisses, seamed with epidiorite intrusions. The knobbly, ice-polished, rocky surface, with its low rounded hills, rock pools and peat bogs, gives us a foretaste of the landscapes of the Outer Hebrides. But, as with those islands, patches of blown sand, shelly strands and raised-beach terraces bring welcome oases of pastoral green to these cliff-girt, storm-lashed western headlands. The curving western coastlines of both Loch Indaal and Loch Gruinart have been carved along an important tear- or wrench-fault, the Loch Gruinart Fault, which cuts off the Archaean rocks of the Rhinns from the Bowmore Sandstone in the east (Figure 32). Since the fault is a sinistral one, the western block having moved south-westwards in relation to the rest of Islay, it has sometimes been correlated with the Great Glen Fault. It is now suggested, however, that such a correlation is unlikely and that the latter fault passes beneath the sea to the west of Islay and Colonsay (see Figure 37).

Before we leave the fascinating scenery of Islay, two other points of geological interest must be noted. The first of these concerns the presence of the so-called 'Boulder Bed' which is exposed between the Islay Quartzite and the underlying limestones near to Port Askaig. This is another example of the Pre-Cambrian boulder clay or tillite which we have already encountered on the slopes of Schichallion in the central Grampians (see p. 185) and which occurs again among the Donegal metamorphic succession in north-west Ireland. The second phenomenon is the suite of raised beaches which occurs around the coasts, but which is especially well developed along the northern shores. The late-glacial shorelines are almost comparable in stature with those of Jura, but the post-glacial beach is much more extensively developed in Islay. During the post-glacial transgression the Rhinns and The Oa may well have become islands as the seas flooded across the isthmuses of Gruinart and Port Ellen respectively. In earlier times, however, the Rhinns themselves must have been turned into an archipelago, for in the vicinity of Loch Gorm the late-glacial raised beaches lie athwart the central tract of this western peninsula.

The same must have been true of the islands of Colonsay and Oronsay farther north, for the highest late-glacial strandlines can be traced in between the low rocky hillocks from coast to coast. It is partly owing to these marine sands and gravels that the isles of Colonsay and Oronsay are amongst the most fertile of the Hebrides. In addition, however, the Lower Torridonian mudstones, flagstones and phyllites have a moderately good lime content, so that the light gravelly soils support extensive tracts of hayfield and arable land. Despite the lowness of its hills (Carnan Eoin, 470 feet, 143 metres), rare examples of natural woodland survive in the sheltered hollows, with birch and oak predominant amongst the scrubland of hazel, rowan, willow and aspen.

One of the most remarkable features on Colonsay is the northern vale of Kiloran, which Tom Weir describes as '... slung like a yellow hammock between hills of purple heather'. Sheltered by its surrounding ridges from the searing Atlantic winds, this fertile, loch-filled valley is one of the most delightful of the Hebridean landscapes. Woodlands, sparkling waters, sheep-dotted hills and flower-decked grasslands lead down to the dunes and the long creamy sands of Kiloran Bay. The tidal island of Oronsay, at the south end of Colonsay, is also a place of dazzling beaches, dunes and rocky skerries. It was here that St Columba first landed in Scotland in A.D. 563, but since he could still see Ireland from the highest hill he sailed on to Iona (see Chaper 12). Nevertheless, the ancient priory ruins survive on the site of his original foundation, although it is the much older Mesolithic remains amongst the dunes which have brought most fame to Oronsay. Here some of Scotland's earliest inhabitants lived some 8,000 years ago by hunting and fishing along these hospitable shorelines. Known as strand-loopers, these Azilian folk (similar to those of the Oban caves, see p. 202) left sand-hill middens of mussel shells and bones, including those of the now extinct great auk.

11. Glen Mor and the Western Highlands

The islands of south Argyll, described in the previous chapter, can almost be likened to an armada of vessels converging north-eastwards into the narrows of the Firth of Lorn. This major inlet of the sea, together with its continuation, Loch Linnhe, marks an important line of transition in the Scottish landscape, for northwards lie the remarkable landforms of the farthest Hebrides; furthermore, once across the Great Glen we have left the Grampians for the Western and Northern Highlands.

Nowhere else in Scotland is a topographical boundary line so linear as that followed by the chain of lochs between Loch Linnhe and the Moray Firth. Here is the gigantic gash known as Glen Mor, or the Great Glen of Albyn, which has been carved along the line of one of Britain's major faults. This remarkable physical feature, where selective erosion by glaciers has transformed the shatter-zone of the fault-line into a steep-sided trough, runs for a distance of almost 100 miles (160 kilometres) between the island of Lismore and Inverness. This line of weakness in the mountainous Highlands has provided a ready-made means of communication from coast to coast, which is partly reflected in the Caledonian Canal, constructed between 1803 and 1822. Originally surveyed by James Watt but later built by Thomas Telford, this artificial channel linked together the existing lochs of Glen Mor, thus allowing small vessels to avoid the stormy northern passage of the Pentland Firth.

Loch Linnhe and Its Region

Where the enclosed waters of Loch Linnhe pass imperceptibly seawards into the Firth of Lorn, the estuary contains a bevy of islands, including Shuna, Lismore and its satellites. These islands, together with the wooded peninsulas of Barcaldine and south Appin, nowhere rise above the 500-

foot (150-metre) contour, thus providing ribbons of coastal lowland with a reasonable agricultural potential in a region which is almost completely dominated by relatively unproductive mountainland. This favoured eastern shoreline of Loch Linnhe, midway between the thriving urban centres of Oban and Fort William, supports a rural way of life not dissimilar to that of other marginal hill country in south Argyll. A contrast is at once apparent, therefore, between these lands of Appin, Benderloch and Lorn and the more rugged, empty mountains and moorlands of Ardgour and Morvern that flank the western shores of Loch Linnhe and will be examined in the latter part of this chapter.

Lismore, with its familiar flashing lighthouse, is a long narrow island which dominates the watery cross-roads where the Sound of Mull and the fjord of Loch Etive meet at right angles the broader marine straits of Linnhe and Lorn. Like neighbouring Shuna, Lismore is composed of a Dalradian metamorphosed limestone which has weathered into a fertile rendzina soil. Since the limestone has also been extensively quarried and burned in the past for use as an agricultural fertilizer, it is not difficult to understand the derivation of the name 'Lismore', which means the 'Great Garden'. Some of the old lime kilns were constructed in caves and notches behind the so-called 'pre-glacial' raised-beach rock platform which makes such a distinctive feature on these islands. This is the same rock platform which occurs around Oban, where the well-known Dog Stone is in fact an uplifted sea-stack, and also on Kerrera, whose 'raised' wave-cut arch below Gylen Castle is another local curiosity. This raised shoreline creates one further noteworthy landform, the so-called Shepherd's Hat on Eilean nan Gamhna, where the unconsumed rock knob of the central 'crown' stands above the wide 'brim' of the platform.

As noted elsewhere (see p. 205), these raised wave-cut platforms of western Scotland are much older than the post-glacial raised-beach shingle which sometimes rests upon them. Even without the signs of glacial deposition and erosion which often underlie the shingle, it is difficult to envisage how post-glacial waves could have fashioned the broad platforms (up to 100 yards, or 90 metres, across in places) in so short a time, before isostatic uplift raised them all beyond the reach of marine erosion. We are therefore forced to conclude that, so far as the lower rock benches and caves are concerned, the sea has occupied the same level on at least two occasions, in both pre- and post-glacial times. In fact, the shoreline in question is more likely to be of inter-glacial than pre-glacial age, even though its freshness appears to belie the fact that it must have been overrun by ice-sheets.

The signs of glacial erosion are manifest everywhere, but are nowhere better illustrated than in the long, curving fjord of Loch Etive (Plate 25). Like its neighbouring fjords of Loch Creran and Loch Leven, Loch Etive was gouged out by glaciers along pre-existing lines of weakness, including a section of the Pass of Brander fault-line. This same fracture, incidentally, was instrumental in down-faulting the Old Red Sandstone lavas of the Lorn Plateau relative to the massif of the Etive granitic complex (see below). If it were possible to view the floor of Loch Etive's overdeepened glacial trough, we should find that, unlike a normal drowned river valley (known as a 'ria'), a fjord does not deepen systematically as it is traced seawards, but is composed of a series of interconnected rock basins. These are a response to selective glacial erosion, whereby glaciers tend to overdeepen when the trough walls are constricted by hard-rock zones (such as the Etive granites). Farther seaward, however, where glaciers may pass into tracts of less resistant rocks and where the valley widens, they have room to expand laterally, thus reducing downcutting and leaving a rock barrier or threshold across the valley mouth. During the post-glacial rise of sea-level, however, many of the coastal troughs would have been drowned, forming a shallow submerged threshold at the newly created sea-loch entrance. Such is the case at the mouth of Loch Etive, for beneath the Connel Bridge the submerged rock barrier has produced the well-known Falls of Lora. At ebb tide the waters of Loch Linnhe fall faster than those of Loch Etive because of the restricting threshold, thus creating a cataract several feet in height. During flood tide the opposite situation prevails, so that sea-water descends noisily *into* Loch Etive at the Falls of Lora.

Having crossed a curving fault-line, whose weakness is picked out by the aligned valleys of the Pass of Brander and Glen Salach, the upper reaches of Loch Etive, above the Bonawe ferry, assume the deeply incised grandeur of a Norwegian fjord. The steep valley walls in places plunge some 2,500 feet (well over 750 metres) into the sea-waters of the narrow loch, especially near Ben Starav (3,451 feet, 1,052 metres), where ice has gouged a breach in the mountain chain more than 4,000 feet (1,200 metres) in depth (Plate 25). The reason for this sudden change in scenic character is

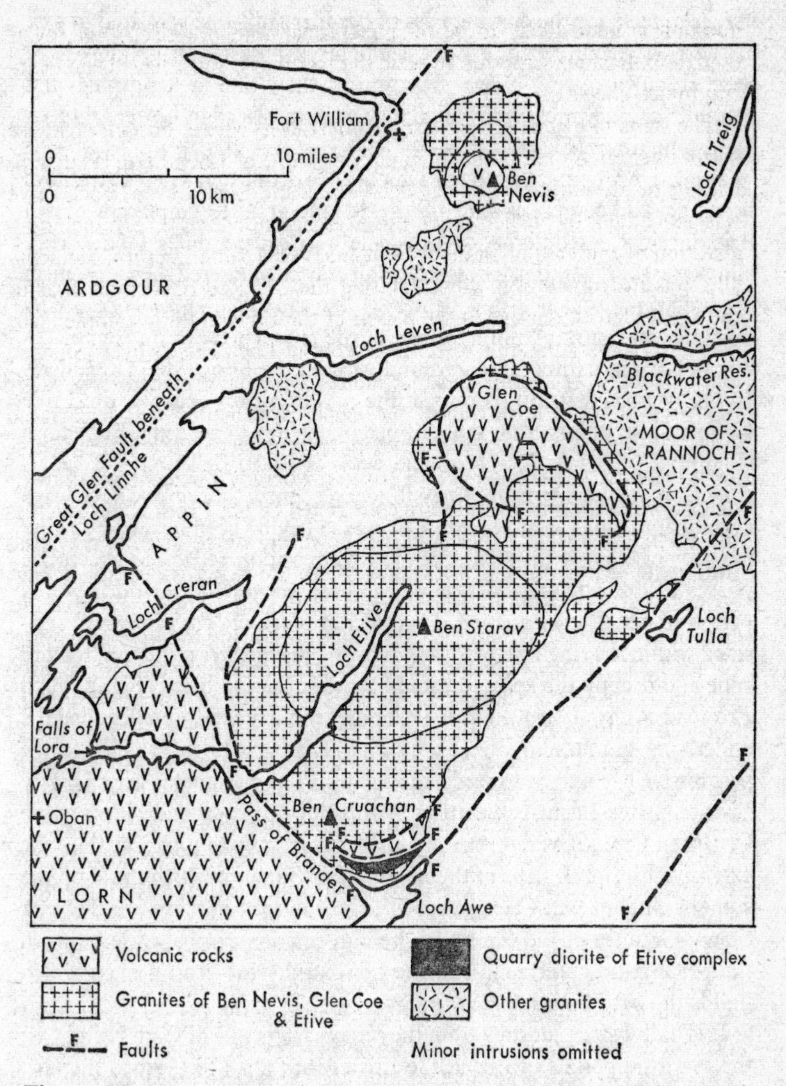

Fig. 35. *The igneous rocks of the Western Grampians (based on material prepared by the Institute of Geological Sciences). See also Figure 36 and Plate 25.*

not far to seek, for the curving Pass of Brander fault-line is related to part of the ring fracture of the Etive Complex, whose tough granites have resisted downwearing more successfully than the metamorphic and volcanic 'country rock' into which they were emplaced in Lower Old Red Sandstone times (Figure 35).

The Etive Complex, through which Loch Etive and Glen Etive have been incised, comprises one of the largest granitic emplacements in Britain, since its outcrop covers some 140 square miles (360 square kilometres) of mountainland. We have already encountered the mechanism of cauldron-subsidence in Chapter 4, but the Etive subsidence differs from that in central Arran by virtue of the fact that the granite emplacement occurred entirely underground without the formation of a caldera. It will be seen that in this respect the cauldron-subsidence of the Etive Complex also differs from those of neighbouring Glencoe and Ben Nevis (Figure 36). Thanks largely to the work of E. B. Bailey and J. G. C. Anderson, it has now proved possible to recognize four or five successive granitic intrusions around Loch Etive, each having been formed by rising magmas of slightly different chemical composition related to successive phases of cauldron-subsidence. It has been suggested that during each period of cauldron-formation a cylindrical block of 'country rock' subsided within the ring fracture, thus allowing magma to ascend the fracture zone and occupy the space vacated by the subsiding block (Figure 36a). Erosion has now destroyed the enveloping 'country rock' and denuded the underlying granites to reveal their ring-dyke structures, which become progressively younger towards the centre. The peripheral grey Quarry Diorite, above Loch Awe station, is the oldest; the second oldest is the Cruachan Granite, which varies in colour from grey to pink; next are the extensive bare pink slabs of the Meall Odhar Granite which make up the summit cliffs of Ben Cruachan itself; the youngest granites are the pink Starav Granites of the centre, where a distinction can be made between the porphyritic outer zone and the non-porphyritic central mass which builds the bare slopes of Ben Starav.

It is only a short journey from the remote fastnesses of Glen Etive, past the towering portals of Buachaille Etive Mor (3,345 feet, 1,020 metres), beloved by climbers, to the overwhelming spectacle of Glencoe. The sudden descent into Glencoe from the apparently endless wastes of Rannoch Moor must rank high among the most spectacular scenic experi-

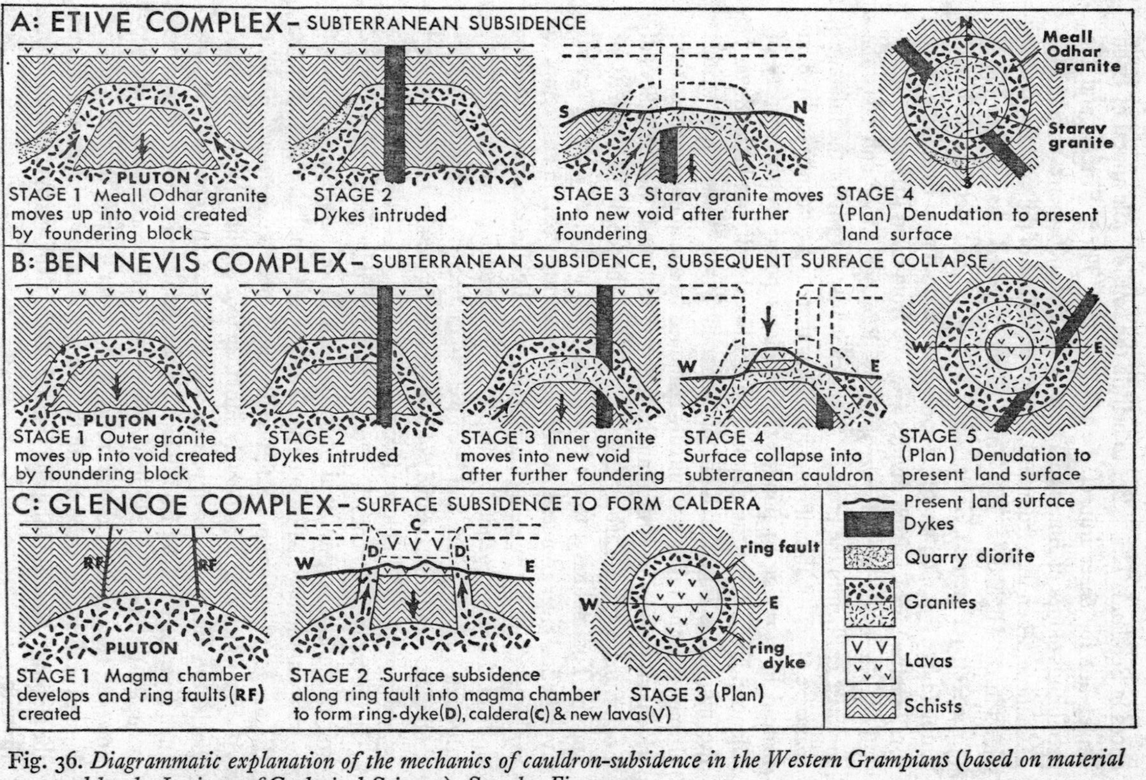

Fig. 36. *Diagrammatic explanation of the mechanics of cauldron-subsidence in the Western Grampians (based on material prepared by the Institute of Geological Sciences). See also Figure 35.*

ences in Scotland. This is not to belittle the considerable appeal of Loch Etive; but for sheer impact the cliffs of the 'Three Sisters' and the precipitous slopes of Aonach Eagach (3,168 feet, 966 metres) at the Pass of Glencoe, when seen from the 'Study', create one of the most impressive views in the Highlands. The Three Sisters referred to are Beinn Fhada (3,120 feet, 951 metres), Gearr Aonach (2,500 feet, 762 metres) and Aonach Dubh (2,849 feet, 869 metres), shoulders of Bidean nam Bian (3,765 feet, 1,148 metres), which itself divides Glencoe from Glen Etive.

Today a modern road sweeps across Rannoch Moor and takes the motorist easily into the great defile which is best known for the infamous massacre of its inhabitants in 1692. Thus there is a tendency to drive on instead of stopping to ponder the problems posed by the scenery of Glencoe. Noting that the glen had no headward wall, but broke through the mountains into the basin of Rannoch Moor, some early writers claimed that the valley was initially fashioned by a major consequent river which, rising near to Ardnamurchan, flowed eastwards via the valleys of Loch Sunart, Glen Tarbert and Glencoe, and thence by Rannoch Moor and the Tummel River to the North Sea. This was merely one of several hypothetical river courses that were once 'reconstructed' in an attempt to explain some of the puzzling drainage lines of Highland Scotland. Modern opinion, however, generally supports the suggestion that the gigantic 'through-valleys' which run from west to east across the primary watershed of Scotland were created not by rivers but by ice-sheets during several episodes of glacial diffluence (see p. 186).

In the case of Glencoe, the breach was initiated by the enormous reservoir of ice that accumulated over Rannoch Moor before finding escape routes via such west-flowing river valleys as Glen Etive and Glencoe (Figure 30). Thus the pre-glacial valley-head cols were destroyed, and the valley floors lowered considerably until they took on the typical U-shape of glacial erosion. Such downwearing was facilitated in Glencoe by an isolated layer of phyllites from which the ice fashioned a basin later to be occupied by Loch Achtriochtan. Wherever harder rock bands appeared, however, they resisted glacial erosion to a greater degree. Examples are to be seen at the valley step beneath the Study and in the precipitous cliffs on the slopes of Aonach Dubh and Gearr Aonach, all of which are made up of rhyolite.

Many fine examples of glacial moulding and polishing can be seen around Glencoe, including *roches moutonnées*, glacial pavements and grooving, whilst morainic debris and perched blocks (including the remarkable Rannoch Moor Granite erratics on the ridge of Aonach Eagach) illustrate the type of load once carried by the ice. As we might expect the major corries face north-eastwards, so that the crags of Aonach Dubh and Gearr Aonach are divided from each other by one of these armchair depressions, out of which a waterfall plunges over the lip of the glacially oversteepened walls of Glencoe.

From a geological standpoint the most interesting corrie is Coire nam Beith, which lies between Aonach Dubh and the peak of An t-Sron, for here can be clearly seen yet another example of cauldron-subsidence (Figure 36c). The ring fracture itself has been picked out by erosion and can be seen as a prominent gully running up the slopes of An t-Sron to its summit. This fault is in fact a boundary between the granite of An t-Sron (which welled up as a magma around the ring fracture) and the existing volcanic succession (of Old Red Sandstone age), which subsided some 4,000 feet (over 1,200 metres) into the cauldron and has thus survived subsequent denudation. The cliffs and crags of the Three Sisters have been carved from these rugged volcanics, as have the summits of Bidean nam Bian, which stands at the back of Coire nam Beith. The cliffs of Aonach Dubh, for example, were built by three massive flows of rhyolitic lava, each about 150 feet (45 metres) thick.

To the north of the notched ridge of Aonach Eagach, another long arm of the sea penetrates deeply into the mountains as Loch Leven. Although General Wade's ancient 'Military Road' takes a direct route over the ridge from Glencoe to Kinlochleven, it is likely that the majority of visitors will prefer to take the more circuitous route through Carnoch to avoid the arduous tramp up the Devil's Staircase. From the shores of Loch Leven the prominent quartzite knob of the Pap of Glencoe (2,430 feet, 741 metres) forms a distinctive feature amongst the gentler surrounding slopes of schist and metamorphosed limestone. The latter gives rise to relatively fertile pastures at the lower end of Glencoe. Near by, the defunct Ballachulish slate quarries disfigure the skyline, but the long-awaited bridge will make unnecessary the lengthy drive around Loch Leven which entails a visit to another sombre industrial village, Kinlochleven. Here the settlement has grown up through the availability of copious hydro-electric power, which is used for the production of aluminium at the Kinlochleven plant. To achieve the high temperatures required to smelt the imported bauxite the British Aluminium Company received permission to create the lengthy Blackwater Reservoir by damming the headwaters of the west-flowing River Leven. Water is now led off from this upland valley and taken through enormous pipes to drive the turbines as it falls to the overdeepened fjord of Loch Leven.

The waters of the Leven, like those of the neighbouring River Spean, may not always have flowed westwards: there is good reason to believe that in the Ben Nevis area deep glacial breaching of Scotland's primary north–south watershed has caused a reversal of drainage in some of the pre-glacial river systems. Thus, Dr Sissons believes that the Leven's sudden 1,000-foot (300-metre) descent from the gently sloping floor of the Blackwater valley down to Kinlochleven '... is clearly a relatively recent development in the history of the drainage and strongly suggests that formerly the Leven flowed eastwards to be continued by the Tummel'. Nevertheless, the Leven's former source probably lay to the east of the Great Glen, and there is no need to consider a hypothesis which seeks to place its pre-glacial source somewhere in the Hebrides (see p. 216). Later expansion of the aluminium company led to the development of a second smelter at Fort William, whose water power is derived from Loch Laggan and Loch Treig via a 15-mile (24-kilometre) tunnel beneath the northern foothills of Ben Nevis.

Glen Mor

Having crossed Loch Leven, we are now in Lochaber, where the tides of Loch Linnhe creep far into the Great Glen until they reach almost to the foot of Britain's highest mountain, Ben Nevis (4,406 feet, 1,343 metres). Before we examine its gigantic bulk, however, a short detour into Glen Nevis will provide a preview of the monolithic scenery hereabouts – scenery which has been compared in stature with that of the Himalayas by W. H. Murray, the well-known mountaineer. He sees in the Nevis gorge a peculiar combination of cliff, woodland and water unparalleled in Britain and reminiscent of the Nepalese valleys.

The hayfields and the oak, birch and pine woods of the lower glen flourish on the more fertile soils derived from an outcrop of the Ballachulish Limestone. A bend in the valley ensures that the full impact of the dramatic Nevis gorge is withheld until the last moment. But then the character of the valley changes immediately, for the main river foams and tumbles through a narrow, twisting cleft, carved partly through a band of steeply dipping Glencoe Quartzite which crosses the valley at Steall. To the left a silvery cascade from a high corrie sweeps down the dark, heathery talus slopes which mantle the pink granite of Ben Nevis, which at this point creates the highest amplitude of relief in Britain: an almost continuous slope of 4,000 feet (over 1,200 metres) descends from its summit to the valley floor. On the right a waterfall leaps 350 feet (107 metres) from a perfect example of a hanging valley. This is Allt Coire a'Mhail, which drains the shapely peaks of the Mamore range (highest point: Binnein Mor, 3,700 feet, or 1,128 metres). Although rising from a base of Leven Schists, the graceful summits of, for example, Stob Ban and Sgurr a'Mhaim owe their form to a cap of white Glencoe Quartzite which is often mistaken for a snow cover.

When viewed from Glen Nevis, Britain's highest mountain is too foreshortened to appear attractive, and even from the western shores of Loch Linnhe it is impossible to appreciate its great height. When seen from the north-west, however, its majestic ice-quarried cliffs are revealed, their frost-riven 2,000-foot (600-metre) precipices unequalled in height even by those of the Cuillins (see Chapter 13). At their base extensive snow patches lie in the gullies and hollows, frequently surviving the summer in their sheltered north-east facing corries and serving to demonstrate how easily corrie glaciers could regenerate in this marginal 'sub-Arctic' environment. Although the U-shaped hanging valley of Coire a'Mhuilinn, like the lower slopes and shoulders of Ben Nevis, has been carved from granite, the cliffs and uppermost parts of the mountain are made of volcanics of Lower Old Red Sandstone age (Figure 36b). It may be seen in Figure 36b that the granites of the pluton were subsequently affected by the collapsed roof of a subterranean cauldron (established during earlier subsidence), thus allowing the former cover of lavas and their underlying schists to sink about 1,500 feet (460 metres) into the Inner Granite while it was still in liquid form. It is these bedded lavas and volcanic agglomerates which today form the summit dome of Ben Nevis.

Although our journey now takes us north-eastwards into the inner recesses of the Great Glen, we must make yet another detour at Spean Bridge to view the remarkable phenomena known as the Parallel Roads of

Glen Roy (Plate 26). At Roybridge a narrow valley winds away into the Lochaber hills, and if we were to follow this upstream it would bring the so-called 'roads' into view. High up on the mountainside three grassy terraces stand out in the darker masses of heather as if to mark contour lines along the slopes. So artificial do they look that local legends claim them as the 'King's Hunting Roads', constructed by ancient monarchs who dwelt at Inverlochy (Fort William). Despite the contention in 1840 by an eminent Swiss glaciologist, Louis Aggasiz, that these parallel terraces could relate to a slowly draining pro-glacial lake (ice-impounded), Sir Charles Darwin thought that they were marine formations. He remained sceptical even when the ice-dammed lake overflow (relating to the middle 'road') was discovered in 1847. Darwin finally admitted that he was wrong in the 1860s, by which time the chronology of the former pro-glacial lake had been fully worked out.

It is now known that a tongue of ice blocked the drainage not only in Glen Roy but also in neighbouring Glen Gloy and Glen Spean, creating marginal lakes similar to the modern Marjelen See at the Aletsch glacier in Switzerland. The Parallel Roads are in fact strand-lines of a lake which, in Glen Roy, stood at a maximum height of about 1,150 feet (350 metres). As the ice-front retreated the lake was progressively lowered first to approximately 1,075 feet (320 metres) and then to some 850 feet (260 metres), as overflow cols were revealed from beneath the ice. Apart from the 'roads' of Glen Gloy and Glen Spean, similar strand-lines have been found elsewhere in Scotland only at Rannoch Moor, but here they are on a smaller scale.

Having returned to the remarkable corridor of the Great Glen, we can move rapidly past Loch Lochy and Loch Oich, where we cross the imperceptible watershed between west-flowing and east-flowing drainage. The divide has been displaced some 15 miles (24 kilometres) to the north-east by glacial breaching and subsequent river capture, for the primary pre-glacial watershed lay near to Ben Nevis. At Fort Augustus we get our first glimpse of one of Scotland's best known lakes, Loch Ness. Even without its legendary 'monster' the Loch would have a leading place in British water bodies because of its gigantic dimensions. At 754 feet (230 metres) maximum depth, it is Scotland's second deepest loch, but since it remains more than 600 feet (180 metres) in depth for the greater part of its 23-mile (37-kilometre) length it holds the largest volume of water. Its capacity of 263,000 million cubic feet (7,443 million cubic metres) is almost three times that of Loch Lomond and more than three times that of the deepest loch, Loch Morar.

Nothing illustrates the efficiency of glacial scouring along a fault shatter-belt better than this set of figures. When one pauses to consider that the valley walls rise abruptly throughout to some 1,500 feet (460 metres), it is clear that Loch Ness occupies one of the greatest glacial troughs in the British Isles (Plate 27). Its shores are so steep that aquatic vegetation is virtually absent, whilst its waters are so dark that it is little wonder that Loch Ness is associated with legends of living 'fossils' which may have survived from earlier geological periods. But no one has yet explained how such creatures survived the bulldozing effects of the glaciers! The intense glacial action was not confined to the trough itself, for glacial roughening is manifest on both shoulders of the Great Glen, judging by the plethora of lochs and lochans on the plateaux to the west and east of Loch Ness.

Two of these rock-basin lakes in Stratherrick, on the eastern shoulder of Loch Ness, have been linked to create Loch Mhor, in order to provide a head of water for the hydro-electricity station at Foyers on the shores of Loch Ness. Reputed to be the first hydro-electric scheme in Britain, the power station at Foyers, like those at Kinlochleven and Fort William, was initially required for the smelting of bauxite. The aluminium factory here was closed down in 1967 to be replaced by a pumped storage scheme, similar to that which operates between Ben Cruachan and Loch Awe. The idea is to utilize off-peak periods (when electricity demand is low) to pump water the 589 feet (179 metres) from Loch Ness up to Loch Mhor, thus allowing the water to descend again and generate 300 megawatts of electricity during periods of peak demand. The location of hydro-electric schemes in Scotland is often related, not surprisingly, to the presence of the over-steepened walls of glacial troughs, and the well-known waterfall at Foyers must have helped the early industrialists to choose this particular site. So spectacular were the falls at Foyers that Burns was moved by them to write one of his few poems of purely natural description. Johnson and Boswell also visited the spectacle, but the eminent Doctor was so fatigued by the climb that he could do little but consider 'the asperities of the rocky bottom' and comment that Loch Ness was 'a remarkable diffusion of water without islands'.

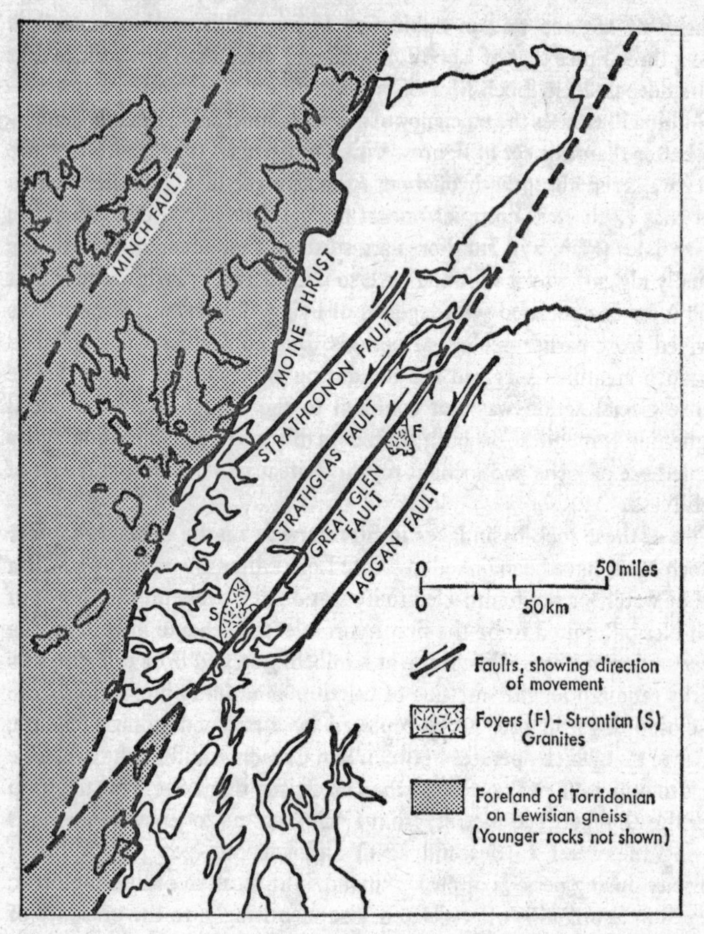

Fig. 37. *The major wrench-faults of northern Scotland and the Moine Thrust.*

The lack of islands in the lochs and the steepness of the flanking slopes of the Great Glen highlight the efficiency of the ice in clearing out not only the fault breccia but also much of the less resistant faulted outlier of Old Red Sandstone which occupies part of the floor of the Great Glen. But ice has also scoured out much of the regolith, so that the valley flanks of these central glens are both too steep and too rocky for agricultural improvement. Instead they have been given over almost entirely to forestry, so that the majestic water vistas of Loch Ness are now framed by endless ranks of dark conifers marching along the valley sides.

Some of Scotland's largest forestry plantations are to be found in this region, especially in the sheltered valleys of Glen Moriston, Glen Urquhart and neighbouring Glen Affric. Here, around the ice-scoured knolls, podzolic soils (see Glossary) have formed on the glacial tills and outwash deposits. Magnificent remnants of the ancient Caledonian pine forests, such as that of Glen Affric, have survived in these sheltered eastern glens where the tree-line is considerably higher than on the exposed western coastlands and islands. Exposure to salt spray and the dessicating effects of prevailing westerlies are two of the most inhibiting factors in tree-growth. Native forest once covered this region but has been largely destroyed by man and his grazing animals, which has often led to soil deterioration and the formation of thin iron pans. However, the growth rate of conifers in Scotland far exceeds that in Scandinavia and in Central Europe, so that afforestation has now become the most important land use in these central glens and has led to the location of a gigantic pulp mill at Corpach, near Fort William, to utilize the forest 'crop'.

Before leaving the confines of Glen Mor at the town of Inverness, it is important to examine the mechanism of the faulting which created the line of weakness from which the corridor was eventually created. We have already encountered the mechanisms of normal faulting and thrust faulting in Chapters 7 and 9 respectively, whilst Figure 37 also illustrates some of the major wrench- or tear-faults which affect the geology of the Central Grampians. The Great Glen Fault falls into the last category; indeed, it is the best example of a wrench-fault in the British Isles. Although following a 'Caledonoid' trend the Great Glen Fault is known to be largely of Hercynian age, though lateral displacement took place at intervals up until post-Carboniferous times. Indeed, minor earth-tremors, recorded on several occasions at Inverness since 1769, demonstrate that movement has not yet ceased.

Despite the knowledge that the sinistral movement has transported the Northern Highlands block bodily south-westwards in relation to the remainder of Scotland, it was not until the work of Professor W. Q. Kennedy that we were able to estimate the amount of that lateral displacement. By demonstrating that two of the Great Glen granites, those of

Foyers and Strontian, were once petrographically and structurally part of the same geological unit, he was able to show that the northern part of the granite pluton has been moved 65 miles (105 kilometres) south-westwards to its present position opposite Lismore Island (Figure 37). Such a reconstruction could also explain how the Great Glen Fault, by displacing the Moine Thrust, allows remnants of the ancient Archaean 'foreland' to survive in Colonsay and western Islay (see Figure 37 and pp. 207–8).

Later work by Dr N. Holgate has suggested that in post-Jurassic times there was another major shift along the Great Glen Fault, but this time a dextral movement (in the opposite direction) of about 5 miles (8 kilometres). Taken together with a postulated dextral move of 13 miles (21 kilometres) along the Firth of Lorn Fault (between Lismore and Appin), it is conjectured that the Northern Highlands block and the Grampian block have changed their position relative to each other by a total of 83 miles (134 kilometres) throughout geological time. Indeed, the impossibility of matching the alignment of the Tertiary dyke swarms of Morvern and Lismore with that of those which occur to the south-east of the Great Glen has pointed to a possible post-Tertiary movement along this major line of faulting.

It is not surprising that an important urban settlement grew up at the more sheltered north-eastern end of this great natural corridor, where the waters of Ness debouch into the Moray Firth. Inverness, the 'capital of the Highlands', commands this nodal position at a bridging-point of the river on the edge of its delta. Once the capital of the Pictish kingdom, the town has remained important throughout Scotland's chequered history, since it dominates the only viable north–south route across the Scottish Highlands at a point where it crosses both Glen Mor and the route to the eastern coastlands. Inverness's Victorian castle and modern commercial buildings have now replaced the thatched roofs, the balconies and the projecting towers and stairs which characterized the old town at the time of Culloden, which seems to highlight the fact that at this point there is a rapid transition from the rugged Highland scene to the more gentle landscapes of the Moray lowlands.

The scenic transition is due in part to the appearance of the Old Red Sandstone with its rolling topography and more fertile soils. It also results from the fact that during the Pleistocene the ice-sheets utilized the basin of the Moray Firth as a dumping ground for the products of their mountainland erosion. Thus, in contrast with the hard rock landforms of the Great Glen proper, the topography around the inner recesses of the Moray Firth has been moulded very largely from relatively unconsolidated materials of Pleistocene and Recent age. These can be classified according to their origin as glacio-fluvial deposits, glacial deposits and marine deposits: a journey from Loch Ness eastwards will enable us to examine examples of each in turn.

Soon after leaving Loch Ness the main road climbs over an undulating tract of kame-and-kettle terrain. In addition, however, the area between Dochgarroch and Tomnahurich has some excellent examples of the glacio-fluvial landforms known as eskers. Not as well developed as the magnificent Flemington esker complex around Croy (to the east of Inverness), those which flank the River Ness are nevertheless excellent illustrations of the principle of sub-glacial deposition. Where the modern river has truncated these narrow, sinuous ridges it is possible to see that they are composed of bedded layers of unconsolidated sand and gravel, orientated in the general direction of the former ice-movement. Although it was once believed that eskers were remnants of the Biblical Flood, it is now generally accepted that they were formed by glacial meltwaters in tunnels beneath the ice. Thus, as the ice-sheet waned the tunnel infills became exposed from beneath the down-wasting ice, finally to be left as steep-sided ridges running indiscriminately across the drift plains. At the Flemington esker it is possible to reconstruct the former sub-glacial tunnel network, for here the serpentine ridges subdivide and rejoin each other in what is termed an 'anastomosing pattern'.

The ice-sheets beneath which the eskers were formed also left a variety of glacial deposits around Inverness, including so-called 'boulder-trains', erratics and extensive till sheets. In addition, their periodic still-stands, during their final decay in this region, have left remnants of moraines in the landscape, none being more important than those at Ardersier, Alturlie and Kessock. At each of these locations the smooth coast of the Moray Firth is broken by a low headland of unsorted morainic drift. The curving moraine of Ardersier, for example, can be identified as recurring at Chanonry Point on the other side of the Firth, despite its subsequent mantle of marine deposits. Such morainic remnants are believed to have been the local equivalent of the so-called Perth Readvance glacial stage but, more importantly, they have since provided the natural breakwaters around which the sea has fashioned the remarkable triangular forelands upon which Fort George and Fortrose now stand.

These terraced forelands are composed of no less than four raised beaches and represent all the various episodes of wave construction, from late-glacial times to the present day. A deep, tide-scoured channel (24 fathoms) now separates the two points, as waves and tidal currents still work upon the unconsolidated drifts of the Moray Firth to maintain the shape of the two forelands. In much the same way the marine processes must have fashioned first the late-glacial and then the post-glacial (Flandrian) beaches before each was isostatically elevated above the reach of the waves. The late-glacial seas succeeded in breaking into Loch Ness, where their strand-lines have now been uplifted to an elevation of some 90 feet (27 metres). The late-glacial rivers formed deltas where they entered this ephemeral arm of the sea, and these uplifted deltas have proved to be the only worthwhile sites for settlement around the fearsomely steep shores of Loch Ness.

The Western Highlands

To the west of the Great Glen lies a little-known mountain massif in which the 3,500-foot peaks (over 1,000 metres) at the head of Glen Affric give way to a series of lower but equally attractive summits as they approach the greatly fragmented coastline of the Western Highlands. Here, in a remote region where the sea-lochs penetrate deeply into the heart of the Highlands, is a land which was well known to Bonny Prince Charlie after his defeat at Culloden but has remained partly uncharted by the Geological Survey for over a century. Few roads traverse the area even today, so that the western peninsulas of Knoydart, Morar, Moidart, Ardnamurchan and Morvern are virtually cut off from the tourist-haunted thoroughfare of Glen Mor by their mountain bastions, lochs and fjords.

From Loch Ness itself there is a choice of two routes to gain access to the western coast: the northern, via Glen Moriston to the delectable Glen Shiel, with its entourage of conical peaks known as the Five Sisters of Kintail, adjudged by W. H. Murray to be the epitome of the West Highland scene; or a more central route past Loch Garry and Loch Quoich to the hidden sea-waters of Loch Hourn. When we consider that the slopes of Glen Moriston and Glen Garry are now thickly clothed with mature plantations of fir, pine and spruce, it is instructive to read how the famous Dr Johnson witnessed a very different scene when passing this way two centuries ago. He sardonically remarked that the mountains to the west of Glen Mor represented '... matter incapable of form or usefulness, dismissed by nature from her care and disinherited of her favours, left in its original elemental state, or quickened only with one sullen power of useless vegetation'. The burning and destruction during the infamous Highland Clearances must only have exacerbated the desolation of the scene in this, the wettest area of the Highlands (with an annual precipitation of more than 120 inches, 3,048 millimetres), but it is salutary to discover that the pine forests above Loch Arkaig were accidentally destroyed by fire during the Second World War. Since Loch Arkaig occupies the only major glen pointing westwards from Glen Mor that lacks a through road to the coast, it has remained virtually untouched for centuries. As a result, the peninsula of Knoydart and its curving moat of Loch Nevis can only be explored on foot or from the sea.

In Knoydart (as in Morar) the Moine metasediments have thin strips of Lewisian schist and gneiss occurring within their alternating sequence of psammitic and pelitic rocks (see Glossary). Although the ancient Lewisian rocks, which crop out extensively in Glenelg, to the north of Loch Hourn, originally formed the cores of the earliest anticlines, later phases of folding and thrusting have deformed the Pre-Cambrian rocks of these western coastlands. Thus, subsequent denudation has worked not only upon a recumbent fold, termed the Knoydart Fold, but also upon later structures known as the Morar Antiform and the Ben Sgriol Synform. The resulting fold complex is, therefore, responsible for the generally north–south alignment of the individual rock types in these western peninsulas, an alignment later to be followed by many of the Tertiary dykes belonging to the Skye 'swarm' (see Chapter 13). Not that this alignment has manifested itself in the scenery hereabouts, for one is much more conscious of the deep glens and fjords which trench the mountains in an east–west direction, at right angles to the regional strike. Furthermore, as if to demonstrate that geological relationships are not always as straightforward as we would suppose, the highest peaks of this area (for example, Ladhar Beinn, 3,343 feet, 1,019 metres, and Meall Buidhe, 3,107 feet, 947 metres) are more often than not built from the reputedly less resistant argillaceous metasediments (pelites) rather than the tougher metamorphosed sandstones (psammites). This is probably to be explained by the fact that many of these high summits coincide with the axis of the Ben Sgriol Synform, so that their joints are tightly closed – yet another example of 'inversion of relief'.

Despite the possibility that many of the discordant east–west valleys in these western peninsulas may have been formed initially by pre-glacial rivers draining westwards from Scotland's primary watershed (see p. 145), there is little doubt that their present overdeepened form is a result of glacial modification. Each of the westward-trending valleys between Glen Mor and the coast (most of which carry the only communications in the region) crosses the primary watershed by means of a deep glacial breach: that which carries both road and rail past Loch Eil to Mallaig is now a mere 60 feet (18 metres) above sea-level. The elevation of the breached cols behind Lochs Hourn and Nevis is, however, considerably greater, thus adding to their inaccessibility. The high peaks and steep flanks of Loch Hourn make it the closest Scottish counterpart of a Norwegian fjord and, like neighbouring Loch Nevis, it exhibits the typical submerged threshold across its entrance. Farther south, however, the post-glacial marine submergence has failed to cross the rocky threshold at the western end of Loch Morar, so that this, the deepest of British lakes (about 170 fathoms), has remained a fresh-water loch. Today it drains seawards across a narrow isthmus, where its once picturesque waterfall has now been reduced to a trickle by a hydro-electric scheme. Although ice is undoubtedly responsible for the prodigious overdeepening of its rock basin, the scouring may have been facilitated by the presence of a submerged fault, similar to that along which the near-by linear valley of Loch Beoraid was excavated. Even though many of the glens and fjords of western Scotland have been carved along fault-lines, there is no need to revert to the views of Professor J. W. Gregory who, some fifty years ago, argued for their tectonic origin. He suggested that all these western valleys were created almost exclusively by faulting or even rifting, glacial processes being relegated to a minor role.

South of the River Morar the Mallaig coast road passes through some of the most magnificent coastal scenery in the whole of Scotland. The harsh bareness of the Knoydart moorlands gives way to the softer oak and birch woods of Arisaig, whilst the narrow outcrop of a micaceous psammite has weathered into the glittering silvery sands which adorn the multitude of tiny bays: 'Their sands flash white across the sea to Skye, Rhum and Eigg . . . Long lines of inshore skerries make a ragged foreground to the Cuillin. Behind some of the bays rise sand-dunes bound by marram, behind others slope level green machair, or the hayfields of croft-land.' This evocative description by W. H. Murray reminds us that we are now in the realms of the Hebrides, which have their own special scenery. For the remainder of our journey through the western peninsulas of Ardnamurchan and Morvern, we cannot fail to be lured by the tantalizing presence of those fabled isles across the sapphire-coloured seas.

Were it not for a narrow isthmus near Salen, the long arm of Ardnamurchan would itself be insular, but instead it remains the westernmost point of mainland Britain. In most other ways, however, it is Hebridean, not least in its remarkable geology and scenery. A new coastal road around the mountainous bulk of Moidart leads us quickly to the outlet of the long, curving trench now occupied by the ribbon-lake of Loch Shiel, which, like Loch Morar, has just failed (by a mere 20 feet, 6 metres) to be reached by the sea and turned into yet another fjord. Like the flanking mountains of Moidart, Ardgour and Sunart, the hills at the base of the Ardnamurchan peninsula have been carved from Moinian metasediments similar to those described earlier. The lithological and structural complexity of the psammites, pelites and metamorphic limestones has been further complicated by the intrusion of the massive Strontian Granite and the later invasion of a Tertiary dyke swarm. Nevertheless, it is possible to distinguish detailed topographic contrasts between the various Moinian metasediments through their differing responses to weathering. Thus, the more evenly grained psammites usually coincide with the smoother outlines in this hilly landscape, whilst the pelites, which represent the metamorphism of shales into gneisses, give rise to a rougher, knobbly land surface.

The western end of Ardnamurchan presents a very different scene. Here we shall renew our acquaintance with the types of landforms produced by the gigantic intrusive centres which dominated the Brito-Icelandic volcanic province in Tertiary times. In Chapter 4 we saw how Arran's mountains had been carved largely from the Tertiary igneous rocks of the Goat Fell Granite and the Central Ring Complex; likewise, in Chapters 12 and 13 the landforms produced by the Tertiary intrusive centres of Mull, Rhum and Skye will be explored and explained.

The central intrusion complex of Ardnamurchan clearly illustrates most of the major examples of Tertiary igneous activity, including ring-dykes, cone-sheets, volcanic vents, sills, dykes and plateau basalts, most of whose formations have been described on pp. 73–7. In contrast with the Central Ring Complex of Arran, however, Ardnamurchan had three

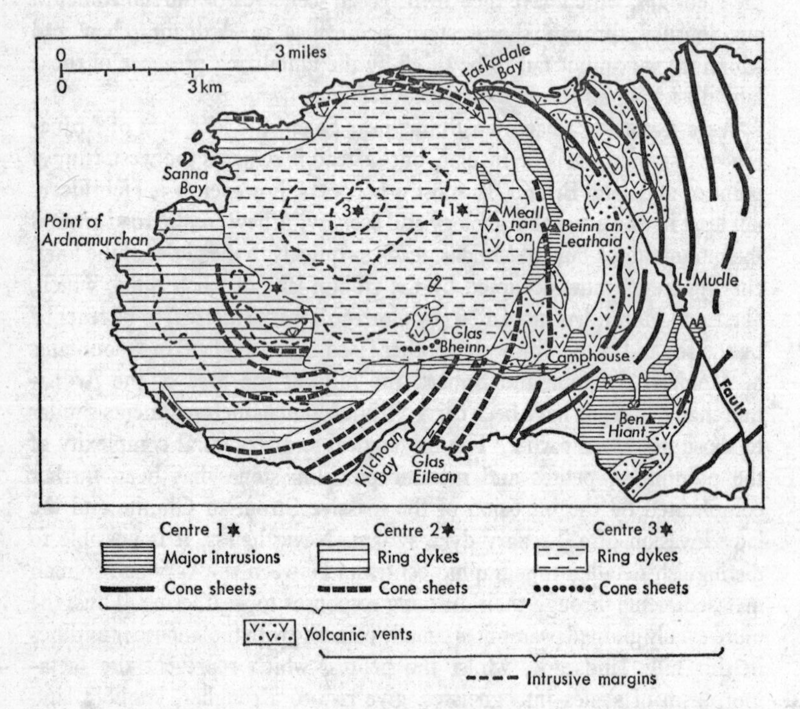

Fig. 38. *The igneous geology of Ardnamurchan (based on material prepared by the Institute of Geological Sciences). Unshaded areas include rocks of Moinian and Jurassic age.*

centres of intrusion (Figure 38), whilst another difference between the two areas becomes apparent when we examine their petrology. If we exclude dykes and other minor intrusions (except cone-sheets), we find that virtually all the Ardnamurchan plutonic rocks are of basic composition (mainly gabbro and dolerite); Arran, on the other hand, is almost exclusively an island of granite and granophyre (that is, acid) composition, so far as its plutonic rocks are concerned.

Although plateau basalts occur on a limited scale in both Ardnamurchan and Morvern (see below), their influence on the landforms is not as marked as it is in Mull, so a full description of basaltic characteristics will be deferred until Chapter 12. In Ardnamurchan the most important igneous activity began with the formation of the first explosion vents, which were punched right through both the plateau lavas and their underlying Moine Schists and Mesozoic sedimentaries. By referring to Figure 38 the reader will be able to follow the succeeding history of the igneous activity as the Ardnamurchan volcanic pile grew steadily in stature. Officers of the Geological Survey have recognized the following time-sequence, as the centre of activity shifted its position: first, the volcanic vents of Centre 1, picked out by their agglomerates, were cut through by the massive cone-sheets and intrusions of the opening phase; secondly, as the activity shifted westwards Centre 2 produced a further series of cone-sheets and a succession of massive ring-dykes now exposed in the western end of the peninsula; finally, the activity shifted back to an intermediate location and Centre 3 was responsible for an innermost set of magnificent ring-dykes which are both basic and acid in composition. Denudation has lowered the massive volcano to nothing more than a 'basal wreck', although the concentric pattern of the structure is still clearly discernible in the landforms.

Ardnamurchan's highest peak, Ben Hiant (1,729 feet, 527 metres), has been carved from a complex of vent agglomerates emanating from Centre 1, bolstered by an intrusive mass of quartz-dolerite which forms the summit. To the west of Kilchoan Bay the ring-dykes and cone-sheets associated with Centre 2 have helped to create a curving rampart of low hills surrounding a semicircular valley, which in turn gives way to the tougher gabbroid ring-dyke eminence of Beinn na Seilg (1,123 feet, 342 metres). But it is the ring-complex of Centre 3 which is most noteworthy, for it is the best preserved example of ring-dyke formation extant in Britain. The most striking of the ring-dykes, as far as the scenery is concerned, is that of the very basic Great Eucrite (see Glossary): its hardness has been responsible for an almost unbroken ring of hills some three miles in diameter. Its inner margin coincides with a circle of steep slopes and crags which in turn give way to a nested series of circular valleys and ridges centred around the settlements of Glendrian and Achnaha. As these inner ring-dykes are traced inwards, their composition becomes progressively more acid, culminating in a tiny central exposure of quartz-monzonite. Marine erosion is currently battering away at the outer ramparts of the igneous complex, and at Sanna Bay has broken through as far as the Great Eucrite ridge. The coasts of western Ardnamurchan, therefore, are characterized by an alternation of dark, forbidding cliffs and headlands with delightful hidden bays backed by beaches of creamy shell sand.

The last of the peninsulas which make up the varied scenery of the Western Highlands is that of Morvern, cut off from Ardnamurchan by the 20-mile (30-kilometre) incursion of Loch Sunart. This broad peninsula can be divided into three contrasting areas: first, the deserted eastern massif, which coincides with the outcrop of the Strontian granite; second, the central hills of Moine Schists, whose scenery is little different from that of neighbouring Moidart and Sunart and whose broken terrain has inspired the title of Taobh Garbh (Gaelic: 'the rough side)', and, finally, the basaltic plateaux of the west, which flank the Sound of Mull and whose forested eastern escarpment overlooks the loch-filled glens of central Morvern.

The Strontian granite (whose southern extension we have already encountered at Foyers, dramatically truncated and shifted by the Great Glen Fault) is composed of three main intrusions. These exhibit evidence of increasing acidity as they are traced inwards from the surrounding tonalite (see p. 59) through a granodiorite to the inner core of biotite-granite, and they are thought to represent successive phases of emplacement. Only the biotite-granite exhibits a resistance to denudation comparable with that of the surrounding Moinian, so that the high granite peaks of Kingairloch are now separated from the Moinian summits of central Morvern by a depression floored with granodiorite and tonalite and followed by the waters of Allt Beitheach.

Following the discovery in 1791 of 'Strontianite', the element of strontium was isolated and named after the tiny village of Strontian at the head of Loch Sunart. This area remained an important lead-mining centre between 1722 and 1872, although production has now ceased. The galena is part of a zone of mineralization occuring in the Moinian rocks that surround the Strontian Granite, which itself is of Caledonian age.

Although there appears to be some genetic connection between the granitic emplacement and the mineralization, isotopic datings show that the Strontian lead was formed in at least two phases between Permo-Triassic times and the Early Tertiary. It is possible that the structural doming caused by the granite pluton resulted in a crustal weakness which was later to act as a focus for rising geochemical solutions at different periods of geological time.

It is now time to turn westwards and follow the road down Gleann Geal to Lochaline. Here one final surprise awaits us. Above this crouching coastal loch the towering cliffs of Tertiary plateau basalts, with their silvery waterfalls, give a foretaste of Mull. If we troubled to climb their forested slopes, however, we should find that entombed beneath the black lava cliffs there are thin layers of sedimentary sandstones, shales and limestones resting unconformably on the Moinian floor, the very pure, friable sandstones being quarried for glass sands. Furthermore, the outlying basal hills of central Morvern (for instance, Beinn Iadain, 1,873 feet, 571 metres) overlie narrow bands of similar rocks.

It comes as something of a surprise to discover that these sediments represent layers of Triassic, Jurassic and Cretaceous rocks, isolated by several hundred miles from their more extensive counterparts in England. Such Mesozoic sedimentary rocks are not uncommon in the Hebrides, for the seas of Keuper, Jurassic and Upper Cretaceous times extended far to the north. The fact that so few remnants now survive, however, indicates the enormous amount of denudation that has subsequently taken place in Scotland. The majority of the Mesozoic rocks in Northern Scotland owe their preservation to the Tertiary lavas which flooded over them and gave them a protective cover. There is little doubt that the Morvern lavas formerly extended farther east, judging by the outliers to the north of Loch Arienas; all of these were once part of the main lava plains of Mull and Morvern, but erosion has cut them off from the main basaltic exposure. We must not think, therefore, that the prominent east-facing scarp of Sithean na Raplaich (1,806 feet, 551 metres) represents the terminal limit of the lava flows, for this scarp was produced by subsequent uplift and denudation of the lava plains. With the gradual wearing back of the escarpment since the lavas were first extruded in Eocene times, small undestroyed outliers have remained to the east of its present limit, representing unconsumed relics of a former continuous basaltic cover.

The same denudation processes, including those caused by Pleistocene ice-sheets, have also isolated the Morvern lavas from their counterparts in Mull, the present Sound of Mull having been created by glacial breaching at the head of a deep trough later to be flooded by the rising post-glacial seas.

12. The Inner Hebrides – the Mull Group

During our examination of the mainland scenery of south Argyll in Chapter 10, it was convenient to look also at the southernmost Hebridean islands of Islay, Jura and Colonsay, because of their structural continuity and geological similarity with the mainland. Once across the Firth of Lorn, however, the Tertiary igneous rocks of North-west Scotland, which were briefly encountered in Arran (see Chapter 4), begin to dominate the Hebridean scenery through their widespread occurrence. It will be seen that the landforms of Mull, Rhum and Skye, for example, differ from those of Islay and Jura just as much as the landscape of Arran was seen to contrast with the neighbouring coasts of Ayrshire and Kintyre. Thus, in the present context, the geographical term 'Inner Hebrides' will be regarded as including only those of the Western Isles which lie to the north of the Firth of Lorn and to the east of the Minch.

Mull

Throughout any journey along the indented coastlines of Ardnamurchan and Morvern it is impossible to ignore the gigantic presence of Mull, that island of contrasts, seemingly only a stone's-throw away. Its cloud-capped southern mountains have formed an impressive backdrop to the oceanic vistas for countless tourists in the mainland resort of Oban, especially when favoured by the unforgettable colours of a Hebridean sunset. Little wonder that one of the most popular sea trips in western Scotland is the circumnavigation of Mull, which includes an opportunity for visiting the fabled isles of Staffa and Iona. To understand fully its scenic complexity, however, it is necessary to land on the island of Mull itself, and no better place exists to commence an excursion than Tobermory.

Few ports can claim so romantic a setting as Tobermory in its 'hilly theatre', for its curve of gaily painted quayside houses is surmounted by a halo of sycamore woodland clinging to the steep lava cliffs. The luxuriance of these deciduous woodlands flatters only to deceive, however: when we have climbed steeply westwards out of the tiny bay it soon becomes apparent how Mull received its name, which may be interpreted as meaning in Gaelic 'high, wide tableland'. So far as the northern part of the island is concerned this is an apt description, for here is a plateau of deeply dissected Tertiary lavas, surpassed in extent only by the basaltic plateaux of Antrim and Skye. Nevertheless, Mull has some compensation in that its 6,000 feet (1,800 metres) of basalt far exceeds the thicknesses found elsewhere in the British Isles. To say that these northern plateaux have a structural simplicity is not to suggest that their scenery is everywhere mundane, for there are few places in Scotland where a volcanic 'trap' landscape is so clearly exemplified.

The term 'trap' is derived from the Swedish *trappa* meaning 'step', and is used internationally to describe the way in which differential erosion of successive lava flows gives a stepped character to the landscape. Where rivers and ice have combined to deepen the valleys and isolate the plateau-remnants into tabular hills, the structural benches created by the lava flows can be traced for many miles along the hillslopes. Each of the 'treads', having weathered into a gently sloping terrace on the more friable upper surface of each lava flow, is generally covered with peat and a heathery vegetation. Separating the 'treads' of this volcanic staircase are the steeper, and sometimes precipitous, 'risers' which represent the more resistant lower layers of each lava flow.

The road westwards from Tobermory zigzags up and down the lava terraces, affording glimpses of the lonely Loch Frisa, whose thickly forested slopes mark the limits of a glacially overdeepened valley. Beyond the conspicuous volcanic plug of S'Airde Beinn (959 feet, 292 metres) stretch the bleak headlands of Mishnish, Quinish and Mornish, which complete the steep, cliff-girt coast of northern Mull. The suffix '-nish' means 'a flat headland', for this is a coastline carved entirely from basaltic lavas. Dr Johnson spoke of these northern plateaux as a 'gloom of desolation' and suggested that they should be planted with trees to give them a 'more cheerful face'. Some two centuries later the Forestry Commission has obligingly clothed the moorlands with widespread conifer plantations, a policy which has created some controversy amongst farmers and conservationists alike. This is because the base-rich basalts weather into an excellent soil which, where adequately drained and maintained, produces 'the most fertile land in the Highlands and some equal to the best in the kingdom' (F. Fraser Darling). It is fair to comment, however, that long before the Forestry Commission turned its attention to Mull the rich cattle-grazing lands had already been abandoned to the ubiquitous sheep. Their more selective grazing has meant the appearance of a profuse growth of bracken and hazel scrub in many of the hollows of northern Mull.

Foremost amongst the glories of these northern headlands are the views which they afford northwards to the mountains of Ardnamurchan and Rhum, although the coastal walker will also find hidden coves whose dazzling white shell-sand beaches are often carpeted with sea-pinks and dotted with aquamarine pools. Beyond the pretty village of Dervaig the road reaches the west coast at Calgary, whose attractive wooded bay affords us our first view of Tiree and Coll on the western horizon. It also gives us a glimpse of a typical Hebridean *machair* landscape, in which shelly sands and coastal dunes combine to support a carpet of flowery grassland on the 'sweeter' soils, although the Calgary *machair* is only a tiny pocket in comparison with those of Tiree and the Outer Hebrides.

The coast road now turns southwards along an intensively farmed shelf which stands more than 100 feet (30 metres) above the sea and is backed by the steep scarped edge of the basalt plateau. In many places this shelf is a structural bench, created by erosion of a lava flow, but as it is traced out onto the headlands it becomes clear that it is also a raised-beach platform whose abandoned wave-cut notch stands some 150 feet (45 metres) above modern sea-level. Since it can be seen that ice has moulded parts of the platform it is now regarded as part of the so-called 'pre-glacial' raised beach already encountered in Islay and Jura (see Chapter 10). More positive proof of its antiquity has been found on the neighbouring island of Ulva, where not far from the summit of A'Chrannog (387 feet, 118 metres), an ancient sea-cave, about 150 feet (45 metres) above present sea-level, has glacial till surviving on its floor, pushed there by former glaciers.

The island of Ulva, composed entirely of basaltic lavas, rises in regular tiers to its flat-topped summit (1,025 feet, 312 metres) and is another excellent example of a trap landscape. But it has also been singled out by the eminent ecologist Sir Frank Fraser Darling as a warning to Hebridean landowners because of the way in which bad land-management on the fertile basaltic soils has led to the profuse growth of bracken in place of the once flourishing natural grassland. He demonstrates that, in the virtually uninhabited isles of Ulva and Gometra, '... the Highland paradox is apparent again, in that here are green acres gone derelict while there is still congestion on the poverty-stricken Archaean gneiss of Lewis.'

Beyond the narrow channel at Ulva Ferry, where a dolerite plug forms the conspicuous conical hill of Dun Mor, the lengthy sweep of Loch na Keal carries an arm of the sea eastwards to the narrow central isthmus of Mull. The name of this sea-loch means 'Loch of the Cliffs', and it is impossible to traverse its shores without being aware of the prodigious Gribun precipices which flank its southern coastline (Plate 28). At Gruline, where the waist of the island is less than three miles across, one enters a very different region, much larger than the northern tract, a great deal more mountainous and more geologically complex. Indeed, a glance at Figure 39 supports the claim made by the officers of the Geological Survey of Scotland, who noted in 1924 that here was '... the most complicated igneous centre as yet accorded detailed examination anywhere in the world'. Although the apparently interminable sequence of basaltic rocks recurs in even greater thicknesses where these form the island's highest summit of Ben More (3,169 feet, 966 metres), the topography of this mountainous area has been carved very largely from a complex sequence of igneous intrusions which break through the extrusive lava flows. Before coming to grips with the somewhat daunting complexity of southern Mull, however, we must turn to the western headland of Ardmeanach, for here are to be found not only some of the most remarkable landforms in Mull but also some geological phenomena unmatched elsewhere in Britain.

At Gribun cliffs the coast road is one of the most spectacular in Scotland, but also one of the most hazardous, owing to the constant rockfalls and occasional landslips (Plate 28). The instability of the 1,000-foot (300-metre) basaltic precipices is due to the underlying Mesozoic sedimentaries, which have been eroded by ice-sheets moving parallel with the cliff face to cause local oversteepening. Such landslipping, due to the so-called 'incompetent strata' beneath the lava flows, will be described in greater detail in Chapter 13. The Gribun area serves to demonstrate not only the detailed stratigraphy of the Mesozoic sedimentary rocks where they are

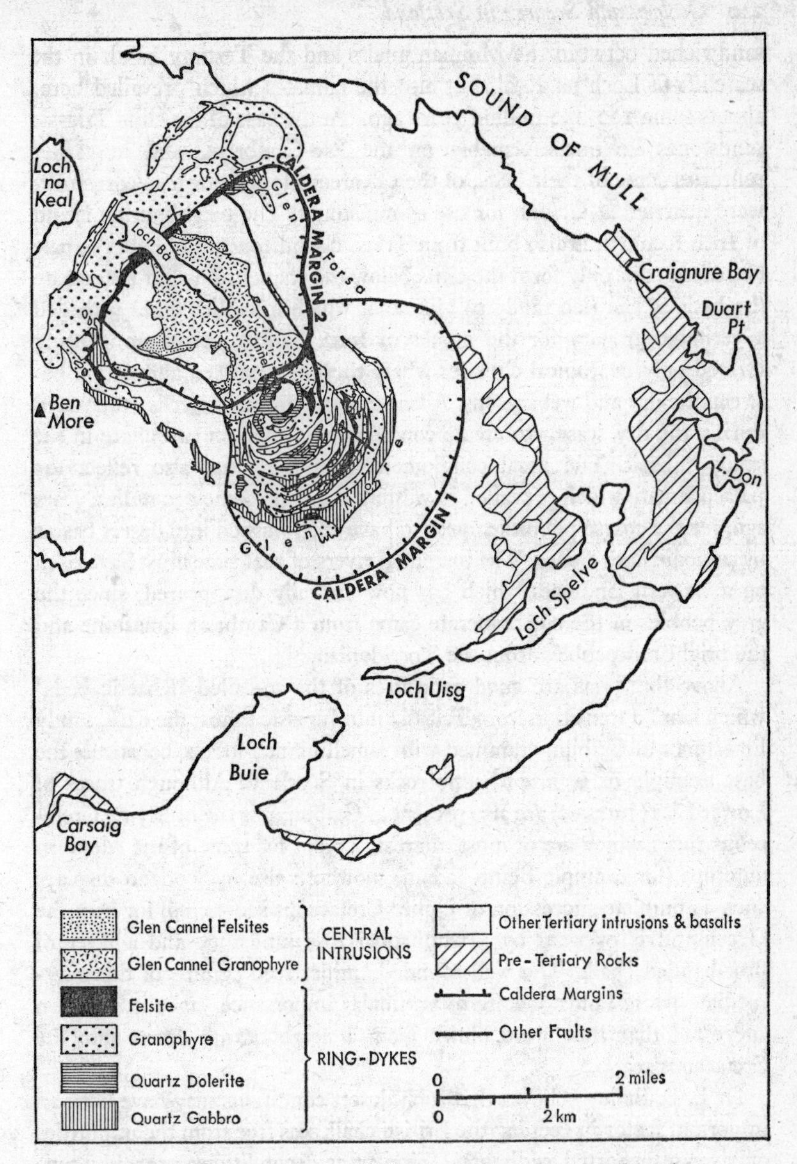

Fig. 39. *The geology of southern Mull (based on material prepared by the Institute of Geological Sciences).*

sandwiched between the Moinian gneiss and the Tertiary basalt in the sea-cliffs of Loch na Keal, but also the climates which prevailed hereabouts some 100–250 million years ago. At the base of the cliffs Triassic sandstones lie unconformably on the Pre-Cambrian, and in earlier centuries some of their beds, of the calcareous type known as cornstone, were quarried at Gribun for use as millstones. The neighbouring island of Inch Kenneth is also built from Triassic sandstones, and here the pale cornstones not only form the cliffs below the chapel ruins but also create the basis of the rich sandy soil like that which in earlier times provided a veritable granary for the monks of Iona. Today, cornstone is being formed only in tropical climates where there is a marked alternation between the dry and wet seasons. Alkali-charged groundwater is evaporated during the dry season to create concretions of calcium carbonate in the sandy subsoil. The basal conglomerates of the Trias also reflect the palaeoclimate which prevailed in western Scotland some 250 millon years ago, for the coarse pebbles appear to have been washed into desert basins by periodic flash-floods. The torrential rivers of that time must have risen on a western landmass which has now virtually disappeared, since the grey pebbles of the conglomerate came from a Cambrian limestone and the bright red pebbles from the Torridonian.

Above the Trias are good exposures of the so-called 'Rhaetic beds' which mark a transition from Triassic into Jurassic times: the dark, sandy limestones of Gribun, crammed with lamellibranch fossils, constitute the best example of *in situ* Rhaetic rocks in Scotland. Although traces of Lower Lias (Jurassic) are also present at Gribun, it is the overlying Cretaceous rocks which are of most interest. Except for some of the Morvern outcrops (for example Beinn Iadain), nowhere else in Scotland displays such a complete succession of Upper Cretaceous sediments, for here the Greensand is overlain by a white quartzose sandstone and a layer of flint-banded chalk. The well-rounded 'millet seed' grains of the intermediate white sandstone are of particular importance, since it has been suggested that they were blown from a neighbouring desert into the Cretaceous sea.

Dr E. B. Bailey believes that such desert conditions may have been an important factor in keeping the British chalk seas free from the impurities of river-transported sediments. Similar arid conditions probably continued into the Palaeocene in western Scotland, for the Gribun chalk exhibits solution cavities infilled with wind-blown sand grains. In places these formed a thin Early Tertiary sandstone which, together with the underlying chalk, became silicified during tectonic uplift before the lavas were extruded. The actual base of the overlying lavas is marked by a narrow band of dark red mudstone, representing the lateritic decay of a volcanic ash and demonstrating that desert conditions had been replaced by a humid, tropical climate during the onset of the Tertiary volcanic episode.

The final rewards of the Ardmeanach headland are reserved for the intrepid walker, since they can only be reached after several miles of scrambling along the shore to the foot of the basaltic cliffs which form the aptly named 'Wilderness'. The northern approach passes the gigantic Mackinnon's Cave, which Dr Johnson thought one of the greatest natural curiosities witnessed by him on his Scottish tour. It is certainly of greater dimensions than the more famous caves of Staffa and Eigg, but it must be remembered that the eminent Doctor never landed on Staffa.

At the farthest extremity of the headland, below the gigantic basaltic cliffs and screes of the Wilderness (near by is Loch Scridain – Loch of the Screes), lie the remains of MacCulloch's Tree. This remarkable geological phenomenon was discovered in 1819, but owing to 'erosion' by fossil-hunters it is now only a 40-foot (12-metre) cast of the original fossil tree which had been buried by the basaltic lavas. Only the lowest part of the trunk can now be seen, a partially silicified cylinder of fossil wood glistening with quartz crystals and surrounded by a sheath of soft, black, charred wood. The 'tree', which stands vertically in the lava cliffs above high-water mark, is rooted in a carbonaceous mud with a film of coal, overlying a bed of red volcanic ash. It is one of only three examples recorded in the Ardmeanach promontory and is clearly related to the organic beds discovered at Ardtun, on the opposite shore of Loch Scridain (see p. 248). However, before examining these deposits, which give evidence of the vegetation prevailing during the volcanic episode, we must turn our attention to the mountains of southern Mull, for these have been carved from a vast central intrusion complex which in its early history probably acted as a feeder for the extensive plateau basalts.

During the nineteenth century there were two schools of thought relating to the formation of the basaltic lavas of this igneous province. On the one hand Professor J. W. Judd believed that the basalts were associated with the gigantic central vent volcanoes of Skye, Rhum, Ardnamurchan and Mull, of which only the 'basal wrecks' remain. Sir Archibald Geikie, on the other hand, supported the contention that the lavas were extruded from linear fissure-eruptions fed from dyke-swarms, by analogy with the modern lava plains of Idaho in North America. In Mull, however, there is evidence to support Judd's hypothesis, since the dykes were created far too late in the Tertiary igneous record for them to have acted as feeders for all the lavas. Furthermore, there is some indication that a massive caldera, or basaltic crater, existed in central Mull during part of the basaltic period (Figure 40a), although almost all traces of it have been obliterated by the subsequent intrusion of the plutonic rocks.

After the eruption of the extrusive lavas, updoming occurred as volcanic activity increased in intensity, thus building an enormous central vent volcano from which acid volcanics were ejected with explosive force (Figure 40b). This second phase of activity was also marked by the intrusion of granophyres along the concentric margin of the early basaltic caldera (Figure 40b), and these hard rocks now form the eminence of Glas Bheinn (1,611 feet, 491 metres) overlooking Loch Spelve. In addition, however, this early intrusion of granophyre gave rise to peripheral folding of the existing rocks (Figure 40b), including those of the Loch Don anticline in the south-east (see p. 245). Slightly later the acid magma was replaced by one of basic composition as olivine-rich gabbro was intruded along the early caldera margin. The gabbro now forms the conical mountain of Ben Buie (2,354 feet, 718 metres), behind Loch Buie, and the lesser ridge of Bein Bheag farther north-east, near Glen Forsa. During this early phase of alternating acid/basic magmatic activity large numbers of cone-sheets were injected, themselves reflecting the differences in mineral composition.

As in the case of Ardnamurchan, we find that there was a progressive shift of igneous activity away from the initial centre. So far as Mull is concerned, this meant a two-stage shift towards the north-west, Centre 1 being Beinn Chaisgidle and Centre 2 Loch Ba (Figure 39). Since the mechanism of ring-dyke formation has already been described in some detail (see p. 214), it is not proposed to dwell for long on these final phases in the history of Mull's important volcano, except to comment on some of the more prominent landforms which now pick out several of the igneous phenomena. The curving shape of Glen More, for example, that

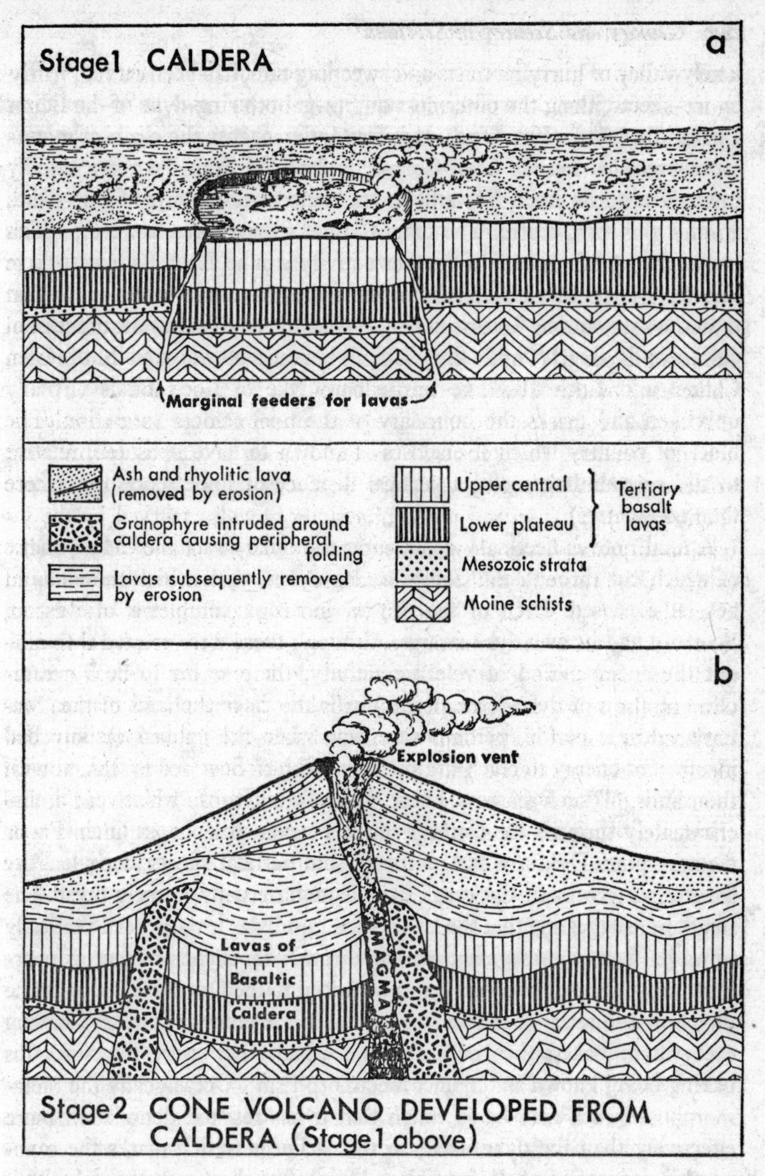

Fig. 40. *Hypothetical reconstruction of* (*a*) *the caldera and* (*b*) *the cone volcano of southern Mull* (*based on material prepared by the Institute of Geological Sciences*).

lonely valley of hurrying mists and sweeping rain, has been carved, partly by ice-sheets, along the outermost quartz–gabbro ring-dyke of the Beinn Chaisgidle centre (Figure 39). It is also apparent that the circle of mountains which surrounds Glen Cannel, and whose drainage runs radially inwards towards Loch Ba, reflects the concentric pattern of ring-dykes around this first centre. The Glen Cannel granophyre, however, which marks the opening phase of igneous activity around Loch Ba (Centre 2 in Figure 39), appears to have been less resistant to the forces of erosion than the surrounding lavas and ring-dykes, for its topographic expression corresponds largely with the glacially overdeepened trough of Glen Cannel and Loch Ba. The felsite ring-dyke of Loch Ba is virtually unbroken and marks the boundary of the final caldera formation. The block of country which it encloses is known to have subsided, relative to the surrounding rocks, a vertical distance of some 3,000 feet (more than 900 metres).

A final note is necessary concerning the dyke-swarms of Mull, many of which cut through the Centre 2 ring-dykes, with which they should not be confused. Each of the central intrusion complexes of western Scotland had its own dyke-swarm. Although these were created throughout the entire period of volcanic activity, there seems to be a greater concentration of dyke-formation towards the closing phases of the Tertiary volcanic period, perhaps at a time when the igneous activity had insufficient energy to reach the surface. Western Scotland is seamed with thousands of such narrow linear igneous outcrops, which cut indiscriminately through the country rocks, be they igneous, sedimentary or metamorphic. Dykes, which occur as thin vertical sheets along lengthy fissures, or cracks in the crust, are characterized by a finely crystalline structure, suggesting that the lava cooled rapidly as it was forced vertically upwards from the magma chamber through thousands of feet of overlying crust. The dyke-swarms can be traced to the Outer Hebrides in one direction and as far as northern England in the other (Figure 41), having everywhere baked the rocks through which they were injected, this baking being known as 'contact metamorphism'. Occasionally the metamorphosed rock may be so tough that it has resisted denudation more effectively than the dyke itself, so that a linear trench marks the topographic expression of the intrusion. More often, however, the dyke itself is harder than the surrounding rocks and stands up from the landscape

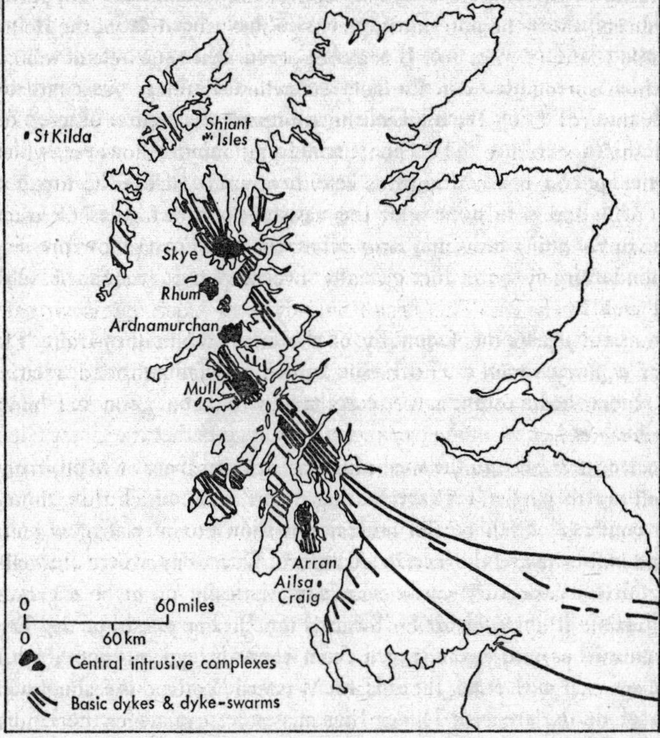

Fig. 41. *The Tertiary dyke-swarms and central intrusive complexes of north-west Scotland (based on material prepared by the Institute of Geological Sciences).*

like an artificial wall (or 'dyke'). Dykes are particularly visible on the seashore all along the south-eastern coast of Mull, to which we must now turn.

The curving shorelines of the south-eastern coast, together with the parallel arcuate trend of Loch Uisg and Loch Spelve (Figure 39), reflect the concentric structures of the island's central intrusion complex. It was noted above (see p. 242) that the early intrusion of a Tertiary granophyre created a series of concentric folds around the perimeter of this igneous complex, and these can best be studied in the Loch Don area. Here a sharply folded anticline has been denuded sufficiently to uncover the

Dalradian schists of its central core, where the overlying Palaeozoic lavas, Mesozoic sedimentaries and Tertiary basalts have been stripped away. The present landforms reflect the cigar-shaped inlier (or 'window') of older rocks where these emerge from beneath the ubiquitous cover of basaltic lavas. The tiny Loch a Ghleannain marks the central outcrop of the Dalradian Schists; the steep surrounding ramparts, such as the forested ridge of Druim Mor Aird, have been carved from the tougher lavas of Old Red Sandstone age; the narrow perimeter vales of Glen Rainich (to the west) and Glen Ardnadrochet (to the east) show the less resistant qualities of the slender outcrop of the Lower Jurassic shales and Inferior Oolite.

From a structural point of view, the occurrence of a blue-grey Dalradian limestone at the centre of the Loch Don anticline is of more than academic interest. Since the limestone can be correlated with that of Lismore Island, at the mouth of Loch Linnhe (see p. 211), it suggests that the Great Glen Fault crosses into Mull to the west of Loch Don and that it was probably bent southwards by the Tertiary folding described above. Furthermore, late movements along this major fault can be shown to have affected both the Mesozoic strata and the Tertiary basalts of Duart Bay, where the well-known thirteenth-century castle stands majestically on its black lava cliffs (Duart is from *Dubh Ard* – 'Black Point'). The neighbouring bay at Craignure has also been carved from an anticlinal structure, with Triassic as well as Lower Jurassic rocks revealed along the shoreline. These include the greenish Lower Lias shales (equivalent to the Pabba Shales of Skye), which can be traced intermittently south-westwards around the coast, where they project from beneath the lava cliffs to form a conspicuous coastal shelf between the mouths of Loch Spelve and Loch Buie. Between Port Donain and Port nam Marbh the shales are accompanied by a thick, white, cliff-forming sandstone (Middle Lias – the Scalpa Sandstone of Skye), together with the Upper Liassic shales and the Inferior Oolite.

The mountain-girt recesses of Loch Spelve and Loch Buie are also of great interest, for the Triassic sandstones are well-represented and the extremely fertile soils of this area have caused it to be called the 'Garden of Mull'. But the chequered hayfields and beautiful hanging woodlands between Loch Buie and Loch Spelve reflect something more than the basic mineralogy of the underlying rocks, although it is true that the basalts, gabbros and dolerites contribute to the fertility of the local soils. In fact, the richness is derived very largely from the calcareous glacial drifts which floor this narrow arcuate corridor.

Glaciers have clearly been responsible for the overdeepening of Loch Spelve (20 fathoms), whose waters now spill seawards over a shallow submarine threshold; the ice may have been generated from the neighbouring mountains of southern Mull, for it is known that Mull had its own ice-cap during the Upper Pleistocene. The marine shells of the glacial drifts have been dated by radiocarbon techniques which indicate an age of some 11,000 years, suggesting that the local glaciers scooped up marine deposits which had been laid down during the preceding interstadial period of the Late Pleistocene, when sea-waters must have flooded into the arcuate corridor. Further research (by analysis of fossil pollen-grains) has finally established that the Loch Spelve moraines are of similar age to those at the Lake of Menteith (see p. 120) in the Forth Valley, which confirms that Mull had its own ice-cap during the Loch Lomond Readvance. Other terminal moraines of this age may be seen at the western ends of Glen More and Loch Ba, whilst Glen Forsa exhibits hummocky lateral moraines and an outwash fan at its northern end. Whether or not it was this late piedmont phase which caused the major glacial breach now followed by the Lussa River, where it breaks out of the eastern end of Glen More, is a matter for conjecture.

Before turning westwards along the narrow peninsula of the Ross of Mull it is important to make a short detour to the south coast of Mull at Carsaig, for here the thick basaltic lavas have again been breached to reveal an interesting succession of Mesozoic rocks. The richness of the soils in this tiny pocket of calcareous rocks introduces an unusual oasis of prosperous farmland and watercress beds into the deserted moorlands of the basaltic trap landscape. Below the grey basaltic beach sands, Carsaig Bay is floored by flat slabs of dark grey shales (Pabba Shales) crammed with fossil ammonites, each glinting with pyrites minerals. It was in this lonely bay that the type-specimen of the fossil, *Uptonia jamesoni*, was collected by no less a personage than the great geologist, R. I. Murchison.

The overlying Scalpa Sandstones, of Middle Lias age, form the cliffs to the west of the bay, where, near to the so-called Nun's Pass, they were once quarried for use in the doorways, pillars and windows of the renowned abbey on Iona. This massive creamy sandstone, much in demand as a freestone in earlier centuries, is unusual in possessing large, sandy, calcareous inclusions known as 'doggers', which are generally darker and harder than the main sandstone, in which they were formed as lime-rich concretions. The overlying Cretaceous greensand and thin Tertiary sandstone are only of academic importance, although the Tertiary lignitic coals and the Nun's Cave sapphires have been of a more widespread interest. The latter occur as tiny impure crystals of corundum associated with the thermal metamorphism created by the doleritic sills, whilst the carbonaceous deposits represent ancient weathered land surfaces which existed during the Tertiary volcanic episode in Mull.

Similar carbonaceous deposits are more clearly exhibited at Ardtun, several miles farther west, where the trap landscape of Brolass is terminated abruptly by the line of the Bunessan fault. Like the lignite bands, the famous 'leaf-beds' of Ardtun Head substantiate the belief that the Tertiary volcanoes of the Inner Hebrides were not continuously in eruption. During dormant periods soils and vegetation became established on the basaltic lava plains, only to be buried by later eruptions. First described by the Duke of Argyll in 1851, the leaf remains demonstrate that the trees growing in the warmer climate of Early Tertiary times had Far Eastern affinities and were not related to modern British varieties. Of these the maidenhair tree (*Ginkgo*) of China, the *Cryptomeria* of Japan and the beautiful *Magnolia* appear to have been the most exotic, although species of oak, hazel and plane were also present. The leaf-beds were deposited in temporary lakes on the lava surface which also acted as receptacles for the gravels and sands that now mark the interbasaltic horizon. Above and below these beds are exposures of beautifully columnar lava, the hexagonal tops of which provided ready-made seats for Johnson's and Boswell's picnic party when they visited this site in 1773. The columnar form of the rocks, which rise in a series of pilasters, is comparable with that of their more famous counterparts in Staffa, and a complete description will be attempted when we examine that fabled isle (see p. 251).

To the west of Bunessan the undulating terrain of the Moinian mica schists soon gives way to the rolling, peat-covered moorlands and bare hills of the Caledonian granite of the Ross of Mull. Varying in colour from pale to deep red, this coarse biotite granite has been extensively quarried as a building stone. The quarries near Fionnphort, at the western tip of the peninsula, demonstrate how the widely spaced jointing and the proximity of natural harbours facilitated the export of the gigantic blocks to all parts of Britain. Following the introduction (from the Mountains of Mourne) of the so-called 'plug and feathers' method of granite-splitting, the Mull granite was utilized in such London structures as Blackfriars Bridge, Holborn Viaduct and the Albert Memorial. Somewhat earlier it had helped to build the notorious Skerryvore lighthouse, which rises from the western ocean some 25 miles (40 kilometres) beyond Iona. Its earliest recorded use, however, was in the seventh-century Abbey of Iona itself, where rounded glacial erratics and other surface blocks were used in the rubble walling.

Iona and Staffa

To visit the tiny island of Iona is to experience a feeling of tranquillity and harmonious well-being that epitomizes the whole atmosphere of the Hebrides. Numerous visitors have commented on this remarkable Hebridean 'experience', which seems to be created from a combination of idyllic scenery, romantic architecture and extraordinary atmospheric effects. From the low hill of Dun I it is possible to view some of the scenic elements that make up this harmonious landscape. To the north, the misty hills of Rhum are framed by the headlands of Mull and Coll; nearer at hand the basaltic specks of Staffa and the Treshnish Isles are outliers of Mull's northern tablelands; at our feet lies the green *machair* whose white shell-sands continue offshore along Iona's northern coast, turning the encircling sea into pools of brilliant turquoise and hyacinth. Westwards the low hills of Tiree are barely visible in the pearly Atlantic haze, but southwards the surf breaks uneasily on the treacherous granite fangs of the Torran Rocks, famed as the shipwreck site in R. L. Stevenson's *Kidnapped*. Far away beyond them the Paps of Jura announce their unmistakable presence, whilst to the east the deep blues and soft purples of the cloud-capped mountains of southern Mull contrast with the shining red granite of the Ross and the sparkling waters of the Sound of Iona itself. The overall effect is completed by the pale shell-like iridescence of the Atlantic light, diffused by the moisture-laden air.

The ancient Lewisian and Torridonian rocks of Iona are so different from the Mull succession that it has been suggested that the Moine Thrust is submerged beneath the Sound of Iona (Figure 37). Few of these rocks have, however, been utilized in the ecclesiastical buildings themselves. The Nunnery walls, for example, are constructed largely from pink Ross granite, local black schists and the creamy Carsaig freestone – although the coloured mosses and the riot of valerian and other wild flowers often distract from the hoary stonework. The red granite and dark schist are also used in the main buildings of the Cathedral, where they form a naturalistic mosaic. Here some of the older roofing and flooring is made of flaggy Moinian schists, quarried on a small scale on the Ross of Mull, but much of the newer restoration has been carried out with imported materials. In addition to the Carsaig freestone used in the mouldings, the interior of the Cathedral has some ornamentation created from the Iona Marble, which is known to have been quarried from the sixteenth century onwards. This attractive green and white rock – a white metamorphosed limestone streaked by a green variety of serpentine – occurs in narrow veins among the Pre-Cambrian rocks of the seashore at the south end of the island. Where the sea has eroded and worked a submarine extension of the marble, one may be fortunate enough to find the beautiful Iona Pebbles on such beaches as that of Port na Churaich, the spot where St Columba landed more than fourteen hundred years ago. Because of its subsequent sanctity no less that forty-eight Scottish kings are reputedly buried on Iona, including the notorious Macbeth. Unhappily, of the 360 crosses said to have been erected in the grounds of the Abbey only three have survived; these include the schistose Cross of Maclean.

When one considers its minute size (smaller than Ailsa Craig), it is surprising to find that Staffa is one of the best-known islands in Britain. Although its geological wonders must have been known to the inhabitants of Mull, Ulva and Iona, it was not until the visit of Sir Joseph Banks in 1772 that this basaltic masterpiece became more generally known. Thereafter it became part of a recognized 'Highland Tour', having qualities sufficient to attract the attention of Queen Victoria and most of the nineteenth-century Romantic poets, including Scott, Keats and Wordsworth, and was recorded for posterity by such eminent artists as Turner and Copley Fielding. But it was the musical genius of Mendelssohn, as displayed in his famous overture, 'Fingal's Cave', that did more than any thing to publicize the scenic grandeur of Staffa.

The aesthetic appeal of the island is based entirely on its geological character; the basalts of its sea cliffs are not only excellent examples of columnar structures, they also have the advantage of remarkable caves. There are taller columns on the Shiant Isles, north of Skye, and a much greater number at the Giant's Causeway in Ulster, but only on Staffa do they display such a natural simplicity in the composition of the lava flows as to fulfil the aesthetic principles of harmony and symmetry. At Fingal's Cave (Plate 29) the structural differences within a single lava flow, reflecting the contrasting rates of cooling after its intrusion, are clearly shown. The slaggy crust cooled rapidly from above to produce an almost structureless capping (now forming the roof of Fingal's Cave); the middle part of the flow also cooled from above, but at a slower and more uniform rate, the loss of water and regularity of contraction leading to the formation of narrow polygonal columns; the lowest part of the flow cooled from below, with an even more uniform rate of contraction, to produce the massive, regularly spaced columns which now form the 'causeway' leading into the cave. The columns are always arranged at right angles to the cooling surface, and most of them are vertical because in the majority of cases this isothermal surface was horizontal. If, however, the lava spilled into an existing hollow or valley, as at the Buchaille Rock and Clamshell Cave, then in some sections the columns will be curved or tilted obliquely.

The affinities of the Staffa lavas with those of the Giant's Causeway are very clear, since both are part of the Tertiary igneous province of North-west Britain. The architectural character of the rocks is such that it is difficult to conceive how nature could have fashioned such a phenomenon; little wonder that our ancestors created the legend of Fingal (Finn MacCool) to account for the apparent 'causeway' link between Ireland and Scotland. As late as 1694 it was thought necessary to devote time at the Royal Society to confirming that the Giant's Causeway was not man-made – because of the lack of mortar between the columns! If the speaker had visited Staffa, however, he might have been less certain in his pronouncements. Here, the pressures of contraction have caused the calcite of the basaltic constituents to recrystallize along the vertical joints, where it makes the columns look as if they have been cemented together.

The neighbouring Treshnish Isles are, like Staffa, parts of the same lava sheets which form the western headlands of northern Mull and the isles of Ulva and Gometra. Their tabular form is most pronounced, especially on the island of Bac Mor (The Great Hump), popularly known as the Dutchman's Cap, where the 'brim' is a marine-eroded platform and the 'crown' a remnant of several overlying lava flows. The wave-cut platform stands at an elevation of some 100 feet (30 metres) and appears to be part of the so-called 'pre-glacial' raised beach of the Western Isles. Today the Treshnish Isles are given over to grazing sheep because of their high-quality grasslands.

Tiree and Coll

Most visitors to these isles will probably note that their geology and scenery are more reminiscent of the Outer Hebrides than of Mull, an impression which the widespread presence of the lime-rich *machair* confirms. It is more convenient to examine them at this point, however, because of their geographical proximity to Mull. The contrast in relief between the latter and these off-shore satellites is most marked, for, while Mull is characterized by the vertical planes of high mountains and precipitous sea-cliffs, Tiree and Coll exhibit a remarkable horizontality, neither of them rising above 450 feet (135 metres).

Despite its low hills of Lewisian gneiss, Tiree gives a persistent impression of flatness. This is because of its widespread deposits of blown shell-sand (*machair*) and its suite of late-glacial and post-glacial raised beaches. The resulting fertile pasturelands have been described thus by G. Scott-Moncrieff: '... a brilliant green flatness contained beneath a great domed bowl of sky ... [giving] a rare and exhilarating quality to the sense of space and light ... light pouring from the sky and bouncing back from the sea, enveloping everything'. The low elevation of the island has meant that although clouds do not persist (making it one of the sunniest spots in Britain) the wind is almost constant, inhibiting virtually all tree-growth.

Wind has also affected the architectural style: the older buildings are squat stone cottages with walls up to 9 feet (nearly 3 metres) thick. The reason for such thickness is that these buildings (known as 'black-houses') were usually double-walled, the space between being filled with sand into which rainwater ran from the thatched or tarred roof. Because of the shortage of local timber the roof-beams were made of driftwood and thus mostly short, making the roofs incongruously low and narrow but less wind-resistant. Today the mortared jointing of the rubble-walled houses has sometimes been picked out in white paint with the centres of the stones left bare, giving a bizarre giraffe-like patterning to the domestic architecture.

Because of its exceptional physical endowment, Tiree supports one of the most successful crofting communities in the Hebrides, one in which the crofter does not have to rely on supplementary incomes from forestry or fishing. (Of course, given the lack of trees and the dearth of natural harbours, the fertility of the land is doubly important for economic survival.) The underlying Lewisian gneiss creates only a handful of low, rocky hills in Tiree, and it seems clear that these must have become insular skerries during the higher sea-levels of raised-beach times.

The ancient tripartite division of land use reflects the three different types of terrain on the island. First there is the lime-rich *machair* of the coasts and the central lowland (known as the Reel), which is given over entirely to the grazing of cattle and sheep. Secondly there are the slightly heavier soils of the raised beaches, where blown shell-sand has alleviated some of the acidity due to impedance of drainage. This land is devoted to arable farming and is termed *achadh*. Finally there are the hill masses of acid peat and bare rock, together with some of the bogs on the raised beaches, which will support only rough grazing, locally known as *sliabh*. Although the majority of the land holders have equal shares in these lands of contrasting quality – *machair*, *achadh* and *sliabh* – over-stocking on the *machair* is tending to break up the sward, generating 'blow-outs' and the encroachment of dune-sand.

Although of approximately equal size to Tiree, the neighbouring island of Coll exhibits a very different landscape because the acid Lewisian gneiss is exposed over three quarters of its surface area. The virtual absence of glacial drift (as found in Lewis) or rich basaltic soils (as found in Mull) means that thin moorland and peat bog occur directly on the stony surface. Nevertheless, about one sixth of Coll is covered by shell-sand, so that *machair* is found along the windward fringes of the western and southern coasts. The very low acreage of land suitable for cultivation has meant that crofting has virtually disappeared on the island, except on the raised beaches of Sorisdale at the northern tip. The more fertile swards of the southern end have been devoted to cattle-grazing ever since the crofters were 'cleared' in the nineteenth century and replaced by Kintyre dairy farmers with their Ayrshire cows. At that time an increasing population was beginning to overtax the natural resources of the island, and it was questionable whether it could sustain the subsistence economy of the crofting townships. As in Tiree, the overgrazing of the fragile *machair* was already causing severe damage and windblow.

13. The Inner Hebrides – the Skye Group

There are few place-names in the British Isles which conjure up such romantic visions as Skye. Even without the literary and historical associations, its scenery alone would distinguish the 'Misty Isle' as something rather special. The Cuillins, for example, are because of their jagged character the only peaks in Britain that pose problems of an Alpine nature, and there are few rock-climbers who do not feel a quickening of the pulse at the mention of their name. The smaller isles of this northern group of the Inner Hebrides, although not possessing quite the magnetism of Skye, have a fascination and remoteness all their own. Their diverse landscapes range from the barren, stony wastes of South Rona, through the oddly fertile lands of southern Raasay, to the majestic rock peaks of Rhum and the verdant crofts of Eigg, Canna and Muck. There seems little doubt that much of the appeal of Skye and its satellites springs from the sudden contrasts in relief, the pronounced changes in landscape texture and colour and the differing cultural patterns. It would be too dogmatic to claim that all of these are a direct result of geology alone, although the curious juxtaposition of ancient Pre-Cambrian rocks with those of Mesozoic and Tertiary age has given rise to some remarkable contrasts in topography and land use.

The Small Isles

Lying between the rugged Ardnamurchan peninsula and the mountain fastnesses of southern Skye is a cluster of islands whose geographical grouping has led to them being dubbed the 'Small Isles of Inverness-shire' as an administrative convenience. Nevertheless, these delightfully named islands of Muck, Eigg, Canna and Rhum each have a distinctive character, different from that of their neighbours.

The smallest island, Muck, is also one of the most fertile of the Hebrides because of its deeply weathered basaltic rocks, its glacial-drift cover and its veneer of calcareous beach sands which have been blown inland. The fertility of the island's soils is reflected in the richness of its pasturelands, which in early summer are carpeted by masses of purple vetches, dazzling blue cornflowers and golden marigolds. Because of the low elevation and relatively low rainfall, peat has never accumulated to any extent, so the few remaining islanders use imported coal now that they have stopped visiting the copious Ardnamurchan bogs for their peat supplies. Basaltic lavas form the land surface, although shales and limestones of Middle Jurassic age are exposed as tidal reefs along the south-west coast. The lavas are virtually horizontal and generally give rise to low moor-covered plateaux, except in the south-west, where the isolated hill of Beinn Airien (451 feet, 137 metres) exhibits a terraced form because of its intrusive sills of dolerite. Among the most distinctive features of Muck are its shore platforms, since these are seamed by a multitude of intrusive igneous dykes. Running into the sea with remarkable regularity, these wall-like doleritic dykes rise 20 to 30 feet (6 to 9 metres) above the wave-cut platforms, having resisted erosion more successfully then the basaltic lavas into which they were injected.

When compared with Muck the island of Eigg presents a very different scene, despite the similarity of their trap landscapes. For a start it is about four times larger; it has a much greater exposure of Mesozoic rocks and its basalts build considerably higher hills; but above all it is the overwhelming presence of the famous Sgurr (to be described opposite) which makes Eigg significantly different. Yet, apart from the Sgurr, the geology of the two islands is quite similar. The thick lava succession rises from a plinth of Mesozoic rocks to form two flat-topped 1,000-foot (300-metre) plateaux in the western and eastern halves of the island, separated by a fault-guided valley along which most of the island's settlement is located. As elsewhere in the Hebrides, the trap landscape is bounded by precipitous cliffs, which are especially marked in the north-eastern corner of the island. This is partly due to the fact that the basalts lie directly on a series of so-called 'incompetent strata' (which they have helped to preserve), including thin Cretaceous sandstones and thick sedimentaries of Middle Jurassic age. The latter are made up of alternating limestones and shales whose erosion has facilitated landslipping (and therefore constant 'freshening' of the overlying lava cliffs), together with some 200 feet (60 metres) of calcareous sandstones which have weathered to produce fertile soils wherever they are exposed.

Some of the best soils are found at the small crofting village of Cleadale, which nestles peacefully on the grassy, south-facing slopes beneath the black basaltic cliffs at the northern tip of the island. Here too are the well-known 'singing sands' of Camas Sgiotaig (the Bay of the Musical Sand). Such phenomena, although rare, occur on several beaches in the British Isles and are due to the uniform size and rounded form of the local sand-grains. In this remote bay of northern Eigg the sand-grains are derived from the weathering of the calcareous Jurassic sandstone, and they emit a curious squeaking note when walked upon.

One final interesting feature, in the eastern part of Eigg, is the so-called Reptile-bed amongst the shales which crop out on the foreshore to the north of Kildonan village. Discovered by the eminent geologist, Hugh Miller (see p. 308), during his Hebridean cruises in the famous *Betsey*, this Jurassic stratum contains some of the earliest described remnants of fossil reptile bones and teeth, together with pterodactyl and dinosaur remains.

All of these phenomena pale into insignificance, however, by comparison with Eigg's most notable landform, the great volcanic skyscraper of An Sgurr (Plate 30). Rising in smooth precipitous cliffs of columnar pitchstone and associated felsite sheets to a height of 1,291 feet (394 metres), the remarkable Sgurr of Eigg is unique in the British Isles. Towering above the landing pier at Galmisdale, not far from the sea-cave where the infamous sixteenth-century massacre took place, the pitchstone ridge presents a profile matched only by that of Suilven in Sutherland (Plate 38). Although of lower elevation and a very different geological composition to Suilven, the Sgurr's summit, with its ancient fortification, matches the Torridonian peaks of Sutherland by virtue of its monolithic character.

Its genesis has been interpreted in several ways, none more intriguing than that of Geikie, who saw it as the remains of a sinuous lava flow which infilled a former valley carved in the surrounding basalts. The glassy pitchstone lava is seen to overlie 'a bed of compacted shingle, in which there is an abundance of coniferous wood, in chips and broken branches', thought by Geikie to mark a former river-course developed after the plateau basalts had solidified. It has been suggested that when the pitchstone lava-flow had overrun the Tertiary vegetation of this valley and itself solidified, its resistance to erosion proved greater than that of the surrounding basalts. By now the latter have been denuded sufficiently to leave the long ridge of the Sgurr rising abruptly above the south-western corner of the island.

Unlike the other members of the Small Isles group, Canna has no Mesozoic rocks exposed from beneath the plateau basalts. Indeed its structure is extremely simple, for the virtually horizontal Tertiary lavas make up most of the geological succession apart from the interbedded agglomerates and tuffs, whilst doleritic sills merely emphasize the terraced appearance of the hill-slopes in this trap landscape. This step-like succession of the various volcanic strata can best be studied in the precipitous sea cliffs of Compass Hill (450 feet, 137 metres) at the north-eastern end of the island. The singular name of this coastal eminence is derived from the fact that the iron content of its lavas is sufficiently high to deflect the magnetic compasses of ships sailing though the sound. Although there are no Mesozoic sedimentary rocks on Canna, there are a number of coarse conglomerates containing pebbles of gneiss, schist and Torridonian sandstone. These suggest not only that the lavas were extruded upon a basement of Pre-Cambrian rocks but also that a large river may have crossed the basaltic plateau in Tertiary times, similar in many respects to that which formerly flowed across Eigg (see p. 257).

This long, narrow island, like neighbouring Sanday (to which it is linked at low water), exhibits the soil fertility which we have come to associate with deeply weathered basalts. The grazing supports a high density of sheep and cattle, although the area near to the village of A'Chill is given over to arable farming on the even richer soils derived from the volcanic tuffs. This sheltered, south-facing slope of eastern Canna, protected by the island of Sanday, can produce the earliest potato crops anywhere in north-west Scotland because of its favoured soils and aspect. Although only a few trees grow there today, fossil pollen-grains within the peat bogs show that pine, hazel, willow, birch, oak and alder flourished on Canna and Sanday during different periods of post-glacial time.

The bright green grasslands of Muck, Eigg and Canna are not repeated on the larger isle of Rhum. The mountainous nature of the island is partly responsible for this, but in addition the widespread occurrences of granitic and felsitic igneous rocks, together with an extensive outcrop of Torridonian sandstone, have combined to produce much more acid soils. When we add to these factors the widespread blanket bogs and the intensive leaching of mineral salts from many of Rhum's soils, owing to its excessive rainfall, we can understand how it is that the island has a vegetation cover and land use very different from those of its smaller neighbours.

It is difficult to land anywhere on Rhum's cliff-bound coast except at the northern outfall of Kilmory Glen or at Kinloch Castle on Loch Scresort. Since the Nature Conservancy has taken over the management of the island, visitors are normally granted permission to land only at Loch Scresort, where the tiny settlement is fringed by the only woodland and farmland on the entire island. From this eastern inlet the glen penetrates deeply into the deserted heart of the island, where it opens out into a curious 'cross-roads' of radiating valleys. To the north lies Kilmory Glen, which is incised deeply into a flat-topped boggy plateau of Torridonian sandstone; southwards, the valley which encloses the ribbon-lake of Loch Long leads into the hidden corries of Rhum's southern mountains. The linearity of these north–south valleys in central Rhum can be explained by the presence of a major fault which slices through all the island's rocks and along whose line of weakness erosion is obviously facilitated. The 'cross-roads' is completed by the west-trending Glen Shellesder, which crosses to an anomalous Triassic exposure and the remote cliffs and waterfalls of the north-western coast. The valley floors and the Torridonian plateau-surfaces of northern Rhum carry a thick cover of blanket bog, whilst the intervening valley-slopes and the better-drained surfaces are more conducive to the growth of *Calluna* moorland, which flourishes on the acid soils and supports large herds of Red Deer.

A contrast in scenery is at once apparent when we turn to explore the more rugged southern sector of the island. Here the conspicuous 1,000-foot (300-metre) surface of the Torridonian sandstone gives way suddenly to the dark, rocky crags of Tertiary plutonic rocks, whose summits rise dramatically from the boggy sandstone tableland to heights of more than 2,500 feet (over 750 metres). Were it not for their relative inaccessibility, the magnificent peaks of southern Rhum would be almost as popular as the Cuillins of Skye among the rock-climbing fraternity. Indeed, their cliff-faces, jagged saw-tooth ridges, pinnacled summits and vast aprons of screes are analogous with those of their more famous counterparts in Skye, and it is no surprise to discover that the geology of the two massifs is extremely similar. We are, in fact, experiencing yet another example of the remarkable scenery created from the igneous rocks of a central intrusion complex of Tertiary age. All the lithological and structural phenomena that we have already seen in Arran, Ardnamurchan and Mull are repeated here: the bounding ring-fault, the cone-sheets, the radial dykes, and the central plutonic complex from which the highest peaks have been carved.

The major differences between Rhum and the other central intrusion complexes are, first, the paucity of its plateau basalts (hence the absence of a trap landscape) and, second, the fact that its plutonic rocks include a greater development of ultrabasic rocks (see Glossary) than can be seen anywhere else in Britain. In addition to the basic gabbro and eucrite, such rocks as, for example, peridotite, harrisite and allivalite help to create the rugged crags, the two latter taking their names from their British type-sites of Glen Harris and the peak of Allival (or Hallival) respectively.

Anyone who is fortunate enough to climb on the bare rock summits of Allival (2,365 feet, 721 metres) and Askival (2,659 feet, 811 metres) will immediately be impressed by their stepped pyramidal form. This has resulted from the differential weathering of the plutonic assemblage, due to the alternating bands of allivalite and peridotite: the small escarpments (or risers) have been carved from the tougher bands of allivalite and the more gentle intervening slopes (or treads) from the peridotite. Since the eucrites and gabbros were later intruded as sheets both beneath and into the ultrabasic rocks noted above, they now form a pronounced plinth from which the high summits rise. One noteworthy feature of these lower slopes is the way in which the stony screes give way downhill to a relatively herb-rich heathland, where the soils have been enriched by such minerals as calcium and magnesium from the underlying basic rocks.

Not all the southern peaks have been carved from basic rocks, however, for Sgurr nan Gillean (2,503 feet, 763 metres) has a summit-capping of quartz-felsite, as does the attractively conical peak of Beinn nan Stac. Over on the west coast a substantial Tertiary intrusion is largely responsible for the more gently domed mountain-group of western Rhum. Here the central microgranite has been attacked by Atlantic waves to produce the spectacular sea cliffs of Schooner Point and Wreck Bay. A basaltic lava capping helps to create the scarped summit plateau on the granite peak of Orval (1,869 feet, 570 metres), but the neighbouring summit of Ard Nev has been carved from an anomalous exposure of Lewisian gneiss.

This, like similar outcrops near Loch Long and on the slopes of Sgurr nan Gillean, has been interpreted as a slice of the ancient basement brought to the surface along steeply inclined fractures during the Tertiary igneous episode.

A final word is necessary concerning the Pleistocene history of Rhum, for much of its distinctive scenery must have been created by glaciers. The scattered patches of glacial drift at all elevations suggest that the island, like the remainder of the Inner Hebrides, was completely overridden by early ice-advances from the mainland. The glacial striations demonstrate, however that at some subsequent stage Rhum carried its own distinct ice-cap, and at that time the island's central glens became greatly overdeepened as the valley glaciers moved radially outwards. Although much of the glacial smoothing and quarrying relates to these earlier advances, the final etching appears to have been the work of the local equivalent of the Loch Lomond Readvance. The corries of the southern peaks contain moraine-dammed lakes, and it seems probable that Glen Dibidil and upper Glen Harris carried small glaciers, judging by their morainic accumulations. Meanwhile the bare rock peaks were sharpened by frost-shattering, thus adding to the cliff-foot screes which abound in Rhum. It also seems likely that during this last glacial phase there was a second tiny ice-cap in the granitic terrain of the north-west, for the northern slopes of Orval have long stretches of glacially plucked cliffs.

Skye

Skye is an island of superlatives: its 50-mile by 20-mile bulk (80 by 32 kilometres) makes it the largest of the Hebrides; it has the grandest mountain-group in the British Isles; it boasts the largest expanse of basaltic plateaux in Britain. Little wonder that its fame is such that for some tourists it is the apogee of Scottish scenery. Few would dispute that it is an island of great diversity and a place of changing mood – from the wind-lashed and cloud-mantled melancholy of winter to the overwhelming beauty of a sunny spring or summer day, when the mountains are sharply etched against the sapphire sky. Skye is a land of bold peninsulas, which dominate its shape to such an extent that there are those who claim that its name is derived from *sgiath*, the Gaelic word for 'wing' – hence its oft-used sobriquet – the Winged Isle. One senses a recurring presence of the ocean, with the sea-lochs biting deeply into the interior to create a majestic picture of high mountains mirrored in restless tidal waters. Such scenic harmony has been eulogized for centuries; where Loch Coruisk and Loch Scavaig are cradled beneath the towering Cuillins (Plate 31) we can witness a view unique in Britain, rhapsodized by Scott, recaptured by Turner's brush and claimed by some writers to be the finest in the British Isles.

It is easiest to commence our exploration, like the majority of tourists, at the island's southern end, where only the narrows of Kyle Akin and Kyle Rhea detach Skye from the mainland. Here is the ribby peninsula of Sleat, resembling more the mainland than the remainder of Skye in its geology and scenery (Figure 42). Although the Torridonian sandstone rises to about 2,000 feet (over 600 metres) in the twin sentinels of Sgurr na Coinnich and Beinn Aslak, which overlook Kyle Rhea, these barren sandstones and shales more often coincide with the lower boggy hills which form the backbone of the peninsula. Few crofts can be found in this peat-covered landscape of northern Sleat, but south of Loch na Dal there is a surprising change of scene, not all of it capable of explanation merely in terms of the geology.

Southern Sleat was considerably involved in the Caledonian thrust-movements, ancient Lewisian gneisses having been carried westwards over younger Torridonian rocks by the Moine Thrust and the Torridonian itself having been thrust bodily across the younger Cambrian near to Loch Eishort. Traditionally, Lewisian gneiss is associated with barren landscapes in which thin soils, rocky outcrops and peat bogs combine to produce a wilderness unconducive to agricultural improvement. Along the Sound of Sleat, however, the Archaean gneisses belie this reputation, so that between Isle Ornsay and Armadale the farming has an air of prosperity missing from the Torridonian country farther north. It is difficult to account for this apparent anomaly, although the presence of basic intrusions and the lighter soils of the raised-beach terraces may have some bearing on the answer.

On the western side of the Sleat peninsula, the prosperous crofts near Ord can be picked out by their lush hayfields, flourishing crops and extensive woodlands, a landscape so different from the remainder of the island that it has been dubbed the 'Garden of Skye'. It is a great deal easier to explain this oasis of fertility, however, because the well-drained

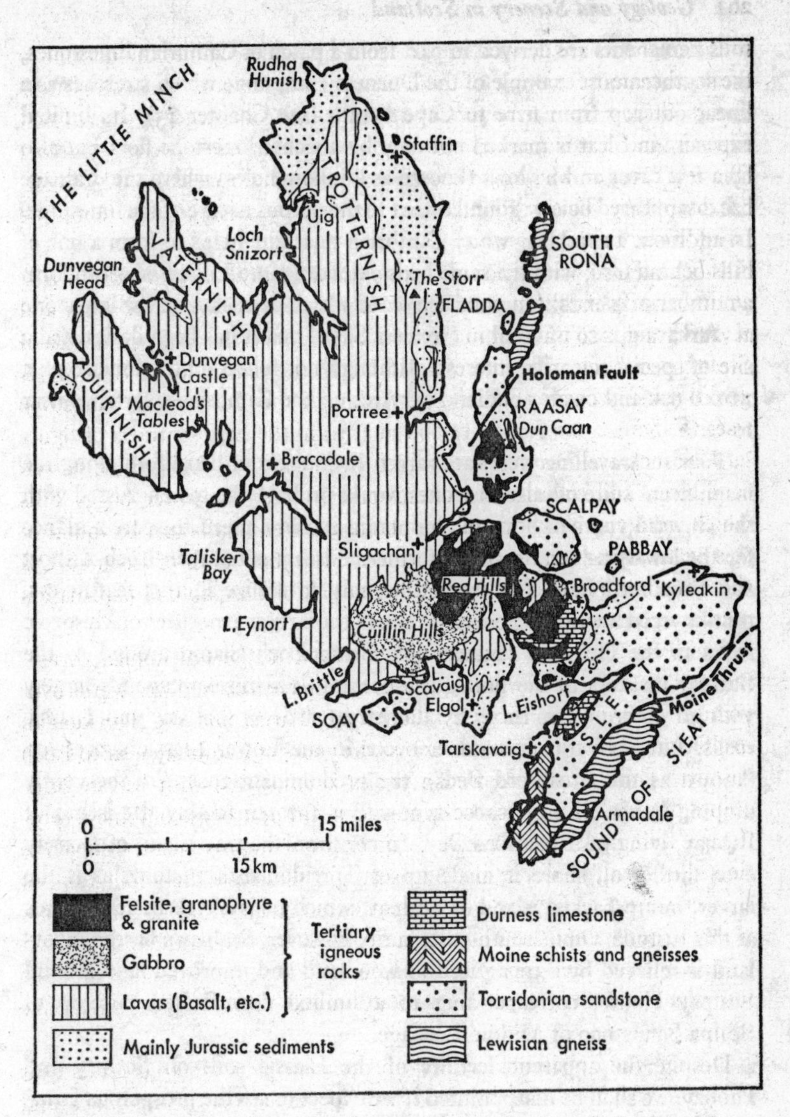

Fig. 42. *The geology of Skye (based on material prepared by the Institute of Geological Sciences).*

soils hereabouts are derived in part from a patch of Cambrian limestones, the southernmost example of the Durness Limestone which stretches as a linear outcrop from here to Cape Wrath (see Chapter 15). Its limited exposure in Sleat is marked not only by a richer limestone flora but also by a few caves and hollows (known as swallow-holes) where the drainage has disappeared below ground owing to the porous nature of the limestone. In addition, a sparkling white Cambrian quartzite helps to form a line of hills behind Ord, with the same beds extending into Loch Eishort to form a number of islands. A natural ash wood which flourishes on the limestone at Tarkavaig is so unusual in 'treeless Skye' that it has been designated a site of special scientific interest. Although not found on the limestone, a mixed oak and birch woodland survives on the Cambrian quartzites near to Ord.

Tourists travelling westwards from Kyle Akin to Broadford along the magnificent suite of raised beaches would probably be too engrossed with the seaward views or with the mountain scenery ahead to spare a glance for the low-lying terrain which forms the isthmus between Loch Eishort and Broadford Bay. Yet, were they only to realize it, this featureless, peat-covered lowland is floored by one of the largest exposures of Mesozoic rocks in the Hebrides (Figure 42). From Loch Slapin around to the Scalpay narrows in the north, the curving lowland corresponds largely with an outcrop 2 miles (3·2 kilometres) wide of Triassic and Liassic rocks. Although thin limestones occur in the Lower Liassic succession (known as the Broadford Beds), the predominant rocks are the gently dipping Pabba Shales, whose type-site is the remarkably flat island of Pabbay, lying off Broadford Bay. In contrast, the tiny island of Longay even farther off shore is made up of Torridonian sandstone, as is the larger, hump-backed island of Scalpay, which hugs Skye's eastern shores at this latitude. On its south-eastern tip, however, Scalpay's barren moorland is relieved by a tiny patch of woodland and improved land around Scalpay House, corresponding to a limited downfaulted outcrop of Scalpa Sandstone of Middle Lias age.

Despite the apparent fertility of the Liassic soils on Scalpay and Pabbay, we shall be disappointed if we expect to see the prosperous farmlands of England's Liassic Vale of Evesham repeated here in Skye. The rainfall is too heavy, the drainage too poor and the felsite intrusions too numerous for the crofting population to make much impact on the peat-covered hills in this district of Strath. The same problems also occur on the Jurassic rocks which build the attractive peninsula of Strathaird, to the east of Loch Scavaig (Figure 42), but before continuing along the Elgol road from Broadford it would be instructive to explore the scenery of Strath Suardal, which curves around the footslopes of Beinn na Caillich's granite peak (2,403 feet, 733 metres).

Because of its thick infilling of glacial drift, there are few pointers to the change of lithology in Strath Suardal apart from the increased woodland in the lake-filled valley. Only the observant botanist will note the appearance of that rare Alpine plant *Dryas octopetala* (mountain avens) opposite Loch Cill Chriosd. The strath, with its richer vegetation, has in fact been carved along the junction of the Tertiary igneous rocks with the less resistant Cambrian limestones which we have already encountered in Sleat. Here, however, the anticlinal limestone outcrop is more extensive, sweeping around the base of the granite mountains in a wide crescent from coast to coast. It is interesting to note that the eastern margin of the Cambrian limestone is terminated by the line of a major thrust-fault, where the Torridonian sandstone can be seen to overlie the Cambrian. Glen Suardal and Loch Lanachan have been etched out along this line of weakness as the thrust-fault curves round to follow the strike of the Cambrian anticlinal axis.

Although the Tertiary granite of Beinn an Dubhaich has invaded the limestone upfold, there are excellent exposures of the limestones, with grike-development (see Glossary) exhibited where they form the low ridge of Ben Suardal. But, as with the neighbouring Liassic limestones, their agricultural potential has not been realized, owing to the drift cover and the igneous intrusions. Nevertheless, in an environment where acid soils predominate, the usefulness of crushed limestone for agricultural purposes has long been apparent. Thus the Torrin quarry, on Loch Slapin, now exports great quantities of stone both for agricultural and for building and decorative purposes. It is locally known as 'Skye Marble', for where the limestone comes into contact with the igneous intrusions of Beinn na Caillich and Beinn an Dubhaich it has been converted into a hard marble by thermal metemorphism. Like their counterparts at Ord, the majority of the Strath Suardaig limestones are of a dolomitic variety – that is, the normal calcium carbonate ($CaCO_3$) has been bolstered by the addition of magnesium to form a dolomite.

From Torrin village it is but a short journey to the promontory of Strathaird, which divides Loch Slapin from Loch Scavaig (Figure 42). Its main attraction lies in the vistas it affords, westwards to the flat, Torridonian sandstone island of Soay, southwards to Rhum and Eigg and, especially, north-westwards to the Cuillins. The view of the Cuillins from the crofting village of Elgol (Plate 31) is famous. In the foreground is the gentler scenery of the Mesozoic rocks, whose limestones nurture a fertile swathe of pastureland amidst the moorland; in the middle distance the movement of the cloud shadows across Loch Scavaig changes the sea from pewter to silver to deepest blue; beyond are the soaring black peaks of the legendary Cuillins, impressive even in cloud and mist.

The Strathaird peninsula has a foundation of Middle to Upper Jurassic rocks, ranging from the Inferior Oolite through the Great Estuarine Series to the Oxford Clay. In the west, however, these Mesozoic sandstones, shales and limestones are capped by thick plateau basalts and dolerite sills; the Elgol road serves to divide the igneous hills, such as Ben Meabost (1,126 feet, 343 metres), from the till- and peat-covered sedimentary platform of the south-eastern shores, where numerous sea-caves have been carved into the well-jointed calcareous sandstones. As in the adjoining district of Strath, the presence of the Mesozoic rocks has for the most part had surprisingly little influence on land use, largely because of the thick drift cover. But it must also be remembered that this favoured corner of Skye carried considerably higher crofting populations until the evictions which heralded the arrival of the Cheviot and Blackface sheep in their stead.

It is now time to turn to the scenery which gives Skye its particular fame and fascination – scenery which has been created from the northernmost example of Tertiary igneous activity in the British Isles. Compared with the complexities of Mull and Ardnamurchan, the plutonic masses of Skye have a simple arrangement of their component parts, partly because of the contrasting topography of the granite and the gabbro hills. Furthermore, the magnificent plateau-lava country of northern Skye adds a third type of distinctive scenery which, taken in combination with the central igneous complexes, has led to the claim in the Geologists' Association Guide that '. . . the broad features of Tertiary igneous geology are shown in a more spectacular way on this island than in any other area.'

A distinction has always been made between the Red Hills and the

Black Cuillins; this reflects not only the differences in colour between the granite and the gabbro but also the contrasts in texture, form and elevation. The highland massif of central Skye can in fact be subdivided into three mountain-groups, each corresponding to a separate plutonic centre: first the gabbro hills of the Black Cuillins, together with the isolated peak of Blaven; second, the granitic Western Red Hills (sometimes called Lord Macdonald's Forest); third, the granitic Eastern Red Hills, which we have already encountered at Beinn na Caillich (Figure 42).

In essence, the Cuillins are an arcuate mass of ultrabasic peridotite which has subsequently been invaded by a large number of olivine-rich gabbro sheets. The whole plutonic complex has been intruded into the earlier basaltic lava plains, so that, as we walk across the streaming peat bogs from Sligachan, the starting point for climbers and geologists alike, we are in fact traversing the surrounding basaltic lavas. Ahead of us the cathedral-like spire of Sgurr nan Gillean (3,167 feet, 966 metres) soars into the sky, its pinnacled northern ridge running out like a gigantic flying buttress. Only skilled climbers should attempt to reach the summit of this splintery gabbro peak; others should content themselves with a scramble around the lower slopes. Here the corrie of Am Bhasteir holds a tiny lochan, backed by the savage cliffs of Am Bhasteir and the pinnacle of the Bhasteir Tooth.

The ravines and vertical gulleys which seam the bare rock-faces have generally been carved from basic dykes, part of the major NW–SE dyke-swarm. Where these intersect the ridges and ice-steepened arêtes they often form notches in the skyline (as in the Waterpipe Gully of Sgurr an Fheadain), which accounts in part for the serrated appearance of the Cuillins themselves. On the Red Hills, however, the basic dykes are not as numerous and are generally more resistant to erosion than the surrounding granite, so that they tend to stand out in relief. This is occasionally true of the Cuillin dykes also, the most famous example being the so-called Inaccessible Pinnacle, which stands up as a narrow wall on the summit of Sgurr Dearg (3,206 feet, 977 metres).

Anyone fortunate enough to ascend to the crest of the main Cuillin ridge will find that it is a sawtooth arête curving away southwards until it plunges directly into the sea at Loch Scavaig. No fewer than twenty of its summits attain heights of 3,000 feet (over 900 metres), the highest being the peak of Sgurr Alisdair (3,309 feet, 1,009 metres), which, like the southernmost peak, Gars-Bheinn, has a cap of basaltic lava surviving on its summit. According to Dr Alfred Harker, the eminent geologist who mapped the Tertiary igneous rocks of Skye, these isolated lava cappings on the high summits represent remnants of the 'roof' of the plutonic intrusion, dating from the time when the plateau basalts were updomed by the ultrabasic and basic magma injections. It was Harker, too, who noted the very marked inclination of the rock structures on the gabbro ridges. These structures take on the form of a pseudo-stratification, although it must be remembered that these are igneous, not sedimentary, rocks with which we are dealing. At Sgurr nan Gillean, at the ridge's northern end, the steep dip is towards the south; along the main Cuillin ridge it is towards the east; at Sgurr na Stri (near Loch Coruisk) it is towards the north, whilst on Blaven it is towards the west. Thus all the structures dip inwards to a central point beneath Glen Sligachan, and Figure 43 shows that they are made up partly of the inward dip of the gabbro banding and partly of the steeply inclined cone-sheets. These last, which form the most classic example in Britain of this type of structure, are restricted to the gabbro and do not invade the granites of the Red Hills farther east.

Needless to say, the final form of the Cuillins owes much to the work of frost and ice. The erosive action of former glaciers is everywhere manifest; the detailed chiselling of the numerous corrie walls and narrow arêtes and the polishing of the massive 'boiler-plate' slabs of Coire a Ghrunnda are small-scale examples of this process; the oversteepening of the valley shoulders, the truncating of the major spurs and the wholesale excavation of the magnificent trough of Coruisk represent the more grandiose aspects of glacial erosion. The ultimate veneer has resulted from frost-shattering on the oversteepened cliffs and pinnacles, so that the main ridge is everywhere draped around with scree slopes. The most spectacular of these is the Great Stone Shoot, which plummets some 1,500 feet (over 450 metres) from Alisdair's summit to the tiny corrie lake of Coire Lagan. Altogether, our impressions may well agree with those of Boswell, who saw the Cuillins as a 'prodigious range of mountains, capped with rocky pinnacles in a strange variety of shapes'. Dr Johnson, however, was ominously silent, no doubt overcome by the inclement weather!

The Western Red Hills are separated from the Black Cuillins by the long funnel of Glen Sligachan, and were it not for the Cuillins' gabbroid pinnacles their graceful domes would take pride of place in Skye's scenic inventory. As it is, their granites have weathered to a paler colour and a smoother, more rounded form than the neighbouring gabbro into which they were intruded at a slightly later date. Their well-known conical summits are crowned with layers of pinkish, frost-shattered detritus

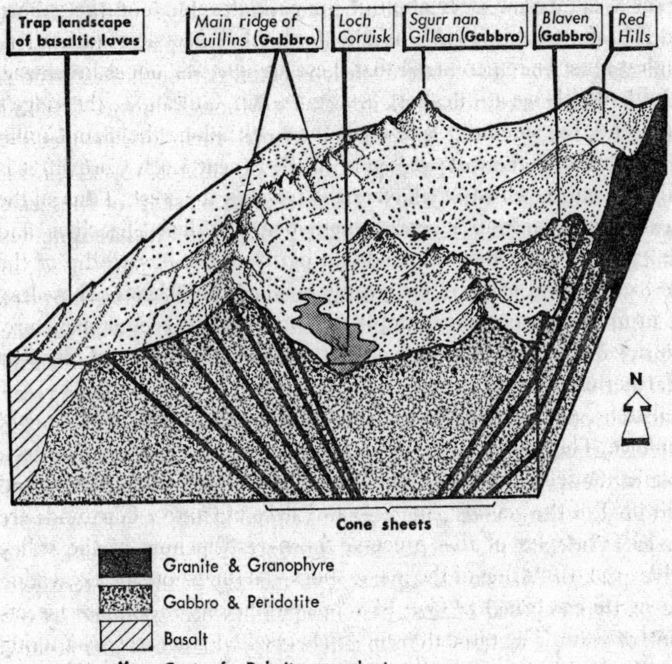

Fig. 43. *The structure of the Cuillins, Skye, demonstrating the principles of cone-sheet injection.*

which can be traced as runnels of scree down their sweeping, uninterrupted slopes and looks just like icing on a cake (Plate 32). It is now known that the Western Red Hills are composed of a number of granitic injections which were intruded concentrically as major ring-structures in a cauldron complex similar to those of Arran (Figure 11), Mull (Figure 40b) and the Mountains of Mourne in Ireland. The fact that the granitic complex caused a local updoming of the basaltic lavas is demonstrated by the steeply tilted lava capping which has survived on the summit and eastern slopes of Glamaig's stately cone, which towers 2,537 feet (773 metres) above Loch Sligachan. Elsewhere the basaltic cover has been almost entirely stripped off by denudation, speeded up by the initial doming of Early Tertiary times.

Similar basaltic remnants can also be seen on the slopes of the Eastern Red Hills, which have been carved from granites emplaced by similar mechanisms but from a separate intrusion centre. Although the splendid summits of Beinn na Caillich, Beinn Dearg Mhor, Beinn na Cro and Glas Beinn Mhor are built solely from granite, their lower slopes are complicated both by faulting and by a mosaic of gabbros, volcanic lavas, vent agglomerates and tuffs which, together with a 'skin' of Mesozoic sedimentary rocks, once formed the 'country rock' into which the granites were injected from below.

Before we leave the spectacular scenery and complicated geology of the Central Ring Complex of Skye, a word must be said about the mechanism known to geologists as 'hybridization', for Skye's central mountains illustrate this phenomenon very clearly. We have already seen in earlier chapters that both acid and basic igneous materials could be generated during the same episode, but we have not yet encountered an example of the way these mineralogically contrasting magmas were able to intermingle when one or both were in a liquid state. Such a mixing is known as hybridization, and its final product is a hybrid rock. The attractive peak of Marsco (2,414 feet, 736 metres) in the Western Red Hills is partly composed of this: basic magma has been acidified by the inclusion of granitic material to form a hybrid rock known as 'marscoite'. This can be distinguished as narrow strips separating the different Marsco granites from each other and from the included gabbro. The prominent gully (known as Harker's Gully) which runs up Marsco's north-western slopes appears to have been picked out along a band of less resistant marscoite. However, the best example of hybridization can be seen nearer to sea-level in the composite sill of Rudh' an Eireannaich, which forms a low headland on the north side of Broadford Bay (Figure 44).

The remainder of our journey on Skye will take us into the plateau-lava country of the north, where the three great peninsulas of Duirinish, Vaternish and Trotternish repeat the trap landscapes of northern Mull, but on a vastly greater scale (Figure 42). Like those of Mull, the Skye basalts are now thought to be related to a number of separate intrusion centres, rather than having been intruded from lengthy fissures, as was formerly suggested (see p. 242). The massive lava pile which has resulted from these basaltic outpourings now forms a series of remarkable scarps and coastal headlands. Professor J. A. Steers has pointed out that many of Skye's sea cliffs cannot be explained merely by marine erosion. He believes that the work of ice-sheets and the mechanism of land-slipping may have been of greater importance in the fashioning of the coastline of

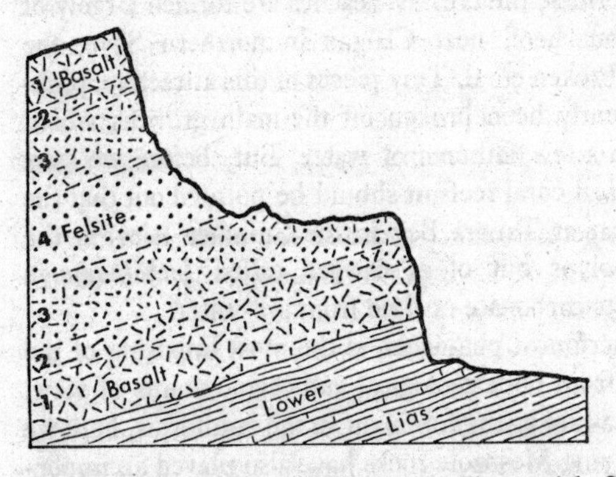

Fig. 44. *The composite sill of Rudh' an Eireannaich, Skye, illustrating the principle of hybridization (based on material prepared by the Institute of Geological Sciences). The numbers 1–4 show degrees of acidification (1 = most basic ; 4 = most acid).*

northern Skye. Moreover, many of the present basaltic cliffs are ostensibly fault-line scarps or even fault-scarps, expecially those in Trotternish (see p. 273). Although Steers states that the linear north-westerly trend of the sea-lochs of northern Skye (for instance, Loch Harport) owes a great deal to fault-control, he is at pains to emphasize that this is not to say their origin lies entirely in faulting.

Dunvegan, with its ancient castle and sheltered deciduous woodlands, is an unusual haven amongst the heather-covered peat bogs which generally blanket the ubiquitous plateau lavas of the north. From here it can be seen that the numerous lava flows are generally flat-lying in the westernmost peninsula of Duirinish. In consequence the plateau basalts have been sculptured into a magnificent example of trap landscape, with the two flat-topped peaks of MacLeod's Tables (Healaval Mhor and Healaval Bheag) epitomizing the characteristically tabular landforms of eroded lava plateaux.

To the north of Dunvegan the sea-loch coastline has tiny coves with dazzling white beaches which, at first sight, remind us of Arisaig or Tiree. Closer examination, however, shows that we are now dealing with a third type of white beach. In Arisaig the beach sands are products of the breakdown of the local Moinian micaceous quartzites; in Tiree, as in all the *machair* lands of the Hebrides, the creamy beaches are formed largely of broken marine shells; but here, near Claigan in northern Skye, the strands are composed of broken coral. Tiny pieces of this attractive, pinkish material have apparently been broken off the main growths which flourish just off shore in a few fathoms of water. But, before we start thinking in terms of tropical coral reefs, it should be pointed out that the Hebridean coral (like that at Tanera Beg in the Summer Isles) is the product not of coral polyps but of a seaweed called *Lithothamnion calcareum*, which abstracts carbonate of lime from sea-water.

Trotternish, the northernmost peninsula, is the most imposing of the three northern 'wings', from both geological and scenic points of view. Although the basaltic lavas are again dominant in the landforms, here we shall find that the underlying Mesozoic rocks have also played an important, if indirect, part in the make-up of the present landscape. Unlike the plateau-lavas farther west the Trotternish basalts are steeply tilted, so that they now form an excellent example of a cuesta, with the gentler dip-slope descending westwards to Loch Snizort and the escarpment cliffs facing eastwards to the Sound of Raasay. This line of fearsome precipices, standing some distance back from the coast, runs for almost 20 miles (32 kilometres) between Portree and Staffin. Seen from the coast road, the wall of frowning black cliffs is one of the most striking landforms in Scotland, for rarely does its elevation fall below 1,000 feet (300 metres).

There are two places of particular interest in this area, for at both The Storr and the Quiraing we can explore the intricate topographic maze created by the largest and most spectacular landslipping in Britain (Figure 45). In front of the escarpment peak of The Storr (2,363 feet, 720 metres) a confusion of screes, boulders and rock pinnacles marks the undercliff zone of slipping. The remarkable 160-foot (49-metre) basaltic

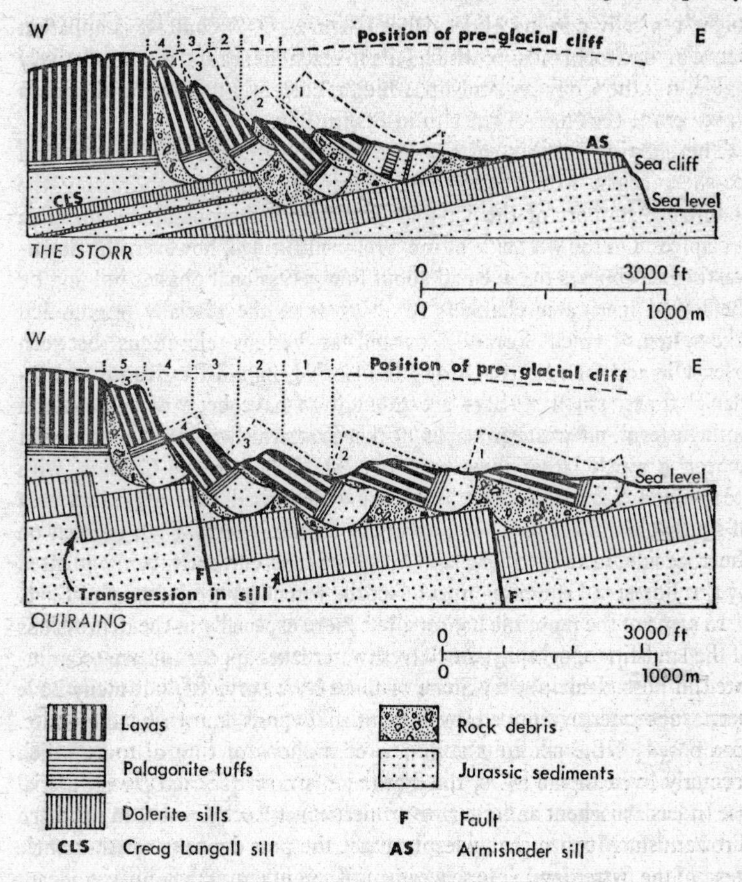

Fig. 45. *Diagrammatic sections of The Storr (top) and the Quiraing (bottom), Skye, to show the mechanics of landslipping. See also Plate 33.*

pinnacle of the Old Man of Storr projects above the jumbled mass which was created when the 'incompetent' underlying Jurassic clays and limestones became incapable of bearing the weight of the overlying lavas, having previously been oversteepened by ice-sheets moving northwards along the scarp face (Plate 33). Farther north, at the Quiraing, similar landforms have been created by the same process of rotational slipping, although in this case major step-faults have also played a part, in addition to the typically curving glide-planes (Figure 45). A contrast is apparent between the Storr landslip, which fails to reach the sea and is now relatively stable, and the Quiraing landslip, which is currently unstable as the ocean waves erode the 'toe' of the slip in Staffin Bay.

The slipping on this eastern face probably started as the mainland ice-sheet began to retreat in late Pleistocene times, withdrawing the stabilizing support of the ice-mass itself from the potentially unstable precipices. On the west side of the Trotternish ridge, however, the down-wasting ice appears to have had about four recessional phases, judging by the linked meltwater channels resulting from the glacially impounded lake-systems which formerly existed at various elevations between Glen Uig and The Storr. During the Loch Lomond Readvance of late-glacial times, small glaciers are thought to have been regenerated on both eastern and western slopes of the Trotternish ridge. As we might expect, a much larger ice-cap was re-established in the Cuillins, with steep-sided kame-and-kettle drift being deposited over a limited area of south–central Skye. This hummocky morainic terrain can be seen on the roadside near to Sligachan. Much of it is currently being planted by the Forestry Commission to offset the legendary treelessness of Skye.

In some of the morainic hollows, but more especially in the depressions of the landslip topography, small freshwater lakes appear to have accumulated in post-glacial times. Some of these have survived, but many have been subsequently infilled by a whitish deposit known as diatomite. Composed of the siliceous skeletons of millions of tiny diatoms which formerly lived in the lakes, the diatomite is now quarried for industrial use (it has absorbent and filtering properties) at Loch Cuither in northern Trotternish. Much more widespread are the peat deposits which mantle most of the flatter land. These were, until recent years, the only significant source of fuel in Skye, despite the sporadic use of local interbasaltic lignites and impure Jurassic coals from the sea cliffs.

Below the lava precipices of Trotternish, the glacial drifts, landslips and peat deposits have mantled the wide shelf of underlying Mesozoic rocks to such an extent that these are rarely exposed at the surface, except in the vertical sea cliffs between Portree and Staffin. Thus the thick Jurassic limestones play little direct part in the inland topography and have had only a localized effect on soils and land use. The sedimentary rocks have here been invaded by numerous 'leaves' of a thick sill of dolerite, which have helped to produce some of the spectacular scenery of this north-eastern coast. There is Kilt Rock, for example, which derives its name from the alternate light and dark banding of the interleaved sedimentary and igneous rocks, and near by, where the cliff-edge Loch Mealt drains directly into the sea, is a 170-foot (52-metre) vertical waterfall. Farther south a similar waterfall has been robbed of its splendour in order to drive the cliff-foot power station below the Loch Leathan reservoir.

Raasay and South Rona

Seen from the sheltered natural harbour of Portree, Skye's capital, which is hidden between flanking basaltic headlands, the neighbouring island of Raasay presents only a featureless shoulder of heath-covered Torridonian sandstone. However, such a viewpoint gives a misleading impression, for, in proportion to its size, Raasay is one of the most geologically diverse of the Hebrides (Figure 42). This diversity is reflected not only in its undulating topography but also in the remarkable contrasts in land use between the northern and southern parts of the island.

A visitor arriving by sea at the village of Inverarish will be astonished to find that the southern end of Raasay is characterized by yellow hay-fields, deciduous woodlands and emerald pastures dotted with dairy cattle. Such fertility is rare amongst the peat bogs and ice-scoured rocks of north-west Scotland, so it is no surprise to discover, as did Boswell, that '... there is plenty of lime-stone in the island.' In fact, southern Raasay has a very extensive exposure of Mesozoic rocks which range in age from Triassic sandstones, through the Liassic shales and limestones, up into the Inferior Oolite, the Great Estuarine Series and even the Cornbrash. Many of these rocks also occur in Skye, but the crucial difference is that in Raasay the Mesozoics are virtually drift-free, so that there is a fair amount of high-quality soil waiting to be farmed. Sir Frank Fraser Darling goes so far as to say that Raasay is '... almost unique in the Highlands and one of the few places where deep ploughing and direct re-seeding of hill land could be applied'.

This soil fertility is due largely to the high calcium content, of course, but also of great importance is the fact that the narrow north–south ridge of Raasay lay parallel with the direction of movement of the mainland ice-sheets, so that pockets of deep soil have managed to survive the general glacial scouring. Nevertheless, the bulldozing effect of the former ice-sheets is illustrated by the remarkable depths of the sea-channels which flank the islands of Raasay and South Rona. In fact the Inner Sound, between Raasay and the mainland, has the deepest submarine hollows in the British sector of the continental shelf. The presence of major faults on Raasay suggests that the entire chain of islands may be fault-controlled. The Holoman fault, running offshore north-eastwards past the ruined castle of Brochel, has clearly played a part in the fashioning of the linear eastern coastlines of both north Raasay and South Rona (Figure 42). Their entire eastern shores may have been carved from fault-line scarps, and the neighbouring ocean deeps excavated along the relatively softer rocks of the downfaulted sedimentary basins.

The highest point of Raasay is the conical basaltic cap of Dun Caan (1,456 feet, 444 metres), which serves as a splendid viewpoint for the entire island. Immediately to the east we can observe a unique geological phenomenon, so far as Scotland is concerned, for here is a 1,000-foot (300-metre) scarp consisting entirely of Jurassic sediments, just as if a section of the Cotswolds had been bodily transferred to the Hebrides (Plate 34). As in northern Skye, the Jurassic clays and shales (in this case those of Liassic age) have collapsed, creating gigantic landslips, because the scarp had earlier been oversteepened by ice-sheets. In the case of the Dun Caan slip, there is evidence to show that the slipping is continuing today, and this instability may have given rise to the newspaper reports of 1934 which claimed that a volcanic eruption had just taken place on Raasay! The reports of steam, showers of stones and rumbling noises could all be explained by a renewed slipping of the unstable mass down to sea-level, where it is currently being attacked by ocean waves.

Turning to the south, we can see the ruins of an ironstone mine in the thickly wooded valley of the Inverarish Burn, for beneath the Inferior Oolite there an 8-foot (2.4-metre) band of Liassic ironstone occurs, similar in character to the well-known Cleveland Ironstone of northern England. Although the Raasay Ironstone has a 30 per cent iron content, its high percentage of lime makes it generally uneconomic to work. Mining ceased after the First World War, despite the 10 million tons of estimated reserves. The final surprise of this Jurassic country in southern Raasay is the occurrence of a bed of oil shale at the base of the Great

Estuarine Series, but unlike the iron ore this mineral deposit has not yet been worked.

Despite the fact that a Tertiary granophyre and some dolerite intrusions introduce a sporadic cover of purple heather, bracken fern and bog cotton into the southern Raasay scene, we could almost forget that we are still in the Tertiary Igneous province of North-west Scotland, were it not for the view of Skye's forbidding basaltic cliffs across the narrow sound. The Holoman Fault acts as a major scenic boundary, however, for it serves to cut off the fertile landscape of the southern Mesozoics from the moorlands and peat bogs of northern Raasay (Figure 42). At Loch Arnish the Torridonian sandstones give way to the knobbly Lewisian gneiss which forms the remainder of the island, as well as the whole of South Rona and the tiny isle of Eilean Tigh. Only the boggy, green pasturelands of Eilean Fladda diversify the northern wilderness of bare, pink, ice-polished slabs and patchy heather, for this remarkably flat island (Gaelic: *Fladday* – flat isle) is made of gently dipping Torridonian sandstones. South Rona is now uninhabited, largely because of its unyielding Lewisian gneiss, mineral-deficient soils, treelessness and inaccessibility. Boswell speaks of it having '. . . so rocky a soil that it appears to be a pavement', which makes it even more difficult to understand how it was able to support crofting villages until the end of the nineteenth century. However, as a final reminder that there are always exceptions to the rule which equates Lewisian gneiss with infertility, the tiny crofts at Arnish, in northern Raasay, stand amidst a thick mantle of birch and rowan trees which make up one of the finest natural broad-leaf woodlands to be found anywhere in the Hebrides.

14. The Outer Hebrides

West of Skye, across the turbulent waters of the Minch, lie the Outer Hebrides, or Western Isles, which, because of their remoteness and dream-like silhouette against the setting sun, have been thought of in romantic terms by countless writers nurtured on the myths of the 'Celtic fringe'. For the most part, however, this 130-mile (208-kilometre) chain of islands, lashed by Atlantic gales and often swathed in low clouds, has a treeless, austere aspect. In general the rural way of life remains difficult here in a land of peat bogs, bare rocks and waterlogged soils. Nevertheless, it cannot be denied that on days of bright sunshine, when the ethereal, rain-washed atmosphere amplifies the colours in the landscape, there are few places in Scotland which can emulate the splendours of this island scenery.

Here is a special kind of Hebridean scenery, for we are no longer concerned with the intricate and often spectacular landforms of the Tertiary igneous rocks, nor with the curious if limited impact made by isolated outliers of Mesozoic sedimetary rocks, as in Mull and Skye; missing also are the bizarre pyramids of the quartzites. Instead, the Western Isles' landforms are governed almost entirely by the intractable Lewisian gneiss: apart from the anomalous Stornoway Beds, the igneous intrusions and a limited exposure of schists in Harris, the gneisses dominate the scene, as they should in the type-locality of these most ancient of British rocks. It is, however, the widespread occurrence of the sandy *machair* which gives the Western Isles some of their most memorable landscapes, composed of flowery greensward, white cottages and far-reaching vistas of sea and sky.

The complex basement of crystalline rocks which is collectively known as the 'Lewisian Gneiss' in all geological literature is not a geological formation in the ordinary sense of the word, since it includes a wide variety of metamorphic rocks of considerably different ages. Although they were formed during early Pre-Cambrian times, the rocks were metamorphosed and deformed during two periods of mountain-building activity associated with pre-Caledonian orogenesis. The older deformation, known as the Scourian (from Scourie in Sutherland), dates from 2,200 to 2,900 million years ago, whilst the younger orogenesis, termed the Laxfordian (from Loch Laxford in Sutherland), has been dated as 1,200 to 1,600 million years old. Not only are these amongst the oldest known rocks in the world, but the interval of time which elapsed between the formation of the oldest and youngest gneisses is a great deal longer than that spanning the period from the Cambrian to the Holocene. Although many of the former Pre-Cambrian rocks have been so altered by heat and pressure that it can be difficult to determine their previous character, two broad classifications of the Lewisian gneiss have been made. Where former sedimentary rocks have been converted into metasediments, such as marbles or mica-schists, they are known as 'paragneisses'; where former igneous rocks have been modified they are termed 'orthogneisses'.

In the Western Isles orthogneiss is more common in the landscape, and a visitor will soon become familiar with the mica-spangled foliation of alternating white and grey or pink granitic material which forms the typical banded gneiss. Richly garnetiferous dark green to black hornblende-gneiss also occurs in pillow-like masses amongst the banded rocks, many of which are laced with white quartz intrusions. Their complex folding can be seen everywhere, ranging in magnitude from gigantic recumbent folds and crumples to minute puckering, so that the individual rocks exhibit much of interest even to the layman. In south-west Harris the paragneisses become more dominant; marbles, quartzites, quartz-schists and graphite-schists build the rugged country between Toe Head and Renish Point. Here too is a massive intrusion of unaltered gabbro-diorite which reminds us that the Lewisian complex was subsequently invaded by plutonic granites and numerous intrusive sheets and dykes of all ages, including the Tertiary.

To complete the apparent confusion of the Lewisian Complex there is' a great belt of crushed rocks, known as mylonites and crush-breccias formed in association with a major thrust-fault which extends along the entire eastern coastline of the Outer Hebrides. The thrusting, from an easterly direction, has caused intensive shearing of the gneisses in a zone several miles wide, while pressure-melting has converted some of the gneisses into a blue-black splintery rock, known as 'flinty-crush' because of its similarity to the better-known Cretaceous flints of considerably younger age.

Minor differences in the mineral constituents of the gneiss mean that

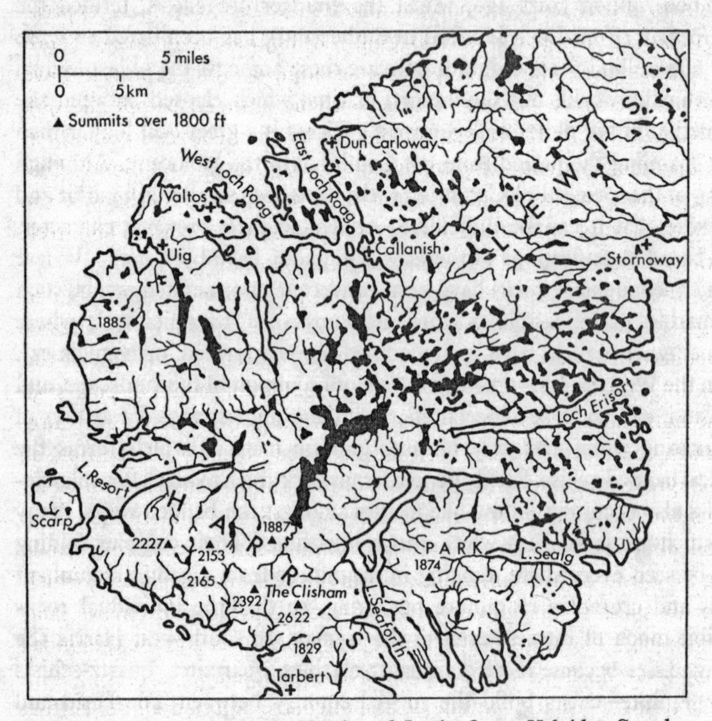

Fig. 46. *Drainage patterns in Harris and Lewis, Outer Hebrides. See also Plate 36.*

the various lenticles, bands and inclusions exhibit differing resistance to erosion, so that one of the most characteristic features of the Lewisian country is the knobbliness of the hard-rock landforms. The topographic complexity has been exaggerated by a complicated fault mosaic criss-crossing the islands which has been picked out by differential erosion, especially by ice, to form a bewildering pattern of hill and valley. A glance at a map of North Uist or Lewis (Figure 46) will reveal the maze of watery depressions, some filled by fresh water and others by the sea.

As on the mainland of Scotland, most of the present-day scenery owes much to the effects of the Pleistocene glaciation. The Western Isles were crossed by mainland ice-sheets which not only scoured the gneisses into a wilderness of bare, rocky knolls but also left boulder clay in the depressions to induce post-glacial waterlogging. The highest mountains of South Uist and northern Harris escaped the invading ice but appear to have nourished their own glaciers. Of the outlying islands (see pp. 289–90), only Rockall and St Kilda were unaffected by the Scottish ice-sheets at their maximum extent. The corries of Lewis, Harris and South Uist appear to have carried small glaciers late in the Ice Age (perhaps in the time of the Loch Lomond Readvance), judging by the moraines and fresh striations which cross the older ones at discordant angles.

The final touches in the make-up of the physical landscape came in post-glacial times, which saw the slow development of three distinct phenomena that were to have very important effects on the life of the succeeding inhabitants. The first of these was the formation of peat bogs in the numerous ice-scoured basins and on the badly drained plateaux. Although these acid blanket bogs have resisted virtually all attempts at land-improvement over wide expanses of the islands, they have provided a constant fuel supply in a region lacking both coal and timber. The second phenomenon was the gradual accumulation of calcareous sands along the western coastal margins to form the well-known *machair*, which has made some of the formerly barren tracts capable of extensive land improvement. The third post-glacial episode has been a gradual inundation of the landmasses by the inexorably rising ocean.

This positive eustatic movement has, of course, affected all the British Isles, but it must be remembered that near the centres of the former ice-sheets an uplift of land has partly off-set the sea-level rise. This uplift was an isostatic recoil from the original crustal downwarping caused by the weight of the ice-mass. The Western Isles, however, were on the perimeter of the ice-sheets and thus suffered little crustal downwarp and no significant post-glacial uplift. This accounts for the lack of raised beaches in the Outer Hebrides, in contrast with the mainland, and also for the gradual drowning of their coastlines since Pleistocene times. Thus, what was formerly one elongated island (still often referred to as the 'Long Island' has now been split into scores of individual islands and skerries by the gradual marine transgression. As a measure of this drowning, Neolithic chambered cairns in North and South Uist are now partly submerged at high tide, whilst numerous examples of submerged peats (containing tree roots in growth position) are known from all the islands' coasts. Local legend even claims that the present ocean floor between the Western Isles and St Kilda was once the hunting ground of a Harris princess!

Sir Frank Fraser Darling has highlighted the way in which the uneven distribution of the fertile *machair* between the northern and southern halves of the Outer Hebrides has created a disparity in the agricultural scene. Thus, while Lewis and Harris have a land area three times that of the southern islands, they possess less than 47 per cent of the improved land. This unequal division is reflected in the human geography; while most of the southern isles are peopled almost exclusively by agricultural communities, the inhabitants of Lewis and Harris have had to find alternative means of livelihood; this achievement of a more diversified economy has led to an increase in population.

Since the *machair* is limited almost entirely to the windward coasts, most of the agricultural settlement of the Outer Hebrides consists of scattered villages along the Atlantic shores (Plate 35). Because these low coastlines have few inlets or safe anchorages (with the major exception of Loch Roag), fishing fleets and their associated harbours are absent. The opposite is the case with the eastern shores. These are generally steeper and rockier than those of the west, so that here agriculture is at a premium. It is, however, this eastern coastline that has the deep-water bays and fjords which provide sheltered harbours, so here are to be found the fishing fleets and all the major urban settlements.

The Southern Isles

The rocky and cliff-fringed islands which surround Barra and Vatersay are now deserted, save for their numerous colonies of sea-birds, but in the nineteenth century seven of these isles carried a total population of over 400 souls. The finely shaped mountain of Heaval (1,260 feet, 384 metres) dominates the Barra scene, for it stands in the middle of this compact little island. The cottages, amidst their clustered fields of hay and potatoes, are arranged around its footslopes just above the fringe of silvery beaches which are such a feature of the island's coastline. The best-known of these is Traigh Mor in the north, the 'Cockle Strand' made famous by Compton Mackenzie's novels and now the location of the island's airport. This popular author's former house is now the headquarters of a company which processes shell-sand from the beach as a basis for the preparation used in the 'harling' (lime-washing) of house-walls. The tiny port of Castlebay nestles around the sheltered haven in which the picturesque Kiesimul Castle stands on an islet of flinty-crush rock.

Beyond the isle of Eriskay, the much larger islands of South Uist, Benbecula and North Uist present some 50 miles (80 kilometres) of virtually unbroken sandy beaches to the Atlantic waves. Although it can boast a group of steep rocky hills along its eastern side, culminating in the rounded summits of Beinn Mhor (2,034 feet, 620 metres) and Hecla (1,988 feet, 606 metres), South Uist is best remembered for its widespread exposure of *machair*. To understand the cultural life of the Western Isles one must visit one of the Uist *machair* landscapes, where geomorphological process, vegetation growth and land use are interwoven into a delicately balanced eco-system in which a change in one of the factors could cause catastrophic changes in the others. Landward of the magnificent Atlantic beaches the sand dunes appear as an undulating golden carpet flecked with green patches of marram grass (*Ammophila arenaria*). Beyond lie the low and flat stretches of the true *machair* (Figure 47).

Research suggests that its sand content was derived largely from the glacial drifts on the shallow off-shore platform and that it represents a symptom of coastal adjustment to excess sand-supply immediately after glaciation. The calcareous component was added from the numerous marine organisms and the final landform produced by erosion and reworking, mainly by the wind. Since there is currently little movement of sand from the modern beaches to the *machair*, the depositional phase is now thought to be over, the major period of building having taken place between 5,700 and 2,000 BP, according to Dr W. Ritchie. The datings are based on radiocarbon readings from buried organic horizons and on first-century archaeological remains. Ninth-century Viking settlements were similarly buried by drifting sand. Today, after many hundreds of years of agricultural activity, the *machair* is a stable, mature surface whose flatness is related to the water-table, which in turn acts as a base-level for wind erosion. Vast areas are flooded in the winter, thus providing wet receptacles for blown sand.

Ecological studies by Sir Frank Fraser Darling and Dr J. M. Boyd have demonstrated that the *machair* formation is primarily dependent on the 'fixing' of the coastal dunes by marram grass and associated species such as sea sedge (*Carex arenaria*). A later stage sees the establishment of a dune pasture based on red fescue (*Festuca rubra*), meadow grass (*Poa pratensis*) and several species of clover and plantain, together with sur-

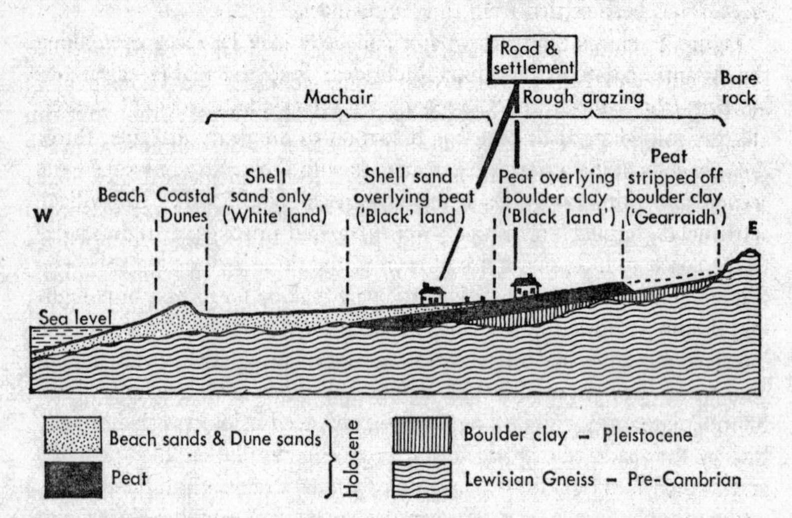

Fig. 47. *Diagrammatic section of a typical* machair *settlement in the Outer Hebrides. See also Plate 35.*

viving clumps of marram. As the nitrogen-fixing plants increase, so the rich calcareous soil becomes more stable and the marram gradually disappears, to be replaced by a wide variety of colourful wild flowers. These, together with the profusion of blue butterflies, make the *machair*, according to Darling and Boyd, '... a brilliant place in July, offset by the blue of sky and sea, the white edge of surf and the cream expanse of shell-sand beaches'.

A final note of warning must be sounded, for historical records show that the delicate turf has been both over-grazed and over-cultivated on several occasions, and once the sward is broken the fierce Atlantic winds cause severe 'blow-outs' and drifting. Great damage was formerly caused by carts carrying heavy loads of seaweed to the kelp factories, the ruts they made falling an easy prey to wind erosion; modern cars and caravans are a present threat. One has only to look at the destruction of the dune rampart behind the Cockle Strand on Barra to see the effects of man and his animals on this finely balanced dune–*machair* system. Nevertheless, with careful husbandry the dark, stoneless loams have long provided first-class crops for the islanders, and it is no surprise to find that the *machair* has been settled from time immemorial.

Figure 47 shows that three major soil-belts may be recognized along the Atlantic coasts of the Outer Hebrides: first, the highly calcareous *machair* (the 'white' land); second, an intermediate zone of darker, alkaline soil where shell-sand has blown on to the peaty margins; third, the acid soils of the waterlogged peatlands with their rocky outcrops – the second and third categories being known as 'black' land. The village settlements, termed 'townships', were arranged in strips, at right angles to the coast, to include a share of each of the three contrasting soil-belts (Plate 35). The true *machair* was particularly valued for its ease of ploughing, despite its lower crop-yields and need for regular fallowing, but the intermediate zone of 'black' land was more productive, even though cultivation was more difficult. Before the advent of machinery this intermediate zone, with its long, narrow, unfenced fields, was traditionally dug by the spade or the 'caschrom' (a crooked spade, or 'foot-plough') and formed into 'lazy beds' separated by open drains. Such heavy work was no problem until twentieth-century emigration caused a shortage of labour. To complete this communal agricultural pattern, known as the 'run-rig' system, the townships were each allotted a proportion of common grazing-land on the poorer acid soils of the bounding hills from which the peat supplies also came. The narrow strips on the arable *machair* once changed hands regularly in an annual ballot, and, although not many remain, the visitor may still discover a few open fields on the highly alkaline *machair*. Until recently the best remaining examples were to be seen in North Uist at Balmartin, Balranald and Kyles Paible, and also at Valtos in west Lewis.

The apparently simple patterns of land-use zoning in the Western Isles are complicated by the remarkably high incidence of water bodies, for the maze of sinuous lakes and sea lochs in Harris and Lewis, shown in Figure 46, is repeated in both North and South Uist and in Benbecula. It was on the latter island that the 1824 chronicler, J. Macculloch, attempted to count the Benbecula lochs from the small prominence of Rueval: 'Great and small, crooked and round, and long, and serpentine . . . my head began to whirl, and at the ninetieth I gave up the point in despair.' These low-lying tracts of lake-dotted terrain include water bodies of different origins: some occupy ice-scoured rock basins, some hollows in the glacial drift; some are coastal lagoons of depositional origin and some are marine-eroded inlets in solid rocks or in unconsolidated superficial deposits; some of the smaller ones are merely pools among the peat-hags. It has been suggested that, whilst the sea is still occasionally breaking through the easily eroded drifts to link salt with fresh water, we must not overemphasize this process, since the tidal inlets of Loch Bee and Loch Balranald, for example, are known to have been joined to the ocean by artificial channels. Nevertheless, the signs of post-glacial drowning are everywhere manifest, and Dr Ritchie has calculated that up until some 6,000 years ago the sea-level of the Western Isles rose at the rate of about one metre (3·28 feet) per century.

Harris and Lewis

Although often referred to as separate islands, Harris and Lewis are in fact a single body of land with the boundary lying almost (but not quite) along the borders of the Harris mountains and the Lewis plateaux. Compared with the remainder of the Western Isles, Harris boasts a much greater proportion of mountainland, culminating in The Clisham (2,622 feet, 779 metres), but it has only a small extent of valuable *machair*. Indeed, no greater contrast may be seen than that which greets us as we cross the Sound of Harris from North Uist; from the relative fertility of the Uists we pass to Harris, where a mere one per cent of the land is cultivated and where 96 per cent remains as peaty moorland or ice-scoured rock. In this inhospitable wilderness the settlements are entirely peripheral; J. B. Caird describes it thus: '. . . no dwelling is beyond the sound of the waves . . . Small paths wind down the ice-etched valleys among the minute lazy-beds to the source of the fertilizing sea-ware (seaweed) and the sheltered anchorages of the small ring-netters and lobster boats.'

As we would expect, the soils of Harris are acid, waterlogged and mineral-deficient, and the absence of raised beaches, with their lighter soils, does not help. It is therefore astonishing to discover that on the tiny pockets of improved land in the so-called Bays area of eastern Harris there is a population density of over 500 per square mile, the result of early-nineteenth-century evictions from the *machair* lands of the western coast. Resettlement has subsequently taken place along what is known as the West Side, but the lazy beds, or *feannagan*, remain along the hostile eastern shores. Sir Frank Fraser Darling gives a moving description of the 'making' of the soil by the careful interlayering of turves and seaweed carried there in creels by the womenfolk: 'One of these lazybeds will yield a sheaf of oats or a bucket of potatoes . . . Spade, mattock and reaping hook are the tools of the lazybeds, whereas on the machairs the ponies plough, the rows are ridged, and scythe and reaper have fuller play.' The lack of level ground and the rockiness of the soils exclude virtually all mechanization from eastern Harris.

In view of this depressing picture of agriculture in Harris, it is not surprising that its population has discovered alternative means of making a livelihood. Apart from the feldspar quarries in the epidiorite intrusion between Northton and Lingarbay, which provided the bulk of British requirements during the Second World War, most of the employment has been found in fishing and tweed manufacture. Since the 1924 collapse of Lord Leverhulme's attempt to expand the fishing industry, however, the small fleets have returned to the traditional harbours of the deeply indented eastern coastline, leaving the 'port' of Leverburgh somewhat forlorn. The world-famous Harris tweed has remained largely a cottage industry, and therefore makes little impact on the landscape itself.

On days of good visibility the high hills of north Harris make splendid viewpoints, not only for the lake-studded boglands of Lewis but also for the remainder of the Western Isles, including the renowned islands of St Kilda, 45 miles (72 kilometres) away to the west. Signs of glacial action are to be seen everywhere in these barren mountains and, although there are few corries, the U-shaped valleys can be studied here to great advantage. Indeed, the lonely valley of Ulladale boasts what can be described as the most unusual truncated spur in the British Isles. Here the notorious Strone rock-face has been so undercut by glacial erosion that its uppermost 750 feet (230 metres) form a considerable overhang.

Like the sea-lochs, the glacially deepened valleys have often been picked out along three predominant alignments, all of which reflect structures within the underlying gneisses (Figure 46). The NW–SE regional strike of the foliation has the greatest influence on the topography of South Harris, whereas in southern Lewis an E–W grain becomes more dominant; the third element is a subordinate SW–NE alignment. Loch Seaforth, the magnificent fjord of the Park district of Lewis, exhibits all three alignments in its 15-mile (24-kilometre) length. This mountainous tract of Park, slashed with radiating fjords, must be one of the largest trackless and uninhabited wilderness areas in the entire United Kingdom.

Leaving behind the mountains of Harris and southern Lewis, we pass into the less interesting scenery of northern Lewis. Here the rolling, loch-studded moorlands show the Lewisian gneiss in all its uncompromising bleakness and austerity. As elsewhere in the Western Isles, the settlements and improved lands are almost entirely peripheral, so that the interior remains a treeless waste of peat-hag, moorland and ice-scraped rocks (Plate 36). Nevertheless, there is widespread evidence to show that at various periods of post-glacial time the Outer Hebrides were a more hospitable place than they are today. Beneath the blanket bogs which flourished in the cooler, wetter conditions which returned just before the Christian era (the so-called Sub-Atlantic period) there are many tree-stumps. These represent remnants of the Sub-Boreal forest which clothed all but the highest Scottish summits around 6,000 to 2,500 years ago.

During this drier climatic period Bronze Age man arrived in the Hebrides, erecting the remarkable Callanish Stones, second only to Stonehenge in archaeological importance, on the shores of East Loch Roag (western Lewis) some 3,000 years ago. When this circle of standing stones was first excavated in 1857, some 5 feet (1·5 metres) of Sub-Atlantic peat had first to be removed. This information provided a valuable yardstick for the reconstruction of the vegetational history of Lewis. Not far to the north Dun Carloway, also on the sea-shore, is the best preserved Hebridean example of the massive Scottish dry-stone circular buildings called 'brochs', whose age and purpose are open to debate. (That at Mousa, in the Shetlands, is the least ruined in all Scotland.)

Before concluding this brief summary of the archaeological heritage of Lewis, mention must be made of the traditional Lewis dwelling known as the 'black house'. This thick-walled, single-room cottage, similar to those on Tiree (see p. 252), was thatched with a motley collection of bracken, turf or bent-grass from the dunes, held down by weighted ropes of straw, and was usually without chimney or windows. The only source of light was from the single door, used by family and livestock alike, the cattle dung being removed once a year. The rough stone corners were characteristically rounded, leaving nothing on the outer face to catch the wind. Not surprisingly, few of these primitive dwellings have survived as habitations, although several are currently in use as byres or barns. The rough stones from which the older houses were constructed were often collected from the boulder clay, laid bare by countless centuries of peat-cutting. Such ground is termed 'the *gearraidh*', meaning 'the skinned land', and it now provides a poor-quality grazing which is nevertheless better than that of the peat bogs themselves.

Apart from the west-coast *machair* lands in the district of Uig, the only areas in Lewis where the soils are of reasonable fertility are Ness, near the Butt of Lewis, and the Eye Peninsula. In both cases the general acidity of the ground has been alleviated by the marine-shell content of the glacial drifts, for it must be remembered that, far removed from the Harris uplands, these northern peninsulas received the unrestricted passage of the mainland ice-sheets, which had in transit dredged up the former sea-floor deposits of the Minch.

In the Stornoway area the calcareous drifts of the Eye Peninsula (in fact a former island now tied to Lewis by a tombolo of superficial material) have become mixed with the sandy soils derived from the underlying Stornoway Beds. Long thought of as Torridonian sandstone, these coarse, reddish-brown conglomerates and finer cornstones, into which Broad Bay is carved, are now reputed to be yet another example of the Hebridean Permo-Triassic rocks. It has been suggested that they were formed as alluvial fans in association with a subsiding basin of sedimentation, itself closely related to tectonic movements within the Minch Fault zone. The red sandstone houses of Stornoway, backed by the thick woodlands of the Castle grounds, single out this urban settlement as an atypical feature in the landscape of the Western Isles, although the gridiron plan of the town merely follows the layout of the old fields and ditches, whilst the fishing fleet reminds us of the long-standing dependence of the islanders on the surrounding seas.

Within these seas lie isolated but well-known islands, such as St Kilda, which merit a brief summary to complete our survey of the Western Isles. The most remote is Rockall, about 200 miles (320 kilometres) to the west of North Uist, which is no more than a 70-foot (21-metre) pinnacle of Tertiary granite rising from a submarine platform of volcanics of similar age to those of the Inner Hebrides.

St Kilda, because of its spectacular coastal scenery and its fascinating cultural history, is better chronicled than many of the less remote Scottish islands. Not only does this group possess the highest sea-stack in the British Isles (Stac an Armin, 627 feet, 191 metres), it also has sea cliffs 1,000 feet (300 metres) in height. The rugged nature of the islands is, as we might expect, due to their having been carved from a Tertiary igneous complex, the gabbroid rocks of which are comparable with those that we have already encountered in Rhum and Skye. Because of the basic mineralogy of its gabbro, the manuring of sea-birds and the lack of acidic mainland glacial drift (see p. 281), St Kilda possesses an alkaline peat (similar to the Fenland peats of England) possibly unique in Scotland. It has been suggested, furthermore, in view of St Kilda's isolation from the mainland ice-sheets, that its flora may have survived from pre-glacial times.

Closer at hand, the Flannan Isles, Sula Sgeir, North Rona, Sule Skerry and Sule Stack are all fragments of gneiss, presumably of Lewisian age. The Shiant Isles, however, are composed of a Tertiary igneous sill overlying Liassic rocks, with the northern cliffs of Eilean Mor exhibiting the tallest examples of columnar basalt anywhere in the British Isles. Although the Shiants lie in the Minch, only a few miles from the shores of Park, their fabric is clearly related to the Tertiary rocks of Skye.

15. The Northern Highlands

Many visitors to Scotland do not travel farther north than the latitude of Inverness – Kyle of Lochalsh, and thus miss some of the finest scenery and some of the most interesting geology to be found north of the Border. This is more true of the western seaboard, because there we can discover, amongst other things, the majestic mountains of Torridonian sandstone, the remarkable structures associated with the Moine Thrust and some of the wildest and least disturbed tracts of the Scottish countryside. Here in the west is a region of ancient rocks, clothed with a natural vegetation of birch, oak and pine forest, heather moor and blanket bog, less habited now than a century ago and with single-track roads which discourage all but the most avid naturalist or those who seek solitude off the beaten track. In contrast, the eastern coastlands of Ross and Cromarty, Sutherland and Caithness exhibit the veneer of the twentieth century, for there are railways, main roads and towns amidst extensive areas of improved agricultural land. There too is a zone which looks eastwards to the offshore oil fields of the North Sea, so that it has already succumbed to the type of major industrial development which was previously confined to the coalfields of the Midland Valley.

The dichotomy between west and east is further emphasized by contrasts in the topography which, as we shall see, reflect both the geological and the geomorphological history of the region. As with the remainder of western Scotland, the west coast is characteristically ragged because of its deeply penetrating sea-lochs, but, unlike those in the southern sector of the western seaboard, the fjords of the north-west take on a markedly NW–SE orientation. This is primarily because to the north of Loch Carron the so-called 'Caledonian grain' (NE–SW), which dominates so many of the Scottish structures and landforms, fails to reach the western coast. Instead, the trend of the Lewisian gneiss (NW–SE), which we have

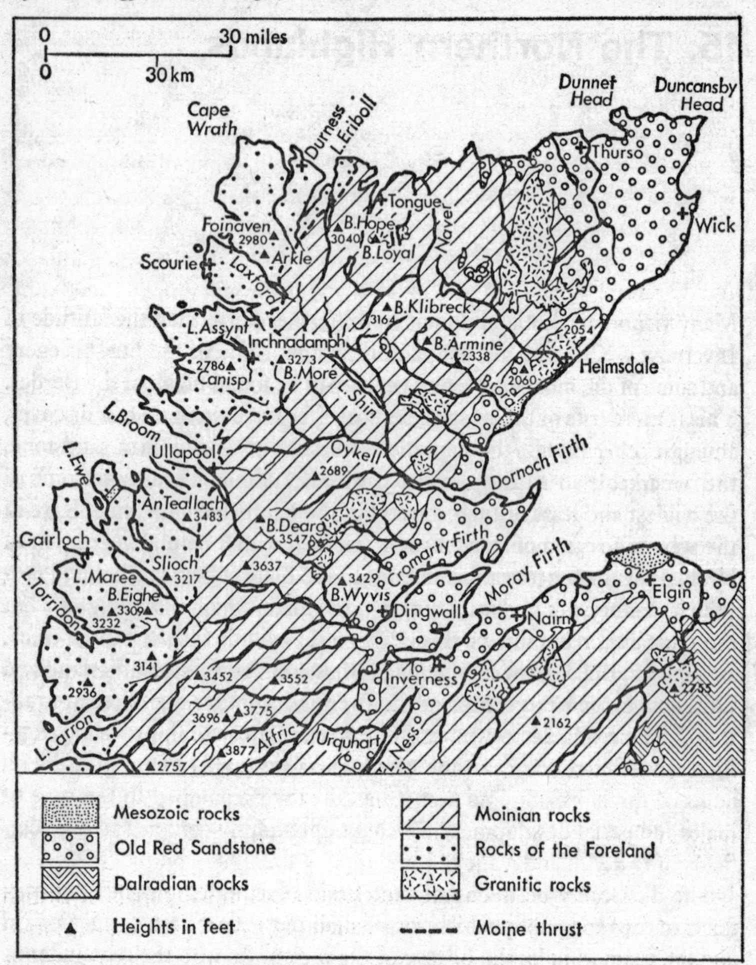

Fig. 48. *The geology and drainage of northern Scotland (after T. N. George).*

already encountered in the Western Isles, has an important influence on the shape of the coastline, which runs discordantly across the regional strike. On the east coast, however, the 'Caledonian' trend still prevails and the coastline itself is broken by fewer sea-lochs. But we must not assume that these morphological differences between west and east are only to be explained in terms of structures, for much of the contrast stems from lithological disparities. In the west, for example, a great diversity of rock types has been picked out by erosion to form a deeply dissected mountainous terrain with many associated linear water bodies. The eastern coastlands, however, have been carved very largely from rocks of uniform lithology, for the Old Red Sandstone extends from the Moray Firth to Caithness and Orkney (Figure 48). There the shorelines are accordant with the 'grain', the relief is more uniform and the lochs less numerous in a landscape of smooth, rolling hills and open valleys.

Professor Linton has demonstrated that the scenic contrasts also owe a great deal to the inequality of glacial erosion during the Ice Age. Corries, troughs and ice-moulded surfaces are far more common in the western tracts than they are in the east, for, as today, the western mountains were the zones of highest precipitation. It follows that the ice-caps lay nearer to the western than to the eastern coast and explains why many of the western sea-lochs have been converted into fjords by periods of intensive glacial overdeepening which most of the eastern sea-lochs failed to experience. Thus, whilst western coasts are often ice-scrubbed and barren (albeit picturesque), eastern coasts sometimes take the form of constructional shorelines in which marine waves have fashioned the enormous volume of glacial outwash into long, sandy beaches backed by prosperous farmland. Since we have highlighted the dichotomy between west and east, it will be convenient to treat the Northern Highlands in two parts: The North-west Highlands and the North-eastern Seaboard.

The North-west Highlands

North-west Scotland is one of the classic areas in the annals of British geology, and its Geological Survey memoir (by B. N. Peach and J. Horne) is often claimed to be the most important ever produced. One reason for this is that some of the major principles of structural geology were established in this region, following the discovery of the Moine Thrust in 1883. This event led Archibald Geikie to conclude: '... the correct explanation of this structure introduced to geologists a new type of displacement in the earth's crust.' Indeed, this was the first place in the world where the normal order of superposition was found to be reversed because of thrusting. The relationships between the Pre-Cambrian rocks and those of Lower Palaeozoic age having been worked out, analysis of the entire rock-succession of Britain was set upon a firm foundation.

Reference to Figure 48 will show that the line of the Moine Thrust separates two contrasting areas of rocks: to the west is the ancient 'foreland' of Lewisian gneisses and Torridonian sandstones, where Moinian schists are virtually unknown; to the east is the main Caledonian fold-belt, where Moinian rocks are extremely deformed, Torridonian is unknown and Lewisian gneiss can be seen only as inliers beneath the contorted schists. It is important for the reader to remember that a thrust will occur if rock deformation is driven beyond the limits that can be accommodated by folding during a period of mountain-building. Thrusts are therefore often developed along the margins of mountain-belts, where the mobile nappes of the orogenic belt (composed of rocks from within the geosyncline) meet a flanking zone of more stable rocks, termed a foreland. Thus the waves (nappes) of Caledonian folding broke against the stable foreland of north-west Scotland, made up largely of the Lewisian and Torridonian. So fierce was the tectonic pressure from the east-south-east, however, that parts of the foreland surface itself were sheared off and carried bodily forwards along these planes of low-angled dislocation.

The marginal belt of Caledonian thrusting can be traced for some 200 miles (320 kilometres) north-north-eastwards from Islay to Loch Eriboll near Cape Wrath (Figure 37). The belt is in fact composed of a few large thrusts and numerous small thrusts; it was the latter which first caused the newly created sedimentary nappes to be driven onto the front of the crystalline western foreland, at the outset producing what is known as an 'imbricate' structure, analagous with overlapping tiles on a roof (Figure 49d). Further compression seems to have caused a gradual pile-up of these minor structures, and as resistance increased, so the whole pile was ultimately driven many miles across the foreland, along several major thrust-planes. The basal thrust is termed the 'sole' of the slide; it is succeeded upwards by the Glencoul Thrust, the Ben More Thrust and finally the Moine Thrust itself, which is the most easterly of these planes of dislocation (Figure 49c).

The curious character of the mountains in the Assynt area is largely the result of the differential weathering of contrasting rocks brought into juxtaposition by major thrusting. Before looking in detail at the scenery of these western coastlands, however, it is important to understand the

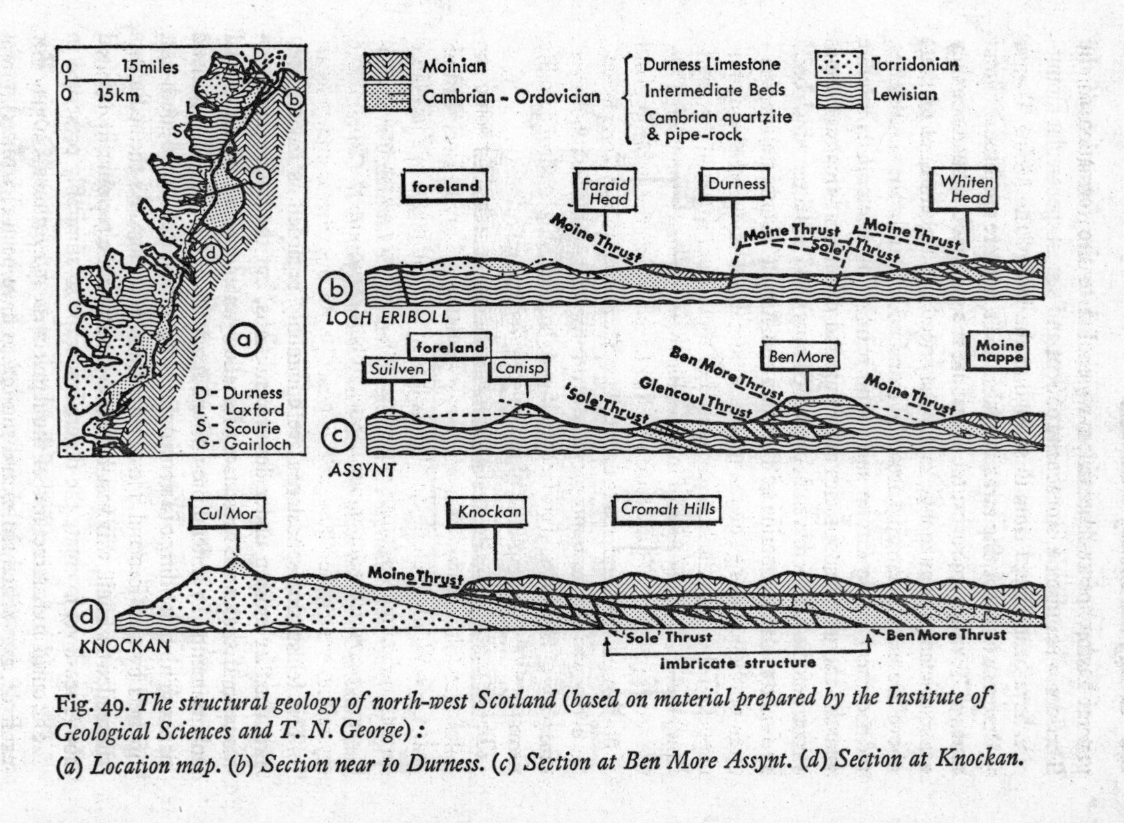

Fig. 49. *The structural geology of north-west Scotland (based on material prepared by the Institute of Geological Sciences and T. N. George):*

(a) Location map. (b) Section near to Durness. (c) Section at Ben More Assynt. (d) Section at Knockan.

general history of sedimentation which led to the formation of the Torridonian sandstone, since many of Scotland's most spectacular mountains have been carved from this tough arenaceous rock.

Research by Dr A. Stewart has established that there are three different types of Torridonian rocks: the oldest are the greywackes, but since they are restricted to Islay and Colonsay they need not concern us at present; above these are great thicknesses of interbedded sandstones and shales, termed the Diabaig Group and found mainly in Skye and its satellite islands; the uppermost facies are composed of feldspar-rich sandstones (known as the Applecross and Aultbea Groups), and it is these which build the spectacular mountains of Torridon and Assynt (Figure 50).

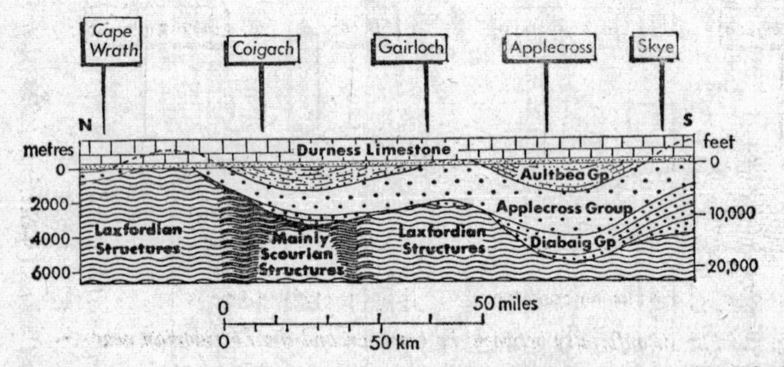

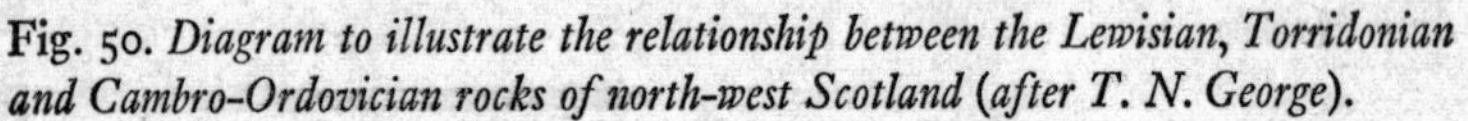

Fig. 50. *Diagram to illustrate the relationship between the Lewisian, Torridonian and Cambro-Ordovician rocks of north-west Scotland* (*after T. N. George*).

The feldspathic sandstones are commonly regarded as having been produced by erosion under desert conditions, but the underlying sandstones and shales are indicative of shallow-water marine or even fluvial environments. In general the upward sequence of facies is characteristic of the gradual infilling of a geosyncline which must have stood south-east of the Lewisian foreland 750–1,000 million years ago. As the sediments succeeded in filling this vast basin, so they became tectonically uplifted above sea-level to create the desert landscape so admirably preserved in the detailed rock structures of the Applecross and Aultbea Groups. For example, rain-pitted shales and sun-cracks have been recognized, whilst the Lewisian gneiss appears to have introduced screes and hillwash into the arid valleys of that time.

Ephemeral water bodies must also have been present, however, and research has shown that streams brought tourmaline-quartz pebbles into the Torridon rocks from sources far beyond the Lewisian of the Outer Hebrides. The mountainous foreland must then have extended as far as southern Greenland, from which major rivers flowed south-eastwards, depositing thick alluvial fans as they debouched from the mountain-front in Wester Ross and Sutherland. Farther east, beyond these primeval estuaries, deltas and land-locked basins, lay the open waters of the geosyncline in which the Moinian rocks were being laid down, so that the Moine schists are probably altered sediments of Torridonian age. Thus the Torridonian and Moinian were deposited under different sedimentary

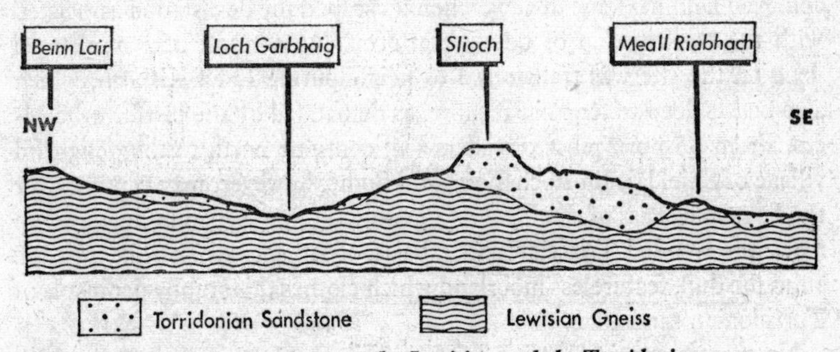

Fig. 51. *The unconformity between the Lewisian and the Torridonian near Loch Maree.*

environments in widely separated areas, but were subsequently brought together by the thrusting of Caledonian times described above.

As the Torridonian sediments accumulated, so they buried an old land surface that had already been sculptured from the Lewisian gneisses of the foreland. This 'fossil' land surface had its own hills and valleys, which are now becoming slowly exposed as the overlying Torridonian is weathered away. The best example of this buried relief may be seen on the slopes of Slioch (3,217 feet, 981 metres) above Loch Maree (Figure 51). It was in this area that the unconformity between the Lewisian and the Torridonian was first remarked upon by John Macculloch as early as 1819. But, before looking at the detailed scenery of this fascinating landscape, we must commence our journey farther south at Loch Alsh, where the Moine Thrust Belt runs out to sea before reappearing in Skye (Figure 48).

The new coast road between Kyle of Lochalsh and Balmacara has not only opened up provocative views of Skye but also enables one to see wonderful sections of the Applecross and Diabaig Groups of the Torridonian in the cliffs which overlook it. The eastward-dipping beds are in fact part of the inverted limb of a gigantic recumbent fold known as the Kishorn Nappe. As we trace the rocks eastwards they get more deformed and crushed until near Balmacara they become mylonites (see Glossary). At near-by Ard Point the equally pulverized Lewisian gneiss can be seen *overlying* the younger Torridonian, driven there by the tangential movements of this thrust-belt.

The crofting village of Drumbuie, near the mouth of Loch Carron, achieved national fame in 1973 when it escaped the devastation associated with the construction of oil-rig platforms. As a result of a ministerial decision the site was transferred to neighbouring Loch Kishorn, which also boasts deep water close inshore, as demanded by the platform-builders. From Kishorn most travellers will continue north to the delightful village of Shieldaig on Loch Torridon. Some, however, may venture over the fearsome 'Pass of the Cattle' (arguably the worst road in Scotland) to reach the wooded Triassic–Liassic basin of Applecross, set like an oasis amid the dull, featureless moorland which clothes this stubby peninsula of Torridonian sandstone.

No greater scenic contrast exists in the North-west than that which differentiates the northern and southern shores of Loch Torridon. To the south lies the drab, unremarkable coastal scenery of Applecross, with the pine-clad isles of Loch Shieldaig as a notable exception; to the north are the gigantic towering peaks of the Torridon Highlands themselves. Torridon village crouches at the foot of a large debris-cone built of material carried down the gullies which seam the precipitous southern face of Liathach (3,456 feet, 1,054 metres). This mountain's name means 'the Grey One', which may refer either to its swirling mists or to the apron of screes below its beetling cliffs. Four of its seven summits, from which much of the frost-shattered debris is derived, are composed of rocks younger than the Torridonian sandstones of the slopes. In fact we shall find that, although the Torridon mountains are built mainly of reddish-brown sandstones (of the Applecross Group), some of the high peaks are capped by a layer of tough Cambrian quartzite which can be seen dipping gently eastwards across the Torridonian rocks until it disappears beneath the Moine Thrust near Kinlochewe. We shall meet this white Cambrian quartzite again in Assynt, where it serves a similar function on some of the mountain summits.

Of the Torridon mountains, the graceful peak of Beinn Alligin (3,232 feet, 985 metres) is composed only of Torridonian rocks, but part of the 7-mile (11-kilometre) summit ridge of Beinn Eighe has its tiers of red sandstone adorned with a shimmering crown of snowy quartzite (Plate 37). To appreciate the grandeur of Beinn Eighe fully, however, one should really ascend Glen Grudie from Loch Maree, for this five-mile (8-kilometre) walk commands a magnificent vista of the stark precipices of the northern face. In the imposing ice-carved amphitheatre, the vertical cliffs of red sandstone and white quartzite soar 1,300 feet (400 metres) from the lonely corrie lake to the summit (3,309 feet, 1,009 metres). This is Coire Mhic Fhearchair, regarded by W. H. Murray as one of the most magnificent corries in the Highlands.

The eastern half of Beinn Eighe is now a nature reserve, the first to be purchased by the Nature Conservancy after its creation in 1949. Apart from rare fauna, such as wild cat, pine marten, golden eagle and ptarmigan, the reserve is best known for its wonderful stand of Scots pine along the southern shores of Loch Maree – a genuine remnant of the historic Wood of Caledon which once clothed most of the Highlands but has since been devastated by fire and axe. The Conservancy's aim is to rehabilitate the native pine forest by fencing against deer and sheep, thus allowing the young pines to regenerate successfully. Mention should also be made of the oak woods at neighbouring Letterewe, for at Loch Maree the native oak reaches its northern limit. On south-west-facing slopes of rich brown soils, derived from basic sills of epidiorite within the acid Lewisian, the fresh green canopies of mature oak woods are reflected in the waters of Loch Maree.

We are fortunate that they have survived, for during the seventeenth century the Wester Ross oak woods were extensively felled to supply charcoal for the local iron-smelting industry. The iron ore was shipped from Cumberland to Poolewe, and thence to the furnace at Letterewe. It is unlikely that the local 'bog-iron' ore ever played an important part in the industry, despite the development of iron pan and iron nodules in the waterlogged mineral soil beneath the peat. After the timber supplies had been virtually exhausted, the introduction of coking coal for smelting caused the transfer of the iron industry to the Midland Valley, where it has remained.

The modern road leaves the fault-guided valley of Loch Maree and winds its way circuitously to the beach-fringed coastline of Gairloch before returning to the Loch Maree fault-line at Poolewe. Here a prominent fault-line scarp of Lewisian gneiss runs north-westwards for a dozen miles (19 kilometres) and overlooks the marine inlet of Loch Ewe. Although this loch is carved essentially from Torridonian rocks, the brighter green fields at the southern end of its island, together with the col leading across to Gruinard Bay, suggest the presence of a different lithology. Sure enough, a narrow outcrop of Triassic and Liassic rocks lies unconformably on the Pre-Cambrian hereabouts, but the well-known 'sub-tropical' gardens of Inverewe owe nothing to these more fertile soils: paradoxically, this 'Oasis of the North' was created entirely on the less tractable soils of the Torridonian. Numerous viewpoints along the coast road of Loch Ewe provide opportunities to view the eastward panorama of Pre-Cambrian mountains. The castellated tops of An Teallach (3,484 feet, 1,062 metres) are Torridonian sandstone, although the Lewisian gneiss of the Fisherfield Forest creates many high peaks, with A'Mhaighdean (the Maiden), above Fionn Loch, taking pride of place.

The longest sea-loch of the Northern Highlands, Loch Broom, penetrates almost 20 miles (32 kilometres) into the mountains from the archipelago of the Summer Isles which clusters across its entrance. So far does the sea-water bore into the uplands that we are forced to make a wide detour across the Moine Thrust and go deep into the Moinian country in order to continue our northern journey (Figure 48). But this provides us with an opportunity to see one of Scotland's finest waterfalls at the head of the mile-long Corrieshalloch Gorge. The river Droma flows down the U-shaped valley of a large glacial breach in the main Highland watershed (see p. 306) until suddenly, at the Falls of Measach, its waters plummet 200 feet (60 metres) into this narrow forested cleft. Since the gorge is incised into the floor of a glacial trough, it has often been taken to indicate the degree of post-glacial river entrenchment. It seems more likely, however, that the incision took place very largely during the Upper Pleistocene, when enormous quantities of glacial meltwaters were coursing around the glacier fronts and cutting similar meltwater channels.

Ahead lies the thriving fishing port and resort of Ullapool, whose line of white houses seems to float in Loch Broom when viewed from afar. Raised strand-lines can be seen almost everywhere around the sea-loch, pointers to the varying amounts of isostatic uplift which have taken place in Wester Ross during post-glacial times. There is no better place for their study than Ullapool itself, for there a late-glacial river delta, created at the mouth of Glen Achall, has been affected by wave-action at various times during its pulsatory uplift. The northern part of the town is located on the highest part of the uplifted delta-terrace, but the bulk of the settlement is perched at an elevation of 40 feet (12 metres) on a flat terrace, also of late-glacial age. No visitor can fail to be impressed by the conspicuous cliff that was subsequently cut into the higher terrace by the waves of the post-glacial sea; Ullapool's waterfront settlement is packed tightly into the notch behind the post-glacial raised beach.

Ullapool is an excellent centre from which to explore the mountains which lie astride the borders of Ross and Sutherland, mountains of such unusual shape and character that some regard them as the most spectacular in the British Isles, despite their lack of absolute height. These are the sundry peaks of Coigach and Assynt, whose soaring buttresses and pyramids are unmatched in verticality anywhere in Britain. It is not merely that their corrie head-walls are precipitous: most of their summits are almost completely ringed by walls of virtually unscaleable cliffs. Some, such as Ben More Coigach (2,438 feet, 743 metres) and Quinag (2,653 feet, 809 metres) are relatively flat-topped domes, but others, like Cul Beag (2,523 feet, 769 metres), Canisp (2,779 feet, 847 metres) and Cul Mor (2,786 feet, 849 metres), are stately pyramids. It is, however, the monolith of Suilven (2,399 feet, 731 metres) which always catches the eye, for it stands alone as a sort of Torridonian lighthouse above an alien sea of Lewisian gneiss (Plate 38).

Like the peaks of the Torridon Highlands, these remarkable hills were carved from thick layers of Torridonian sandstone which must once have completely obliterated the Lewisian basement. But erosion has now laid this bare, so that the surviving mountain residuals rise abruptly from the lake-dotted and ice-scoured terrain of gneissic rocks. The sandstone peaks of Quinag and Cul Mor have isolated layers of white Cambrian quartzite remaining on their summits, while from Canisp's peak the quartzite capping declines eastwards in the form of a far-spreading dip-slope until it passes beneath the Moine Thrust Complex to the south of Inchnadamph.

However, there is merely a scatter of white quartzite boulders surviving on the summit of Suilven and the peak of Stac Polly, farther south-west, has long since lost its tough cap-stone, so that its summit ridge (2,009 feet, 612 metres) is now a splintered cock's comb of red sandstone.

Midway between Ullapool and Inchnadamph the Nature Conservancy has constructed an information centre at Knockan, where the western edge of the Cromalt Hills overlooks the main road. Not only is this an excellent viewpoint for Stac Polly, Cul Beag and Cul Mor, but it also provides an opportunity to understand something of the complex stratigraphy associated with the Moine Thrust Belt. We have already seen how the several large thrusts found in this zone of dislocation have helped to create the complicated geology around Ben More Assynt (3,273 feet, 998 metres) depicted in Figure 49c. In the latter district the distance between the Moine Thrust and the 'sole' of the thrust complex at Loch Assynt is no less than 8 miles (13 kilometres). At Knockan, however, the Moine schists have been transported such a distance along the Moine Thrust that they have completely overridden the slightly older planes of thrusting (Figure 49d). The final result can be viewed with comparative ease at Knockan Cliff, where the undisturbed Cambrian basal quartzite and its overlying Lower Palaeozoic sedimentary rocks are capped by the Moine schists of the Cromalt Hills (Figure 49d). The junction between the creamy-coloured Durness Limestone and the darker overlying Moinian psammite is the Moine Thrust itself (Point 11 on the Nature Trail here), although both rocks have been crushed into mylonites by the thrust movements.

We have several times witnessed the effect of a limestone outcrop on the vegetation of the Scottish Highlands. None is more remarkable than that which results from the appearance of a narrow exposure of Durness Limestone amongst the acid gneisses, sandstones and quartzites of Assynt. Unlike that occurring in Skye (see p. 265), this calcareous Cambro-Ordovician rock is largely drift-free. Thus near Elphin and Inchnadamph significant tracts of limestone are exposed, bringing swathes of brighter green into the dark-coloured moorlands (Plate 39). Such plants as holly fern, bladder fern and stone bramble thrive on the more fertile soils, whilst mountain avens (*Dryas octopetala*) is everywhere abundant. As J. Raven and M. Walters write: 'There are few botanical centres in Britain more rewarding than Inchnadamph itself... within a quarter of a mile, on the long limestone cliff running southwards, not only purple saxifrage and the rare grass *Agropyron doniarium* but dark red helleborine and the sedge *Carex rupestris* are in unusual plenty.'

As we follow the road northwards from Inchnadamph, past the picturesque ruins of Ardvreck Castle on the shores of Loch Assynt, the dipslope of Cambrian quartzite can be seen rising westwards to the summit of Quinag. Along this mountain's northern face the irregular nature of the unconformity between the knobbly Lewisian and the horizontally bedded Torridonian is clearly visible. In contrast with the irregular base of the Torridonian (see also p. 297), the base of the Cambrian everywhere forms an even plane (Figure 50). To the east of the road the gneissic hump of Glas Bheinn hides Britain's highest waterfall, where the Eas a Chual Aluinn leaps 658 feet (200 metres), about four times the height of Niagara, but this can only be reached by a rough walk of about 3 miles (5 kilometres) from the road.

Once across the Kylesku ferry we enter that corner of Sutherland where the Scourian and Laxfordian gneisses create a Lewisian wilderness of lumpy crags and small lochans, comparable with the naked landscapes of South Rona, central Lewis and the Inverpolly Nature Reserve in Western Assynt. Tiny lochans, bestrewn with water lilies and water lobelias, seem a bizarre extravagance in this treeless expanse of ice-scraped rock, where the term 'mamillated surface' seems extremely apt. Bare rock surfaces abound because heather does not flourish on gneiss; its place is usually taken by coarse grass, bog myrtle and deer's hair sedge. The mountains of Cambrian quartzite stand some miles back from the coast, so that there are few viewpoints on the tedious narrow road across this undulating gneissic platform.

The hinterland is uninhabited; the few tiny settlements here hug the rock-bound coast. Like their Viking forerunners, they look to the sea for their livelihood and communication in a land which offers little inducement to agricultural improvement. Along the fretted Atlantic coastline of Sutherland the gneissic sea-cliffs are nowhere impressive until one reaches the remote fastnesses of Cape Wrath, where they rise to 450 feet (135 metres). Nor do the occasional patches of Torridonian make much impact on the cliff scenery, with the notable exception of the island of Handa, near Scourie. Here the horizontal stratification of the sandstone in the 400-foot (120-metre) cliffs provides countless nesting sites in what is one of Britain's most important bird sanctuaries.

Cliffs of a different origin can be seen farther inland: the peaks which lie to the east of Laxford Bridge and Rhiconich are deeply notched by a dozen corries. Although Ben Stack (2,364 feet, 721 metres) is a shapely cone of Lewisian gneiss, the neighbouring summits of Arkle (2,580 feet, 787 metres), Foinaven (2,980 feet, 909 metres) and Cranstackie (2,630 feet, 802 metres) stand resplendent in snowy caps of Cambrian quartzite (Figure 48). One interesting difference between these picturesque hills and their Assynt counterparts is that here the quartzite rests directly on the gneiss, the Torridonian having been destroyed in a period of erosion prior to the formation of the Cambrian cover rocks. The juxtaposition of the gneiss and the quartzite has had an important effect on corrie morphology: where glacial erosion has scooped out a corrie in gneissic rocks only, the depression is generally shallow and dish-shaped with no steep walls; when a corrie is located in the quartzite, however, the cliffs are steeper because this material has better jointing properties. Usually the corries are composite, each possessing a semicircular quartzite wall and a heavily polished gneissic floor, as in the immense northern amphitheatre of Arkle, where Loch an Easain Uaine is constantly supplied by the waterfalls from four hanging corries.

Because the quartzite is a well-jointed, brittle rock it yielded to frost-shattering throughout the Pleistocene, so that today the summit-ridges of Arkle and Foinaven (*Foinne Bheinn* – White Mountain) are a confusion of broken rocks, while their scree slopes are so unstable that they have been likened to glittering spoil-heaps by generations of weary climbers. Mass movement of the mountain-top detritus evidently continued into the post-glacial period, for a buried soil profile on Arkle shows that a vegetation layer flourished there more than 5,000 years ago before being inundated by solifluxion. Today, however, these almost sub-Arctic mountains are bare, and their litter of boulders provides ideal breeding-grounds for the rarely seen snow bunting and pine marten.

Separating Foinaven and Cranstackie is the river Dionard, which meanders northwards down a broad strath before emptying into the tidal waters of the Kyle of Durness. At last the landscape becomes less rugged, suggesting that we have crossed an important geological boundary. Closer inspection of the greyish-white rock exposures will reveal that it is not the barren Cambrian quartzite but the more yielding Durness Limestone of Cambro-Ordovician age. Since the Lewisian gneiss is not so prevalent on this northern section of the Scottish coastline, the barren, rock-bound fjords of the west coast are now superseded by the flowing curves of sand-dunes and the sweeping lines of yellow sand-banks along the shallow estuary. The 'softer' outlines of the topography are matched by a change in the colour of the landscape: the dark, peaty moorlands disappear as if by magic, to be replaced by the emerald pastures characteristic of calcareous soils. The featureless boglands give way to a mosaic of stone-walled fields crowded with Cheviot sheep, whilst the few bare limestone 'pavements' remind us of the Northern Pennines or the Irish Burren, albeit on a more restricted scale. Although the windswept coastal location precludes any significant tree-growth, a rich ground flora exists, including an unusual *Dryas* heathland at Borralie, near Durness.

Of greatest scenic interest, however, are the sinkholes, caves and underground water-courses of this well-jointed limestone terrain. Chemical solution in the highly acid waters of the local streams has produced a small-scale development of karstic scenery (see Glossary). Because of the porous nature of the limestone the surface streams disappear at sinkholes once they cross from the surrounding impervious rocks, their underground courses often being marked by lines of hollows or actual shafts where the roof has collapsed. We can view this phenomenon near the coast road at Durness: here the waters of the Allt Smoo plunge 80 feet (24 metres) vertically down a shaft before reappearing at sea-level from the mouth of a gigantic cave. This is the famous Smoo Cave (from the Norse: *smuga* – cleft), whose cliff-face aperture is 120 feet (37 metres) wide (Plate 40). Inspection shows that it has been created partly by the underground river and partly by marine erosion; it is easy to imagine that the sea-filled gorge outside the cave entrance was once roofed by a layer of limestone, long since collapsed. In the Durness Limestone at Traligill, near Inchnadamph, a cave some 700 feet (213 metres) long has been explored, whilst at Knockan Cliff the caves include a vertical pothole 130 feet (40 metres) deep. Excavations in some of the Inchnadamph caves have revealed that during the Pleistocene they gave shelter to animals, such as lynx and bear, no longer native to Scotland.

As we follow the road around the bare flanks of Cranstackie, the fjord of Loch Eriboll looms into view, reputedly of a depth and size sufficient to harbour the entire wartime strength of the Royal Navy. Although patches of Durness Limestone flank its shore, tough gneisses and quartzites bolster its surrounding hills and its imposing portal of Whiten Head (Figure 49b). We have now reached the northern limit of the Moine Thrust Belt; eastwards lie the seemingly endless hills and plateaux of Moinian metamorphic rocks, named from the near-by district of A Mhoine. It is time, therefore, to return southwards and make our way through the little-known mountainlands of central Sutherland before examining the long eastern and northern seaboard of northern Scotland. Any of three routes can be followed south-eastwards to Bonar Bridge on the Dornoch Firth: one goes by way of the newly bridged Kyle of Tongue, Loch Loyal and Strath Vagastie, and has the most varied scenery; another follows Loch More and the almost interminable Loch Shin, offering the finest waterscapes; the southernmost, via Strath Oykel, has the disadvantage of making us retrace our footsteps, but gives those whose appetite for geological treasures is insatiable a second chance to see the highlights of Assynt.

We have already encountered in earlier chapters the capacity of ice-sheets for overriding the pre-glacial watersheds, thereby creating deep glacial valleys where none existed before. The primary watershed of the Northern Highlands is no exception: in its 100-mile (160-kilometre) extent between Ben Fhada (Kintail) and Foinaven it displays no less than 14 major breaches. The road to Loch Shin follows one, that to Strath Oykel follows another, and so do all the main roads between Inverness and the west-coast towns; one of the lowest breaches, at Achnasheen, is also followed by the railway to Kyle of Lochalsh. At the beginning of this century Peach and Horne observed that during the Pleistocene '... what may conveniently be described as ice-cauldrons were set up in Central Sutherlandshire.' These were later termed 'basins of impeded outflow' by Professor Linton. Thus in areas of heaviest snowfall (in such localities as the present-day Loch Naver, Loch Shin and Strath Bran) ice-caps accumulated, and it was these which were responsible for the glacial 'transfluence' described above. In one of these glacial breaches, at Loch Droma, not far from the Corrieshalloch Gorge (see p. 300), a buried organic deposit of late-glacial age has thrown fresh light on the history of deglaciation in the Northern Highlands. Investigation of this important site has indicated that the ice-sheet which left such conspicuous frontal deposits on both western and eastern coasts (at Loch Broom and Cromarty Firth) had practically disappeared by 12,800 BP. Consequently, the fresh-looking terminal moraines which occur farther inland must belong to the local equivalent of the Loch Lomond Readvance, when small ice-caps re-formed for the last time in the higher glens.

Today, the glens of the Northern Highlands are occupied by a river pattern which can be divided into three distinct categories: first, the west-flowing streams, which are short, fast-flowing and in part structurally adjusted to the rocks of the foreland; second, the much longer east-flowing streams, which cut discordantly across all the major structures of the Moinian Series; finally, the short streams of the north coast, which, being adjusted to the regional strike, have succeeded in capturing some of the headwaters of the eastern rivers (Figure 48). Professor T. N. George believes that the drainage system is unequivocally superimposed, although, unlike Linton, he does not feel the need to invoke ephemeral Cretaceous 'cover' rocks. Instead he suggests that the drainage was created upon a primary upland surface of late-Tertiary age; into this, '... present-day valleys have become deeply incised by a rejuvenation partly brought about in adjustment to emergence on the one hand supplemented by an erosively contracting coastline on the other.' Such dissection by rivers (and also by ice) has succeeded in isolating the high residuals of Ben Hope (3,042 feet, 927 metres), Ben Loyal (2,504 feet, 763 metres), Ben Klibreck (3,154 feet, 962 metres) and Ben Wyvis (3,433 feet, 1,047 metres) from the main massif of the Northern Highlands (Figure 48).

The North-eastern Seaboard

The summit of Ben Wyvis is as good a place as any to get a bird's-eye view of the indented coastline of Easter Ross. When we pause to consider the general linearity of Scotland's eastern coasts, we realize that the complex inter-fingering of land and sea represented by the firths of Beauly, Cromarty and Dornoch is something quite exceptional. We are led to the conclusion that these major inlets probably represent nothing other than the drowned estuaries of three major east-flowing rivers, superimposed from the late-Tertiary upland surface (noted above) and deeply incised into the underlying structures by rejuvenation. That these former river courses were discordant to the structure is demonstrated by the way the mouth of the Cromarty Firth breaks through a line of hard Pre-Cambrian rocks (see p. 310). But since most of the incision took place into the reletively uniform lithology of the Old Red Sandstone, especially during the phase of glacial overdeepening, these eastern firths exhibit nothing of the rugged irregularities of the western sea-lochs whose geology is so complex. Instead we shall find a landscape of low, sandy shorelines with extensive raised-beach remnants cut into the thick mantle of glacio-fluvial deposits and reddish-brown boulder clay.

In general, these sandy drifts provide good loamy soils although some of the glacio-fluvial outwash fans yield such light soils that they are given over to patches of lowland heath, as at Muir of Ord. Nevertheless, the general prosperity of the farming scene and the large acreage of arable remind us of the fertile landscapes to the south of the Moray Firth, described in Chapter 8. The contrast with the subsistence crofting of the rocky western coast is only too apparent, for the low rainfall and prolonged sunshine of this sheltered eastern seaboard encourage not only a wide variety of root crops but also excellent yields of barley and wheat. Consequently the farms here are large, with manicured hedgerows, shelter-belts and well-maintained buildings, in keeping with the better-quality soils and the low, flat terrain. As soon as we cross the western limits of the Old Red Sandstone basin, however, the relief becomes higher and more rugged, the vegetation changes to heather moor and peat bog and the farms get smaller and more sporadic – in fact, the scenery becomes more typically Highland.

The oddly named peninsula of the Black Isle has a fringe of farmlands along its thickly forested central ridge, which attains an elevation of 841 feet (256 metres) before descending north-eastwards to the small town of Cromarty. The ridge is formed from the red and yellow layers of the Millbuie Sandstone, whose pebbles of andesite and porphyritic basalt are probably derived from erosion of Middle Old Red Sandstone lavas no longer exposed. The peninsula is aligned along a broad synclinal structure in the Old Red Sandstone, the axis of which runs SW–NE along the Millbuie ridge (Figure 52). It was the presence of the fossil-fish-bearing beds at the base of the Millbuie Sandstone which gave early inspiration to Hugh Miller, the Cromarty stonemason who became one of the most famous in the long line of eminent Scottish geologists. Born in a thatched stone cottage in Cromarty, now well preserved by the National Trust for Scotland, Hugh Miller wrote a classic volume on the Old Red Sandstone which J. Challinor evaluates thus: 'Its significance in the progress of geology is that, by the charm of its personal narrative and description, it awoke a widespread interest in the methods and results of geological inquiry.' Coincidentally, another eminent geologist, R. I. Murchison, made his home at Tarradale House, at the western end of the Black Isle.

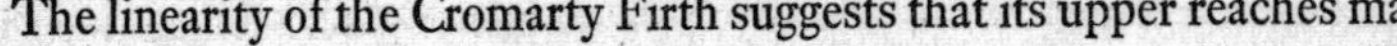

The linearity of the Cromarty Firth suggests that its upper reaches may

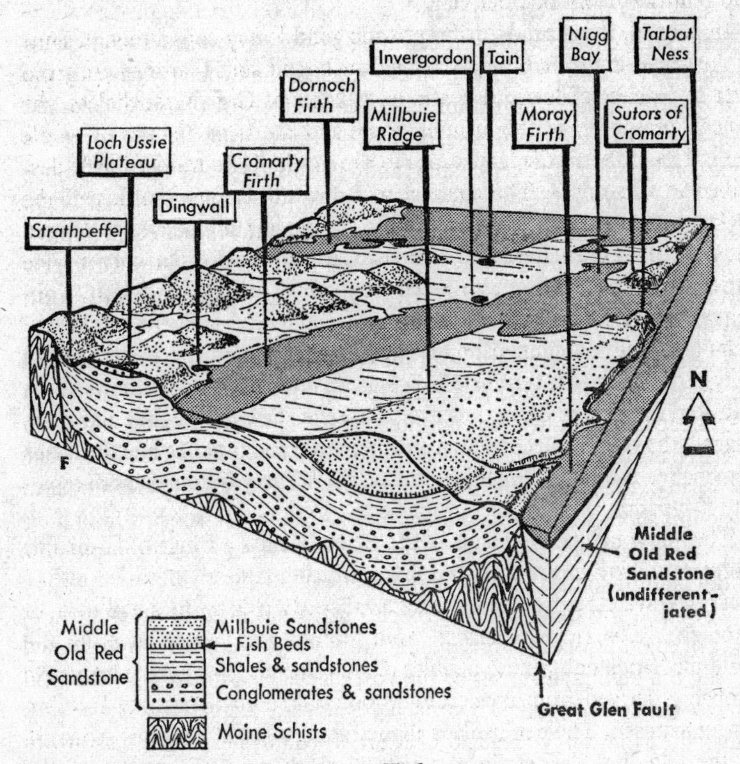

Fig. 52. *The structure of the Cromarty Firth area.*

be structurally controlled. It is therefore no surprise to discover that it is aligned along an anticline parallel to the Millbuie syncline. Thus the firth and its neighbouring Black Isle provide an excellent illustration of the principles of 'inversion of relief' (Figure 52; see also Figure 33). Although it has been suggested that the pre-glacial drainage flowed generally north-eastwards along the regional 'grain', with the 'Cromarty River' debouch-ing somewhere near to present-day Tain, there is evidence to show that ice-sheets not only overdeepened the Cromarty valley near to Invergordon but also succeeded in breaking through the rim of the Cromarty basin at the Sutors of Cromarty. These prominent hills, which flank the present entrance to the firth, are composed of a narrow outcrop of Moinian rocks so resistant to glacial erosion that they caused a constriction of the valley-glacier and a consequent overdeepening of the narrow entrance near to Cromarty (Figure 52). Subsequent drowning of the Cromarty basin by the post-glacial marine transgression led to the ultimate use of this land-locked deep-water firth as a naval base, centred on Invergordon. In the 1970s the same combination of a deep channel and a flat coastal shelf saw the development of a major oil-rig-platform construction yard at the Bay of Nigg. Neighbouring Alness, with its aluminium smelter, has also helped to place the stamp of industry firmly on the pastoral landscape of the Cromarty Firth, whose rural charm will slowly disappear under the impact of North Sea oil (Plate 41).

The remarkable straightness of the coastline between Tarbat Ness and Fortrose (on the Black Isle) is clearly related to the submarine extension of the Great Glen Fault. The enormous lateral movement of the fault (see p. 224) has caused a great displacement of the Mesozoic rocks which fringe the basin of the Moray Firth. Fragments of Jurassic sediments can be seen along this shore, both north and south of the Sutors of Cromarty, and their disposition has given rise to the claim that there was a post-Mesozoic shift of the fault of some 18 miles (29 kilometres).

The shallow waters of the Dornoch Firth are flanked by an interesting suite of coastal spits and forelands, some of which were initially fashioned during the higher sea-levels of the post-glacial transgression, when tidal waters temporarily linked Dornoch Firth with Nigg Bay (Figure 52). Since then the sandy foreland of Morrich Mhor, near Tain, has grown outwards into the firth as the sea-level has slowly fallen. This has happened through the successive addition of sandy barriers thrown up by wave action on Whiteness Sands, the end result of which has been the silting-up of Tain harbour and the end of its function as a port.

The final marine inlet on this eastern coastline is Loch Fleet, which causes a deviation of the coast road between Dornoch, with its squat sandstone cathedral, and Golspie, the county town of Sutherland. Originally an open bay at the seaward end of the deeply incised Strath Fleet, the loch has been almost cut off from the North Sea by the gradual growth of spits from both the northern and southern shores. Like most of the other constructional coastal landforms around the Moray Firth, these beach ridges appear to have been fashioned largely by waves of post-glacial raised-beach age. Professor J. A. Steers has compared the Loch Fleet opposing spits with similar phenomena at Poole Harbour in England and at the western end of the Menai Straits in North Wales, although the explanations of their genesis may be different – the complex configuration of the Dornoch Firth coastline would certainly have affected the angles of wave approach in the case of Loch Fleet. Today its pine woods, shingle ridges and sheltered waters constitute an important nature reserve.

To the north of the trim town of Golspie and its fairy-tale castle of Dunrobin, seat of the Duke of Sutherland, the breccias and conglomerates of Old Red Sandstone age form a west-facing scarp at Ben Horn (1,706 feet, 520 metres), but their fault-guided eastern slopes are even more steeply marked where the coastal plain widens around Brora. In this coastal strip a series of surprises awaits us, for here, in place of the ubiquitous chocolate sandstones of the Old Red, a downfaulted strip of Mesozoic sediments fringes the upland massif. Its effect on the topography is innocuous enough – the prosperous farming scene owes as much to the superficial deposits as it does to the underlying Triassic and Jurassic limestones, clays and sandstones. But it is the Brora colliery and its associated brick pits which halt us in our tracks, for are we not in the Highlands, a long way from any Carboniferous rocks? The same carbonaceous deposits which gave Raasay its oil shales (see p. 276) have given to Brora a metre-thick seam of good-quality coal, since the Estuarine Series (of Lower Oolite age) are well-represented here. In the associated clays, petrified wood and plant impressions have also been discovered, although it is the overlying blue-grey shaly clays of Lower Oxfordian age which have been extensively exploited for brick-making. Prior to its modern use as a local fuel and to raise steam at the Brora woollen mill, the Jurassic coal was traditionally used in the manufacture of salt from saltpans at the river mouth and, somewhat later, to fire the locally excavated bricks. Symptomatic of the impact of North Sea oil on this coastline is the fact that this unique colliery is now closed down after four centuries of production, and its work-force lured to more remunerative jobs in Easter Ross, despite the millions of tons of untapped reserves.

As we approach the tiny port of Helmsdale, nestling in its deeply incised river gorge, the coastal plain peters out where the bounding fault of the Mesozoic sedimentary basin runs out to sea. Granite hills now hug the shore and cliffs replace the sandy beaches of the Brora lowland, while the coast road zigzags into the numerous re-entrants carved in the steep coastal slope by the short, torrential streams. So precipitous is this tract of coastline between Helmsdale and Berriedale that the nineteenth-century engineers took their railway line inland on a remarkably circuitous route before regaining the Caithness coast at Thomas Telford's planned town of Wick. The railway surveyors utilized the Strath of Kildonan, where the Helmsdale river breaks out of the mountainous desolation of central Sutherland on its way to the North Sea.

We must journey inland too, leaving the picturesque landmarks of Scaraben (2,054 feet, 626 metres) and Morven (2,313 feet, 705 metres) to stand sentinel at the junction of the Moinian hills of Sutherland and the Old Red Sandstone plateau of Caithness. It is noteworthy that, whilst the sharp ridge of Scaraben faithfully reflects the morphology of a Pre-Cambrian quartzite, the symmetrical, tor-capped and frost-shattered cone of Morven is built from a patch of Old Red Sandstone, which stands unconformably amongst the granites and Moinian metasediments of these parts and owes its upstanding relief to the synclinal structure of the tough, conglomeratic sandstone.

The Strath of Kildonan received considerable publicity a century ago, when the discovery of alluvial gold in tributaries of the Helmsdale river led to a two-year gold rush. The mineralization was probably associated with the thermal metamorphism induced by the emplacement of a granite pluton, but the veins were subsequently eroded, so that all the discoveries to date have been in stream beds, not in the solid rock. The Geological Survey estimates that the reserves are greater than the amount won in 1868–9; the Kildonan 'goldfield' may prove to be the largest in Britain. Examination of a map of the Helmsdale river suggests that its natural headwaters should be sought in the river Mudale, far to the west of Loch Naver. The deeply embowered waters of Loch Coire, beneath the heights of Ben Klibreck, must also have contributed to the pre-glacial catchment of the river Helmsdale. However, the piratical, north-flowing river Naver has beheaded the Helmsdale, taking all the former headwaters away down the structurally adjusted valley of Strath Naver. One suspects that glacial interference may also have had some part to play in this change of direction. Few of the Scottish glens exhibit such a melancholy air as that of Strath Naver, for here is a zone whose population was utterly decimated by the Highland Clearances. Where once the black cattle of the crofters grazed on the broad valley floor and crofting villages dotted the hillside, nothing remains but the grass-grown ruins, the transient sheep and the interminable ranks of Forestry Commission conifers.

During the melting phase of the Pleistocene ice-sheets, Strath Naver carried a great volume of meltwater, judging by the terraces of glacio-fluvial material on the valley sides. The outwash deposits have undoubtedly been responsible for the great expanse of inter-tidal sandbanks in Torrisdale Bay, and these in turn have contributed to the extensive dune-formation of these parts. Strong northerly winds have blown the beach sands high onto the rocky ridges of Moinian metasediments to create a hummocky terrain unusually rich in flora. The Bettyhill Nature Reserve has been established to preserve some of the rare mountain flowers which here descend to sea-level. Raven and Walters describe how one small hillock exhibits '. . . a veritable carpet of mountain avens interspersed with other such calcicoles as globe-flower, yellow mountain saxifrage, purple saxifrage and dark red helleborine'. In addition, one may also find the Scottish bird's-eye primrose (*Primula scotica*) and the purple oxytropis (*Oxytropis halleri*), which is widely regarded as one of Britain's loveliest plants.

It is not only the vegetation which makes this lonely northern coast attractive, but the landforms too. Part of its beauty springs from the fact that the great variety of Moinian rocks which make up the Northern Highlands are here stripped of their peat cover and seen in all their intricate detail, for the coastline cuts discordantly across the regional strike. Since the narrow bands of Moinian metamorphic rocks and their igneous intrusions exhibit varying degrees of resistance to erosion, the coast is broken up into a series of rocky headlands and picturesque bays. Faults, dykes and jointing patterns have been etched out by erosion to create a detailed mosaic of stacks, arches, caves and geos (see Glossary), to say nothing of the constructional features created by the relentless waves along this exposed Sutherland coastline.

Of particular significance are the raised beaches which can be traced between the Kyle of Durness and the borders of Caithness. Studies have shown that two late-glacial and two post-glacial strand-lines exist, and that the former become higher in elevation towards the east. It thus seems likely that the North Sea ice-sheets depressed Caithness more than Sutherland, causing greater isostatic recoil there. It is suggested that the two late-glacial beaches (60–65 feet, 18–20 metres; 31–39 feet, 9–12 metres) may be equivalent to the so-called '100-foot' and '50-foot' raised beaches found elsewhere in Scotland.

At Melvich Bay the coast road leaves the Moinian rocks for the last time, and soon we cross the boundary into the county of Caithness. Few counties exhibit such a geological uniformity as Caithness: its western hills are of granite and its bordering mountains of quartzite and basal conglomerate (see p. 312), but its rolling plains are all floored by an enormous thickness of Old Red Sandstone. This is part of the structural basin known as the Orcadian Basin, which extends northwards into the Orkneys and contains some 16,000 feet (4,880 metres) of Middle and Upper Old Red Sandstone. Since, however, volcanic rocks are virtually absent, we shall find that crags and high relief, such as are to be found in the Sidlaws and Ochils, are not a feature of the Caithness scene, whose drama lies mainly in its coastline. Only there can the horizontally bedded yellow and red sandstones and the grey flagstones be appreciated in full, for elsewhere the pre-glacial topography is buried beneath glacial drifts and post-glacial peat.

Today the gently billowing outlines of the hills and peat bogs of central Caithness contribute to a landscape more reminiscent of central Ireland than of the Scottish Highlands. However, unlike the successful Bord na Mona schemes of Eire, the attempts by the North of Scotland Hydro-electric Board to win milled peat for a peat-burning power station have proved a costly failure. Amidst the 21,000 acres (8,502 hectares) of Altnabreac bog, for example, the gaunt ruins of the power station overlook the forlorn wastes of the partly skimmed bogland: as in the case of the Brora coal, the Caithness peat reserves remain ignored in the face of North Sea oil production. Thanks to the foresight of eighteenth-century landowners, however, many of the Caithness bogs were drained and reclaimed for agriculture, so that the present landscape is one of prosperous dairy farms among rolling green pasturelands. The older, thatched cottages can still be seen, although many have been relegated to function as barns and cattle-byres.

Another surviving feature in the landscape comes from the use of the renowned Caithness flagstone as a means of field enclosure. Set on end, these slabs of grey stone form excellent 'fences', and they criss-cross the treeless and hedgeless pastures like interminable rows of gravestones (Plate 42). The flagstones have also been used locally for both wall construction and roofing purposes in the older buildings, but production at the Thurso quarries has now ceased. When the quarries were fully operative the 'Caithness flags' were sent to all parts of the United Kingdom, where, being highly durable, they have withstood the tramping of countless millions of pedestrians in British towns and cities. It is a matter of regret that, like the flagstone fences in the countryside, the flagstone paving of Thurso itself is being replaced by more modern materials.

The flagstones belong to the Middle Old Red Sandstone and are part of a series of alternating sandstones, mudstones and limestones which together comprise a further example of a cyclothem (see p. 333). It has been suggested that these different sediments represent the contrasting environmental conditions which operated when vast, ephemeral desert lakes (known as 'playas') were being infilled and subjected to intense evaporation in the slowly subsiding Orcadian Basin, some 370 million years ago.

Although most of Caithness is a flat, monotonous plateau, it is rimmed by a girdle of magnificent coastal scenery. It is here that James Hutton, the founder of modern geology, noted that the affinities of the strata in Caithness and Orkney pointed to a former land connection between the two. He was led to this conclusion by viewing the 'perpendicular cliff of sandstone, lying in a horizontal position', and by observing that '... there are small islands, pillars and peninsulas of the same strata, corresponding perfectly with that which forms the greater mass.' Of the pillars, the gigantic sea-stacks near Duncansby Head (210 feet, 64 metres) are the most spectacular, although a fault-controlled outcrop of scarlet and gold Upper Old Red Sandstone creates even higher cliffs at Dunnet Head (300 feet, 91 metres), the northernmost point of mainland Britain. Between these two imposing headlands lies the renowned settlement of John o'Groats, a disappointment scenically and because it is in fact neither the northernmost nor even the north-easternmost point of the mainland. But John o'Groats is the end of the road and the former location of the Orkney ferry. Ahead of us lies Stroma, set like a stepping-stone in the treacherous waves and tides of the Pentland Firth, and, even farther beyond, the whale-backed shapes of the Orkneys tempt us to explore their fascinating archipelago.

16. Orkney and Shetland

The northern archipelagos of Orkney and Shetland, isolated in the restless Atlantic, convey a variety of moods to the people who know their landscapes. To some their scenery is hauntingly beautiful, a jewel-like mosaic of rocks and skerries dominated by the immensity of the sky, the shifting dapple of sun and cloud shadow and the ever-present fringe of creamy surf on cliff-girt shores. Others view the islands more prosaically: for them the wind-seared, treeless hills are merely stark and gaunt, ravaged by Atlantic waves and Arctic storms in a land where farmer and fisherman alike wage constant war against the elemental forces of Nature. We shall discover, however, that despite their northerly latitude, which means that short growing seasons inhibit certain farming practices, these remote island clusters are by no means barren and infertile. In fact the Orkneys display an ordered pastoral scene which belies their peripheral isolation, and we shall now discover how it is that their fertility owes as much to the geology and soils as it does to their remarkable cultural history.

The Orkney Isles

To appreciate fully the character of the Orkney environment, one should aim to arrive by sea from the Caithness port of Scrabster. Only this way can the immensity of the western cliffs be properly grasped, as the Orkney ferry crosses the troubled waters of the Pentland Firth and creeps past the overwhelming precipices of Hoy. Here the Old Man of Hoy raises its renowned pinnacle of sandstone from a sea-washed base of contemporaneous lavas to create one of the most memorable of Scottish scenes (Plate 43). The neighbouring cliffs of St John's Head are even more imposing in some ways, for their red and yellow sandstones rise vertically tier upon tier to a height of 1,140 feet (348 metres), making one of the highest *vertical* sea cliffs in the British Isles (see p. 290). Like the high cliffs of Dunnet Head in Caithness, viewed on our journey from Scrabster, the mural precipices of Hoy are carved from thick layers of Upper Old Red Sandstone, known as the Hoy Sandstone. These tough, pebbly sandstones, which often stand on a pedestal of dark basalts, are found nowhere else in Orkney and give to Hoy a distinctive form and elevation that is missing from the other islands.

As the Orkney ferry turns eastwards into Hoy Sound, however, we begin to see a different, more typical, Orkney landscape of low whale-backed hills and burrowing silver waters. Our explorations will reveal that beyond the awesome western portals the Orkney flagstones and the less resistant sandstones of Middle Old Red age have produced a more subdued landscape of gentle slopes and favourable soils. Here the land-locked waters and low coastlines have attracted seafarers throughout historic time, their sheltered firths and sounds offering havens from the stormy ocean. In the course of time the seamen settled on the islands, having discovered that the Old Red Sandstone soils could be tilled with comparative ease and that their pastures could support large numbers of livestock. A contrast is immediately apparent, therefore, between prosperous Orkney and the poorer agricultural economies that we encountered in the equally remote regions of Harris, Lewis and Sutherland, where the intractable gneiss holds sway.

To understand the subtle textures of Orkney's scenic fabric, the visitor would be well-advised to begin his journey by scaling the modest hill of Brinkies Brae, which rises steeply behind the fascinating port of Stromness. At first the character of the hilltop will remind him of the Scottish mainland, for this is the only Orcadian locality where Pre-Cambrian granites and schists of the 'basement' are exposed at the surface. Not only do they give a knobbly aspect to the landforms, but their rounded boulders have been cleared from the fields and used in local boundary-wall construction. Such walling is unique here, for elsewhere in Orkney the ubiquitous Orcadian flagstones (including the Eday Beds, Rousay Beds and Stromness Beds – Figure 53) have been built into splendid drystone walls. These well-jointed and laminated flags have also been extensively used for domestic building stone, as a glance at the Stromness houses will verify. The ochre-coloured walls of the older dwellings have been constructed from the Stromness Flags, whose thinner beds were successfully

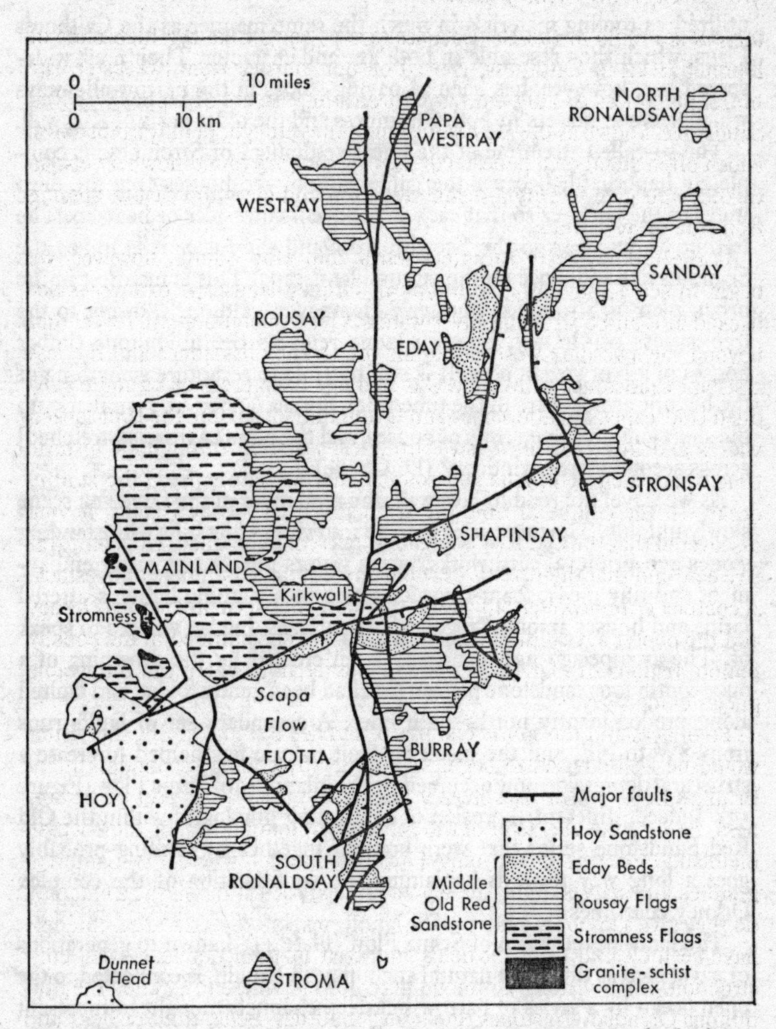

Fig. 53. *The geology of the Orkneys (based on material prepared by the Institute of Geological Sciences).*

utilized as roofing materials in much the same manner as the Caithness Flags, which they resemble in both age and character. Their most widespread use, however, has been as paving-stones in the narrow alleyways which serve as streets in both Stromness and the older parts of Kirkwall.

The so-called streetline of the linear settlement of Stromness is completely haphazard, because the gable-ends of the houses face the deep water of the harbour so that each may have its own pier or boatslip. The farther we move from the Scottish mainland the easier it is to see the Scandinavian influence on the cultural landscape. This is manifest in the urban plan of Stromness (formerly Hamnavoe), although, owing to the treelessness of Orkney, Scottish stone replaces the ubiquitous timber houses of a Norwegian port. It is still possible to recapture something of the historic atmosphere of the time when the sea lapped '... familiarly up the short lanes between rows of houses, and the bows of vessels [stretched] across second storey windows' (D. Gorrie).

As we travel the road to Kirkwall, the remainder of the Orcadian scene slowly unfolds. The outer isles emerge into view – low hills with standing stones and primeval earthworks, ocean waters penetrating every embayment and tiny creek, foam-ringed skerries with warning lights, scattered farms and houses amongst regular field patterns, but no villages to speak of. The archipelago has manifestly been created by the drowning of a once continuous sandstone plateau that had been gently folded and faulted along predominantly north–south lines. A secondary set of faults runs from SW to NE, and the resulting fault mosaic has helped to create a structural depression now occupied by the inland sea of Scapa Flow (Figure 53). Indeed, differential erosion of contrasting lithologies within the Old Red Sandstone series that were brought together by faulting probably goes a long way towards explaining the configuration of the complex Orkney coastlines.

The land-locked basin of Scapa Flow (Plate 44), known to generations of naval men as the finest natural anchorage in Britain, is connected to the open ocean by a series of narrow but deep channels thought to represent pre-glacial valleys drowned by the post-glacial transgression. That the entire archipelago is still continuing to be inundated is demonstrated by the occurrence of submerged peats and tree stumps along the island strands and by the absence of raised beaches. It is clear that Orkney, like Shetland, must have been far enough removed from the centres of the

Pleistocene ice-sheets to have been virtually unaffected by glacial 'loading'; thus, isostatic 'recoil' seems to have played little part in the post-glacial tectonic history of either group.

The northern isle of Sanday and the tiny isle of Flotta at the southern entrance to Scapa Flow have erratic boulders of Scandinavian origin in their glacial drifts, which supports the hypothesis that the earliest ice-sheets crossed Orkney from east to west. Further proof of this are the numerous broken shells and Mesozoic erratics which have been glacially bulldozed from the North Sea floor and then incorporated in the Orcadian drifts. These calcareous boulder clays have contributed greatly to the soil fertility of some of the eastern islands, which thus contrast with those to the west whose poorer, acid soils are reflected in a greater percentage of unimproved land.

The most widespread tract of moorland and rough grazing is to be found in the mountainous terrain of Hoy (Norse: High Island), the only island of sufficient elevation to nurture local glaciers during the Ice Age. Both Ward Hill (1,565 feet, 477 metres) and Cuilags (1,420 feet, 433 metres) have been glacially modified by corries, that of the Kame of Hoy, which is spanned by a conspicuous moraine and located above the northern sea cliffs, being the most spectacular. A few valley moraines and glacio-fluvial deposits are strewn in the ice-deepened valleys to the north of the picturesque deserted settlement of Rackwick. However, a very different type of geomorphological phenomenon can be found high up on the slopes of Ward Hill, for here are widespread examples of 'patterned ground' (see Glossary). Freeze–thaw processes in fact appear to have been active here in relatively recent times, as if to demonstrate that this is a marginally sub-Arctic environment.

We have already seen that Pleistocene frost-shattering played a significant part in the gradual disintegration of the Scottish Highlands' cliffs and summits, but few people realize that even in post-glacial times there have been periods cold enough to regenerate frost-action on some of the more exposed hilltops. Ward Hill is one such place, since, like the Cairngorms (see p. 179) and Ronas Hill in Shetland (see p. 328), it is one of the southern outliers of the Icelandic/Scandinavian terrain known as 'fell-field'. Drs R. Goodier and D. F. Ball have identified three major categories of patterned ground on Ward Hill – turf-banked terraces, wind stripes and hill dunes. Most of the terrace surfaces are devoid of vegetation, whilst the plant-growth along the terrace fronts is gradually being overwhelmed by currently mobile soil and rock waste (solifluxion material) which has been loosened by frost and transported downslope by the action of frost-heaving and rain-wash. The wind stripes and hill dunes owe their origins to the interaction of frost and wind after breaching of the turf cover (by natural or artificial means) had created an exposure of bare earth. Frost action has broken up the exposed mineral soil and allowed the fierce Atlantic winds to blow out the loose material, thereby undercutting the turf. The stripe pattern is thought to be a result of differential erosion caused by the action of wind eddies in alternate wind-break and exposure situations. Such a process of wind erosion is known as deflation, and it has resulted in the almost complete destruction of the turf cover on parts of Ward Hill.

It is a striking fact that as we have journeyed northwards in Scotland the altitudinal zoning of vegetation has decreased in absolute elevation. Thus, the montane flora which we saw only on mountain-tops in the Southern Uplands is found in Orkney at much lower levels because of the more severe climatic restrictions. Wind exposure, which is a recurrent theme in any Orcadian description, is undoubtedly the most important factor in determining the upper limits of agricultural improvement. It is not surprising to find that on the exposed windward coasts of Hoy and Mainland, for example, the montane moorland extends virtually down to sea-level. This is also true of certain parts of the northern isles, such as Rousay and Westray, although in some cases (for instance, Eday) it is the impervious character of the till-covered bedrock which has led to waterlogging and the subsequent growth of peat. Altogether about half the total land area of the archipelago is farmed, however, and in these parts the sheltered valleys and slopes are devoted to intensive agriculture up to heights of about 300 feet (90 metres). Because of the incessant wind Orcadian woodlands are rare; a well-screened deciduous stand near Finstown and a couple of forlorn, wind-trained conifer plantations in Hoy and Eday are exceptional. A fitting commentary is provided by Patrick Bailey: 'One usually thinks of trees sheltering houses; in Orkney houses shelter trees and the largest trees on the islands are to be found in central Kirkwall.'

Apart from the famous tree in the centre of Kirkwall's flagstoned main street, the majority of its trees are found around the splendid sandstone ruins of the Earl's and Bishop's Palaces. But overshadowing these is the redoubtable St Magnus's Cathedral, whose imposing Norman architecture brings to these remote northern isles something of the majesty of Durham. Constructed from blocks of red Eday Sandstone from the neighbouring coastal quarry at Head of Holland, the edifice recalls the days when Orkney came under Scandinavian rule.

Despite its windiness Orkney exhibits no extensive dune systems comparable with, for example, those of the Moray Firth, although there are wide stretches of *machair*, especially on the northern group of islands. A great deal of the improved land of North Ronaldsay, Sanday, Stronsay and Westray is related to the calcareous soils of the fixed dunes and *machair*; the shell-sands of Westray are so thick that they were once exported to the less well-endowed islands for agricultural purposes. Smaller patches of blown sand occur on Mainland at Deerness and Birsay, although the most celebrated example is that at the Bay of Skaill. Here, on the exposed Atlantic shoreline, one of the most remarkable of British prehistoric sites has been laid bare of its blanket of coastal dunes. The cluster of flagstone-built huts at Skara Brae, excavated by the eminent archaeologist Professor Gordon Childe, is now known to have been built some 4,000 years ago by Stone Age inhabitants. Long before these earliest Orcadians had discovered the use of metal, they were capable of erecting quite sophisticated domestic dwellings of the drystone construction favoured by an environment containing plentiful flagstones.

It was these same Neolithic folk, gathering shellfish on the Orkney strands and pasturing their flocks and herds on the fertile grasslands, who were responsible for the erection of the renowned stone-built chambered tombs. Although the gigantic mound of Maeshowe, near the Stromness–Kirkwall road, is the best-known, other important 'cairns', as they are called, occur on Wideford Hill and at Unston, near the Loch of Stenness. The funeral pottery recovered from the latter tomb has a great affinity with that from Windmill Hill, near Avebury in Southern England, and makes Orkney one of the most important Neolithic sites in Britain. When we add to these the Bronze Age standing stones and stone-circles, the Iron Age fortified towers or 'brochs' (the largest drystone structures in Britain), Celtic structures, such as the Broch of Birsay, and the later churches of Eynhallow and Egilsay, to say nothing of the Viking heritage, it is easy to understand why some regard Orkney as one of the archaeological treasures of the British Isles. There seems little doubt that its sheltered harbours, relative fertility and excellent building stone combined to offer an environment which appealed to the earliest settlers sailing the northern seas.

The Shetland Isles

There is a well-known Scottish dictum that the Orcadian is a farmer with a boat, while the Shetlander is a fisherman with a croft. Nothing highlights the contrasting environments of the two groups of islands better than these different cultural responses. Indeed, a closer examination of their land use demonstrates that despite Shetland's greater land area it has only about a quarter of the arable and grassland acreage of Orkney. Tracts of peat bog and moorland occupy about two thirds of the Shetland landscape, while its scattered plots of arable include few of the golden fields of oats and barley which do so much to diversify the prevalent Orkney greensward. The Shetland scene is a tapestry of more muted colours: against a background of dun-coloured grasslands, the russet browns and olive greens of its moorlands are slashed with black peat-cuttings, stitched about and fringed with bare grey and red rocks and threaded with steely blue water. Nowhere in Shetland is more than 3 miles (4·8 kilometres) from the sea, and when we come to examine a map of this tattered archipelago one of the most surprising discoveries is that the incredibly ragged Mainland has only five significant subsidiary islands (excluding Foula) – the remainder being little more than stacks and skerries. In addition, just to the north-east of Mainland, there are the so-called Northern Isles, which consist of three substantial islands: Yell, Fetlar and Unst.

The geology of Shetland is as bewildering as the complexity of its coastline. Besides a great variety of severely folded and faulted Dalradian metasediments and acid igneous intrusives, the Mainland exhibits an extensive unconformable cover of Old Red Sandstone along both its western and eastern fringes, while its north-western limits (in North Roe) have been carved from Lewisian gneiss. Yell, Fetlar and Unst, however, whilst possessing many of the Mainland's schistose rocks, lack the numerous bands of Dalradian limestone which characterize the central Mainland area from Tingwall to Delting (Figure 54). Unst and Fetlar have instead wide expanses of basic gabbro and serpentine. These have contributed to the relative fertility of the Fetlar soils, causing this island to be known as the 'Garden of Shetland'. However, although part of an archipelago where

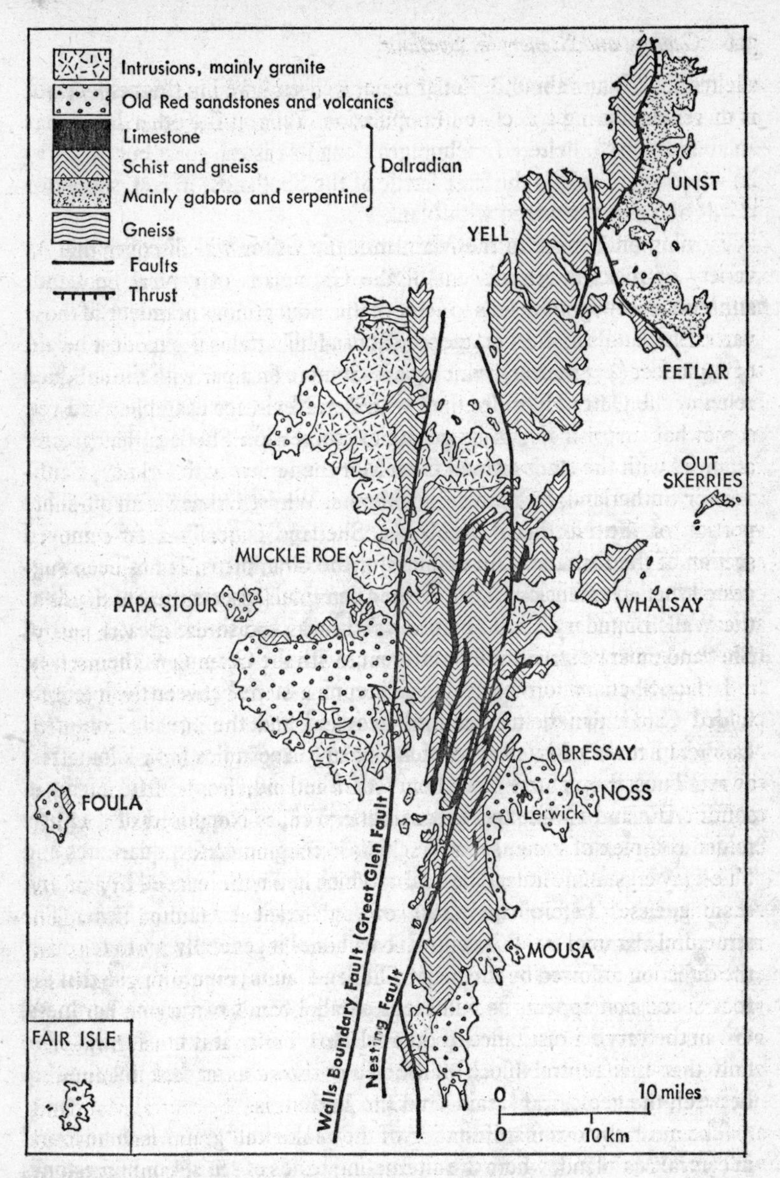

Fig. 54. *The geology of the Shetlands (based on material prepared by the Institute of Geological Sciences).*

sheltered harbours abound, Fetlar is not well-endowed in this respect and is thereby suffering a decline in population. Yell, on the other hand, has an abundance of sheltered anchorages along its eastern coast but owing to its waterlogged soils is the least fertile of the Shetlands; in fact, some two thirds of Yell are covered with blanket bog.

Turning once more to the Mainland, the visitor will discover that its variety of rocks is reminiscent of the Grampians of central Scotland, although its topography fails to achieve the mountainous grandeur of those parts. Shetland's highest eminence, Ronas Hill, attains the modest height of 1,486 feet (453 metres), which places it more on a par with the subdued relief of the Outer Hebrides than with Ben Nevis, for example. And yet, somewhat surprisingly, in its geological succession Shetland has greater affinities with the lands around Glen Mor than it has with Orkney, Caithness or Sutherland, its nearest neighbours. Whilst Orkney is an off-shore portion of the Northern Highlands, Shetland is really a far-removed section of the Dalradian Assemblage of the Grampians. It has been suggested that an extension of the Great Glen Fault bisects the Shetlands as the Walls Boundary Fault (Figure 54), in which case the greater part of Shetland must be structurally continuous with the Grampians themselves.

In fact, Shetland forms a geological stepping-stone between the intensely folded Caledonian belt of central Scotland and the equally contorted Dalradian rocks of Norway, which lies only 190 miles (304 kilometres) away. Thus the central 'backbone' of Shetland, from Fitful Head in south Mainland to Gloup Ness in north Yell, is composed of a tightly folded complex of green schists, schistose conglomerates, quartzites and blue-grey crystalline limestones, all of which had been invaded by acid and basic gneisses before the phase of high-grade metamorphism. The structural alignment of this central 'backbone' is generally north to south, the direction followed by the major Shetland faults (Figure 54), so that the rock succession appears as numerous parallel bands of varying hardness. Given the varying resistance of the rocks to erosion, it is not surprising to find that this central block of Shetland shows a marked relationship between the geological 'grain' and the landforms.

The most obvious manifestation of the geological 'grain' is to be found in central Mainland, where the alternating bands of schists and limestones have produced remarkably linear landforms. The long ridges of West Kame, Mid Kame and East Kame are separated by north–south limestone valleys whose greenness reminds us of the agricultural value of calcareous rocks amongst the acid moorland soils. The narrow limestone outcrops can be picked out by the lines of ruined lime kilns, in use until a century ago for the manufacture of agricultural lime, whose burning was facilitated by the abundance of local peat. Today, however, the limestone is quarried only for road metal, and Shetland's major extractive industries are now confined to Unst. There the broad exposure of serpentine around Balta Sound contains important deposits of chromite and soapstone; the former is no longer being exploited, but the soapstone is the basis of a small-scale talc industry.

There seems little doubt that the general configuration of the coastline reflects the geological structure of Shetland (Figure 54), and this is especially true of the location and alignment of many of the lengthy marine inlets (or rias) known here as 'voes'. But Professor Steers has warned us that the correspondence between the voes and the bands of less resistant rocks may be more apparent than real. Nevertheless, it is difficult to refute the exact alignments of Whiteness Voe, Weisdale Voe, Dales Voe and Colla Firth (all in central Mainland) with bands of Dalradian limestone, whilst the deep and narrow channel of Clift Sound, between the isle of East Burra and Mainland, must owe its form to erosion of a limestone band to the south of Scalloway. Yet when we turn to examine the voes of Unst, Yell and Northmavine there is no corresponding relationship between them and the Dalradian rock structures. Again, in the Old Red Sandstone terrain of Walls the relationship is no longer apparent, with Gruting Voe and Bixter Voe cutting indiscriminately through sandstone and granite alike.

The ever-present glint of sea-water infiltrating deeply into the heart of the Shetlands is, as with Orkney, the outcome of a lengthy period of marine submergence. The gradual drowning of the Shetland hills to form the island cluster, with the inundation of the valleys creating the voes themselves, is the overriding theme in any study of landscape evolution in these northern isles. Recent studies by Dr D. Flinn demonstrate that, like the faraway Scilly Isles, the Shetlands are merely the summits of hills rising from the flat plains of the ocean floor, which is here situated some 400 feet (120 metres) below sea-level. Isostatic depression due to ice-sheet-loading must have been minimal hereabouts, so that post-glacial tectonic uplift has been insignificant. Raised beaches are therefore absent, and the marine transgression has succeeded in drowning a post-glacial peat which must have accumulated on a land surface more than 5,500 years ago.

During the earlier glacial phases of the Pleistocene Shetland seems to have been overridden by Scandinavian ice-sheets, although its later glacial history appears to have been governed by a local ice-cap which probably calved from an ice-shelf into the surrounding ocean. Such an interpretation is based not only on the numerous glacial striae but also on the presence of Scandinavian erratics, such as the 2-ton (2,000-kilogram) boulder at Dalsetter in southern Mainland. Radiocarbon datings of various organic deposits have further suggested that the local ice-cap disappeared around 12,000 years ago.

Although the major Shetland valleys were probably created in pre-glacial times by sub-aerial agencies working along fault shatter-belts and bands of less resistant rocks, there seems little doubt that glacier ice deepened the valleys by clearing out debris, if nothing more. Despite the steepness of the terrain, however, Shetland is not characterized by U-shaped troughs, corries or fjords. Only on isolated Foula is there a semblance of a glacially excavated corrie on the Old Red Sandstone hill of The Sneug (1,373 feet, 419 metres) – the granite dome of Ronas Hill (see p. 326) exhibits no such phenomenon. Instead Shetland's highest summit has, despite its modest elevation, a wealth of 'patterned ground' features which, according to Ball and Goodier, are '... seen more clearly here than on any other British site'. We have already encountered similar features on Ward Hill in the Orkneys (see p. 321), but frost action is thought to be more important in the case of Ronas Hill. Nevertheless, just as frost and wind are both important agents of erosion at Ward Hill, so they are on this exposed, gale-torn summit in Shetland. Turf-banked terraces, wind stripes and hill dunes are all present on this granite blockfield, whilst farther north, on Unst, the gentle slopes of serpentine gravel on the Keen of Hamar (200 feet, 61 metres) constitute the lowest post-glacial frost-patterned ground recorded in Britain, at a latitude of almost 61 degrees North.

Because of its periglacial features and its arctic–alpine flora, Ronas Hill has been designated as a Site of Special Scientific Interest. So too has the Keen of Hamar, for this belt of serpentine rock has particular interest to botanists, as does the neighbouring islet of Haaf Grunay ('green island in the deep sea'), which became a National Nature Reserve in 1959. Shetland's other Nature Reserves, however, have been created to preserve two of the most prolific cliff-breeding colonies of seabirds in Britain. The largest of these is at Herma Ness and includes Britain's northernmost isles of Muckle Flugga and Out Stack. The high, almost vertical cliffs of Herma Ness, facing out to the Atlantic storms, are stained white with the guano of thousands of guillemots, gannets and kittiwakes (Plate 45). The other colony is on the Isle of Noss, which provides a complete contrast to the dark Dalradian gneissic rocks of Herma Ness. Here, off the south-eastern coast, the Noup of Noss cliffs are built from bright red sandstone of Old Red age. As with the Torridonian sandstones of Handa (see p. 303) and the Old Red Sandstone of Foula and Fair Isle, the gently dipping strata provide countless nesting ledges for the vast numbers of seabirds.

The Old Red Sandstone has also been used for building: as in Orkney, the thinly bedded series provides roofing materials and paving stones. The flagstones which paved the streets of Lerwick came from the tiny, grassy island of Mousa, although the most renowned use of these flags was in the construction of the famous broch of Mousa. The flatness of the Mousa flagstones aided the meticulous creation of the drystone structure in this, the best-preserved of the Scottish brochs. On the larger island of Bressay the Old Red Sandstone has also been quarried as a domestic building stone, and the abandoned quarries at Bard Head and Ord Head bear mute testimony to the former demand for it. Most of the older buildings in Lerwick were constructed from this bright red sandstone, but, as a result of the great expansion of harbour facilities to accommodate the North Sea oil bonanza, most buildings in Shetland are now constructed from imported materials.

Oil is, perhaps, an appropriate note on which to end our survey of Scottish geology and scenery, for nothing has made such an impact on specific parts of the Scottish landscape since the Highland Clearances and the Coming of the Sheep. The lonely Sullam Voe, carved in primeval granites and schists in Shetland's northern Mainland, has been chosen as Britain's newest oil port, meeting as it does all the oilmen's site requirements. It has a sheltered, deep-water anchorage; it is close to the oilfields of the east Shetland basin, such as Brent, Thistle, Ninian, Dunlin and Hutton (the last named in tribute to the great geologist); it is also in a tract of underdeveloped and unpopulated moorland. Experts are forecasting that this remote corner of Shetland will soon be housing a vast complex of oil-based industry whose operations could ultimately exceed even those of Holland's massive Europort. Scapa Flow too is to be the site of an oil terminal. Development of the deep-water rias of Milford Haven and Bantry Bay has already brought modern industrial blight into some of the most beautiful scenery in Wales and Ireland, and now we see the same process at work in Shetland and Orkney.

Glossary of Some Technical Terms

Note: A large number of the technical terms used refer to the 'periods' of geological time, e.g. Carboniferous. These are best understood by referring to Figure 2.

Agglomerate: A rock of volcanic origin, composed of irregular blocks of various sizes, comprising solidified lava and fragments of the rocks through which the volcano has broken. Met with mostly in the pipes or necks of volcanoes.

Allivalite: A type of gabbro, of coarse grain and with a mineral assemblage dominated by plagioclase feldspar and olivine.

Amphibolite: A metamorphic rock composed mainly of amphiboles, e.g. hornblende schist.

Anticline: An arch or upfold in the rocks, generally produced by the bending upwards of the beds under lateral pressure. Anticlines are structurally weak, and the upper part rapidly becomes worn away.

Anticlinorium: A system of folds which produce a complex arch or upfold.

Arenaceous: A term used to describe the sandstone group of sedimentary rocks.

Arête: A sharp mountain ridge, usually produced by the headward recession of two opposing corrie walls.

Argillaceous: A term used to describe the fine-grained group of sedimentary rocks, e.g. clays, shales, mudstones and marls.

Armorican: (see *Hercynian*).

Augite: A common mineral found in basic volcanic and plutonic rocks.

Aureole: The zone around an igneous mass within which metamorphic changes, mainly thermal effects, have been produced.

Barytes: A common mineral; barium sulphate ($BaSO_4$).

Basalt: A dark lava of basic composition, containing the minerals feldspar and augite, and sometimes olivine. Mostly crystalline but of fine grain.

Base-level: The lowest level to which a river system can lower a land-surface, usually the sea-level of the time, but occasionally and temporarily a hard band of resistant rock.

Beds, bedding: Sedimentary rocks tend to break along planes parallel to that of their deposition, and thus form the beds or bedding of the rock. See also *Joints*, *Current-bedding* and *Graded bedding*.

Biotite: One of the mica group of minerals.

Boss: A large mass of igneous rock, intruded into and disrupting other rocks. Generally nearly circular in plan; often of coarse-grained rock such as granite.

Boulder clay: A deposit laid down under an ice-sheet. Frequently a mixture of clay with boulders, but may contain very little of either, and therefore preferably called 'till'. May be tens of feet thick and spread over hundreds of square miles.

Breccia: A composite rock consisting of angular fragments of older rocks cemented together with various minerals such as lime.

Caldera: A very large crater formed by the coalescence of smaller craters or by the collapse of surface rocks into a large underground magma chamber (see Figure 40a).

Caledonian: The mountain-building period which dates from Middle Ordovician to Middle Devonian times.

Cauldron subsidence: A process whereby a cylindrical portion of the crust is thought to founder, allowing magma to well up around the sides to fill the resulting cavity (see Figure 36). This is often associated with the structure known as a ring-complex where magma infills the ring cavity as a dyke or pours out at the surface as a lava (see Figure 38). Cone-sheets are also ring structures (see Figure 43).

Cementstone: A type of dolomite (sometimes termed Magnesian Limestone) of Lower Carboniferous age, in which some of the magnesia has been replaced by iron: part of the Scottish Calciferous Sandstone Series.

Chert: A hard, flint-like rock, originally a sand, in which the sand-grains have been cemented together by the deposition of silica from solution.

Cirque (*Corrie or Cwm*)*:* An amphitheatre or armchair-shaped hollow, usually excavated on a mountainside; the slopes are precipitous, the floor nearly level. Due to glacial erosion. In Britain they are commonest on north-east-facing slopes, especially in Wales, the Lake District and Scotland.

Cleavage: A term used to describe a plane of breakage in a mineral and also the re-orientation of mineral particles within a rock due to intense deformation. It represents a partial recrystallization of a fine-grained rock without necessarily destroying the traces of bedding (e.g. slate), although further pressure may produce a foliated rock (e.g. schist) with all traces of bedding destroyed.

Cone-sheets: (see *Cauldron subsidence*).

Confluence: Junction of two streams, or of a tributary to a main river.

Conglomerate: A rock which consists of an aggregate of pebbles or boulders in a matrix of finer material: a pudding stone. Formed by rapid streams and powerful currents.

Consequent (or dip) streams: A stream which flows in the direction of the dip of the rocks on which it was initiated; the direction of the stream is a 'consequence' of the original inclination of the surface.

Corrie: (see *Cirque*).

Cross-bedding (or Current-bedding): A term used to describe inclined bedding planes which reflect the direction of current, the rate of supply and the angle of rest of the sediments during deposition.

Cuesta: A landform consisting of an inclined surface parallel to the dip of the bedding planes (dip slope), and an escarpment steeply inclined in the opposite direction.

Current-bedding: (See *Cross-bedding*).

Cyclothem: A rhythmic sequence of sedimentation, repeated several times, which illustrates a slow spasmodic subsidence. Often found in the Coal Measures.

Denudation: The combined action of weathering, erosion and transportation, to reduce or dissect the existing land-surface.

Dip: The tilt of a bed of rock along its direction of steepest inclination, measured in degrees from the horizontal; also the direction of this dip.

Dip slope: The long, gentle slope of an escarpment, following in general the dip of the rocks.

Dolerite: A dark-coloured igneous rock, basic in composition and resembling a basalt, but composed of rather larger crystals. Generally occurs in a small intrusion such as a dyke or a sill, where it crystallized from a molten state.

Dolomite: Sometimes known as Magnesian Limestone; composed of the double carbonite of magnesium and calcium: $(Ca, Mg)CO_3$.

Drumlin: A low, whale-backed ridge of glacial boulder clay, thought to have been fashioned beneath an ice-sheet. The long axis of the ridge is generally parallel with the direction of former ice-movement (see Figure 3).

Dyke: A sheet-like body of igneous rocks which cuts discordantly through the structures of the bed rock. They occasionally occur as dyke-swarms (see Figure 41).

Epidiorite: A metamorphic rock which has been changed from a basic igneous rock by thermal metamorphism but retains the mineral assemblage of a diorite.

Erratic: A rock which has been transported by ice a great distance from its source, and was left stranded when the ice melted.

Escarpment: A unit of scarp and dip slope. 'Cuesta' is an alternative name.

Esker: A long, sinuous ridge of sand and gravel, formed by a sub-glacial stream, but after melting of the ice-sheet left unrelated to the surrounding topography.

Eucrite: A type of gabbro, similar to allivalite but with more pyroxene.

Fault: A dislocation in the rocks, where one side has moved relatively to the other. Many fault-planes are nearly vertical, others inclined at small angles. Faults have been produced by a variety of causes: some are due to intense lateral pressure, others to differential uplift or to tension (see Figure 37).

Fault-line scarp: A scarp arising along the line of a fault, caused by the differing resistance of the rocks on either side of the fault, the more resistant rock being eroded more slowly, and hence standing up (see Figure 15).

Feldspar: A mineral group comprising various complex silicates of alumina and potassium, sodium and/or calcium. Abundant constituent of most igneous rocks.

Felsite: A fine-grained intrusive igneous rock.

Ferruginous: Containing iron in some form, usually as an oxide; descriptive of rock.

Fjord (*or fiord*)*:* A long, steep-sided, ice-eroded valley in mountainous terrain, now flooded by the sea. The overdeepening of the valley by glaciers has resulted in a submarine threshold or rock bar at the seaward end.

Flagstone: A rock that splits readily into slabs suitable for flagging (as in pavement construction).

Flint: A siliceous rock, grey or black; composed of very minutely crystalline silica. Occurs mostly as nodules in the Chalk.

Freestone: A quarrying term to describe a stone easily quarried and which can be dressed or cut in any direction into blocks of any size.

Gabbro: A coarsely crystalline rock of plutonic origin (see *Plutonic*). It differs from granite insofar as its constituent minerals are more basic than acid.

Geo: A narrow cleft created by the sea, often along a structural weakness in a rocky coastline.

Gneiss: A metamorphic rock, crystalline and coarse-grained, somewhat resembling a granite but showing a more or less banded arrangement of its constituents.

Graben: (See *Rift valley*).

Graded bedding: A term used to describe the change of grain-size in a sedimentary rock in which water sorting during deposition has placed the coarsest material at the base and the finest material at the top of the bed.

Granite: An igneous rock, composed of crystals of quartz, feldspar and mica, of coarse grain. Results from the slow cooling of a large molten mass.

Granophyre: An igneous rock of medium-sized grain in which the quartz and feldspar crystals are intergrown.

Greywacke: A type of sandstone in which the angular particles are poorly sorted.

Grike: A cleft in the bare limestone pavements of karst scenery.

Grit: A rock much like a sandstone, composed mainly of grains of quartz. The grains are either more angular or larger than those in sandstone.

Gypsum: A mineral commonly formed under arid climatic conditions with strong evaporation. Also known as alabaster ($CaSO_4 2H_2O$).

Head: A mixture of clayey material and frost-shattered stones, formed in an earlier periglacial climate, and which under these conditions moved down quite gentle slopes, thereby causing some rounding off of slopes and cliffs.

Hercynian: The mountain-building period of Carbo-Permian times.

Horst: An upthrown block of rocks between two parallel faults.

Igneous rock: A rock formed by the consolidation of molten material. The character of the rock depends mainly on the composition of this material, and on the conditions under which it cooled. Material ejected as lava generally cools quickly, material intruded in small masses less quickly, and material in large masses very slowly: the quickly cooled material may be glassy or very finely crystalline (e.g. basalt); the slowly cooled material is coarsely crystalline (e.g. granite). Thus most igneous rocks are crystalline.

Incised meander: Part of an old meander which has become deepened by river-incision due to uplift of the land or a fall of sea-level.

Inlier: A mass of older stratified rocks showing through the surrounding newer strata.

Ironstone: A sedimentary rock characterized by its ferrous carbonate content. Common in the Lower Jurassic and Coal Measures.

Isostatic: Describing the concept of isostasy, whereby the earth's crust is maintained in a state of near equilibrium. The weight of a continental ice-sheet may cause a temporary downsag, but this will recover when the ice melts. Such vertical crustal movements will result in regional tilting of raised shorelines formed in glacial or early post-glacial times (see Figure 21).

Joints: Divisional planes which traverse rocks perpendicular to the plane of their deposition, cutting them in different but regular directions and allowing their separation into blocks. They are due to movements which have affected the rocks, and to shrinkage or contraction on consolidation. See also *Beds, bedding*.

Kame: A steep-sided ridge or conical hill of stratified sands and gravels formed by meltwaters marginal to a down-wasting ice-sheet.

Karst: Irregular limestone topography characterized by underground drainage; a streamless, fretted surface of bare rock which is honeycombed with tunnels; created by solution of the limestone by ground-water.

Kettle-hole: A depression in glacial drift, often enclosed, made by the melting out of a formerly buried mass of ice and the subsequent collapse of the over-lying drift.

Keratophyre: A fine-grained volcanic rock rich in sodium, potassium and feldspar.

Laccolith: A dome-like intrusive igneous mass which arches the overlying sedimentary rocks but retains a flat floor; it is mushroom-shaped.

Laterite: A residual deposit formed by deep chemical weathering under tropical conditions. It is composed of hydrated iron oxide, although aluminous laterites and ferruginous bauxites are common.

Limestone: A sedimentary rock consisting mainly or almost entirely of calcium or magnesium carbonate, derived from the shells and fragments of former organisms and deposited originally in water.

Lopolith: A saucer-shaped igneous intrusion which is concave upwards, its shape often controlled by the existing folding.

Magma: A molten fluid generated at great depth below the surface and thought to be the source of igneous rocks.

Marl: A calcareous clay; marlstone.

Meltwater channel: A deep gorge or valley, now frequently streamless, which was carved out by glacial meltwaters below or marginal to an ice-sheet. (See *Overflow channel.*)

Metamorphic rock: A rock which may originally have been either igneous (metadolerite) or sedimentary (metasediment) but which has undergone such changes since the time of its formation that its character has been considerably altered: in extreme cases it may be difficult to ascertain its original nature. Heat and pressure are the chief agencies of metamorphism. The commonest rocks of this class are gneiss and schist. Metamorphosed.

Mica: A mineral group comprising complex silicates of iron, magnesium, alumina and alkalis. Several different types can be recognized. Mica can be split into exceedingly thin, flexible plates. An important constituent of granite and other igneous rocks, and of some sedimentary rocks.

Migmatite: A mixed rock formed from the invasion by granitic material, through hydrothermal solution or diffusion, of a pre-existing host rock, usually of metamorphic character.

Mineral: A constituent of a rock, either an element or a compound of definite chemical composition. Mineral as used here has not the same meaning as in the 'Mineral Kingdom', which principally includes rocks. Rocks are aggregates of one or more minerals and usually occur in large masses or extend over wide areas.

Moraine: An accumulation of gravel and blocky material deposited at the margins of an ice-body.

Muscovite: One of the mica group of minerals.

Mylonite: A fine-grained 'crush-rock' formed by fracturing of the individual crystals during intense faulting and tectonic dislocation.

Nappe: An overturned fold in which the axial plane is horizontal (see Figures 28 and 32).

Nunatak: An area of high land or an isolated peak which protrudes above an ice-sheet; it often exhibits widespread periglacial phenomena.

Olivine: An olive-green mineral common in basic igneous rocks.

Oolite, oolitic: A limestone of marine origin, composed of more or less spherical grains, each with concentric layers of calcium carbonate; usually formed in shallow water; probably the grains are of chemical origin.

Outcrop: The area where a particular rock appears on the surface. The rock may continue beyond its outcrop, concealed at depth beneath the surface.

Outlier: A mass of newer stratified rocks detached from their main outcrop by subsequent denudation and separated from it by an area of older rocks.

Overflow channel (*or Spillway*)*:* A deep gorge or valley carved out by the overflow of a pro-glacial lake at the margin of an ice-sheet. These channels are generally streamless.

Patterned Ground: The surface phenomena produced by freeze–thaw processes during a periglacial phase.

Pelitic: A term describing metamorphosed argillaceous rocks, or pelites.

Peneplain: A land-surface of low relief, worn down by prolonged weathering and river erosion. An extensive peneplain needs for its formation a considerable time during which sea-level remains practically stationary.

Peridotite: A class of ultrabasic rocks consisting mainly of olivine.

Periglacial: A term used to describe those of the cold-climate processes and landforms which result from frost action. This term covers frost-shattering, frost-heaving and solifluxion. It is usually most severe at ice-sheet margins.

Permafrost: Permanently frozen ground.

Phyllite: A metamorphic rock with a grain-size and cleavage midway between slate and mica schist.

Piedmont: The plain below a mountain front. Usually formed of outwash materials from the mountain ice-cap and/or alluvial river deposits.

Pillow lava: Lava extruded under water to form pillow-like masses.

Plagioclase: A series of sodium/calcium feldspars.

Platform: An erosion surface produced by marine agencies.

Plutonic: A term referring to rocks of igneous origin formed at great depth, by consolidation from magma; the rocks are generally coarse-grained.

Podzol (*or Podsol*)*:* A soil with a strongly developed horizon of leaching which appears ashen-grey in colour; occurs typically under coniferous forest or heath vegetation.

Pro-glacial lake: A lake impounded by ice-sheets against an area of higher unglaciated ground. The lake disappears either by overflow through the lowest gap in the watershed or by melting of the ice-sheet.

Psammitic: A term describing metamorphosed arenaceous rocks or psammites.

Pyroclastic: A term used to describe rocks formed from the fragmental ejected products of a volcanic explosion.

Quartz: One of the commonest minerals in the earth's crust. Composed of silica (oxide of silicon). A constituent of granite and sandstones. Crystals are hexagonal prisms terminated by pyramids.

Quartzite: A highly siliceous metamorphosed sandstone. A rock composed mainly of quartz. Breaks with rather a smooth fracture. Usually very tough.

Regolith: The loose, incoherent mantle of rock fragments, soil, etc., which rests upon the solid bedrock.

Rejuvenated stream: A stream which has received added power to cut vertically into its bed, usually owing to the uplift of the land.

Residual: An isolated hill which stands above the surrounding country because it has resisted erosion to a greater degree. It is the remnant of an older surface which has been almost totally destroyed by the action of streams or marine agencies. Such a hill rising from a sub-aerial peneplain is known as a monadnock.

Ria: A long, narrow inlet of the sea, caused by subsidence of the land or rise in the sea level. It is generally regarded as a drowned river valley which deepens gradually seawards.

Rift valley (*or Graben*)*:* A valley which has been formed by the sinking of land between two roughly parallel normal faults.

Ring-complex: (see *Cauldron subsidence*).

Ring dyke: (see *Cauldron subsidence*).

Roche moutonnée: An asymmetrical rock outcrop owing its distinctive shape to glacial smoothing. The smoother slopes face the direction of former ice advance, whilst the steeper, ice-plucked slope is on the lee side.

Rock: The geologist does not necessarily confine this term to hard, resistant rocks; he may refer to anything from poorly consolidated sands, gravels and clays to massive, hard stone beds as 'rock'.

Sandstone: A sedimentary rock composed mainly of sand-grains, chiefly grains of quartz, cemented together by some material such as calcium carbonate.

Schist: A metamorphosed rock characterized by the parallel alignment of its constituent minerals.

Serpentine: A clearly banded rock of varying colours produced by metamorphic hydrothermal action on ultrabasic rocks.

Sill: An intrusive sheet of igneous rock, more or less following the bedding of the rocks it invades, but occasionally changing its position among the beds.

Silt: A very fine-grained sediment with particles between $\frac{1}{16}$ and $\frac{1}{256}$ mm in diameter.

Slate: A metamorphosed argillaceous or clay rock which breaks along planes (cleavage) produced by pressure. The cleavage planes may coincide with the original bedding or may obliterate it almost completely.

Solifluxion: The process of slow flowage of the surface rock waste downslope when saturated with ground-water, frequently under a periglacial climate.

Spit: A point of land, generally composed of sand or shingle, which projects from the shore of a water body, frequently at a river mouth or bay entrance.

Stack: A mass or pinnacle of rock left isolated by retreat of coastal cliffs.

Stratum (*plural: strata*)*:* A layer or layers of rock, usually referring to bedded sedimentary rocks.

Strike: The direction of a horizontal line on a dipping stratum; level-course of the miner. The strike is at right angles to the dip. The strike direction is generally the trend of the outcrop of the stratum.

Subsequent (*or Strike*) *stream:* A stream tributary to a consequent stream or originally so, following the strike direction. Generally along the outcrop of a less resistant bed.

Superimposed drainage: A drainage system that has been established on underlying rocks, independently of their structure, from a cover rock which may have disappeared entirely.

Swallet: A funnel-shaped depression in the surface of a pervious rock (usually limestone), linking with a subterranean stream passage developed by solution.

Syncline: A trough-like fold in the rocks generally resulting from lateral pressure. In some cases the surface reflects the structure but, owing to the wearing of the rocks on either side, many synclines ultimately stand out as hills or mountains (e.g. Figure 33).

Synclinorium: A system of folds which produce a complex downfold; the antithesis of an anticlinorium.

Tectonic: An adjective used to describe structural phenomena as opposed to stratigraphic phenomena.

Till: (See *Boulder clay*.)

Tombolo: A sand or shingle bar connecting an island with the mainland.

Tor: A castle-like rock pile which crowns summits of certain well-jointed rock landscapes, or occasionally occurs at the crest of valley slopes; frequently found in granite but can occur in other well-jointed rocks; its origin is uncertain, with deep-surface weathering, normal sub-aerial weathering or periglacial processes all being suggested by various writers.

Transgression: An advance of the sea over a former land area, due to a change of relative land- and sea-level.

Tufa: Material deposited by calcareous springs; usually white or yellowish. Sometimes soft, sometimes hard enough to form a building stone.

Tuff: A rock composed of the fine-grain material ('ash') ejected by volcanoes. Often arranged in beds or layers which have accumulated under water.

Ultrabasic: Igneous rocks, frequently of plutonic derivation, deficient in quartz and feldspar but rich in ferromagnesian minerals.

Unconformity: The relation of rocks in which a sedimentary rock or group of rocks rests on a worn surface of other rocks (sedimentary, igneous or metamorphic). Frequently the plane of unconformity separates rocks of vastly different ages and the rocks beneath the unconformity are of more complex structure than those above. 'Unconformably' and 'unconformable' are the descriptive words derived from this term.

Watershed: The divide between the drainage basins of two rivers (in English usage). American usage defines 'watershed' as the whole of a river catchment basin.

Water table: The level below which the rocks are saturated by ground-water. If the water table intersects the surface a spring will result.

Weathering: The general result of atmospheric action on the exposed parts of rocks. Rain, frost, wind and changes of temperature tend generally to help in the disintegration of the surface rocks.

On Maps and Books

Despite the many text diagrams of the present volume, the reader may wish to seek further information from published topographical and geological maps.

So far as topographical maps are concerned, the publications of the Ordnance Survey remain the most comprehensive. The new 1:50,000 (metric) maps have now replaced the well-established 1-inch series (1:63,360), in which the contour interval was shown in feet. Of particular appeal are the special 1-inch Tourist Maps, which cover such well-known regions as the Cairngorms, Loch Lomond and The Trossachs, and Ben Nevis. A new series of metric Outdoor Leisure Maps (1:25,000) has recently appeared: the first two in the series are entitled *The High Tops of the Cairngorms* and *The Cuillin and Torridon Hills.*

The most up-to-date geological map of Scotland is that published by the Institute of Geological Sciences, London, at a scale of 1:158,4000 (the current, fifth, edition was published in 1969). The same publishers also produce a larger-scale geological map of Northern Britain at a scale of 1:625,000, although the most recent edition was printed in 1957. In addition, the Institute of Geological Sciences publishes a 1-inch (1:63,360) series of detailed geology maps (Solid and Drift editions), although some of these are currently out of print.

Those readers wishing to become more familiar with the general principles of geology may find the following books of interest:

H. H. Read and J. Watson, *Beginning Geology*, Macmillan, 1971
A. Holmes, *Elements of Physical Geology*, Nelson, 1969
J. R. L. Allen, *Physical Geology*, Allen & Unwin, 1975

For more detailed aspects of Scottish geology, geomorphology and natural history the following books are important:

G. Y. Craig (ed.), *The Geology of Scotland*, Oliver & Boyd, 1965
F. F. Darling and J. M. Boyd, *The Highlands and Islands*, Collins (and Fontana paperback), 1969

J. A. Steers, *The Coastline of Scotland*, Cambridge University Press, 1973
J. B. Sissons, *The Evolution of Scotland's Scenery*, Oliver & Boyd, 1967

Of more general interest are the books published by David & Charles in their *Island* series. Currently available are volumes on Arran, Bute, Mull, Skye, Harris and Lewis, the Uists and Barra, Orkney, Shetland, and St Kilda. Some interesting aerial views of Scottish scenery are contained in A. Glen and M. Williams, *Scotland from the Air*, Heinemann, 1972.

Index

Page numbers in italics refer to text figures

More about Penguins and Pelicans

Penguinews, which appears every month, contains details of all the new books issued by Penguins as they are published. From time to time it is supplemented by *Penguins in Print*, which is our complete list of almost 5,000 titles.

A specimen copy of *Penguinews* will be sent to you free on request. Please write to Dept EP, Penguin Books Ltd, Harmondsworth, Middlesex, for your copy.

In the U.S.A.: For a complete list of books available from Penguins in the United States write to Dept CS, Penguin Books, 625 Madison Avenue, New York, New York 10022.

In Canada: For a complete list of books available from Penguins in Canada write to Penguin Books Canada Ltd, 2801 John Street, Markham, Ontario L3R 1B4.

The Environmental Revolution

Max Nicholson

'Here are passages, pictures, tremendous themes which no reader is ever likely to forget. Mr Nicholson is an ecological expert, a practical administrator, a prophet, a man with a splendid sense of history and, best of all perhaps, a flaming, pamphleteering propagandist. All conservationists will pay due honour to Mr Nicholson's pioneering encyclopaedia' – Michael Foot in the *Evening Standard*

Man and Environment

Robert Arvill

This is a book about man – about the devastating impact of his numbers on the environment and the ways in which he can attack the problem. The author is an expert on conservation and planning and his approach is farsighted and informed.

Derelict Britain

John Barr

Land in Britain is scarce and socially precious. Yet large areas of the North and West are now useless, scarred by a century of of industrial plundering. This book is a vigorous indictment of ruthless, profit-seeking industry, of central and local government indifference, and of public apathy.

The Making of the English Landscape

W. G. Hoskins

The Making of the English Landscape can claim to be the only book to deal with the historical evolution of the English landscape as we know it. It dispels the popular belief that the pattern of the land is a result of eighteenth-century parliamentary enclosures, and attributes it instead to a much longer and more fascinating evolution. This book is a pioneer study which traces the chronological development of the English landscape from pre-Roman days to the eve of the Black Death, onwards to the Industrial Revolution and up to the present day. With the help of photographs and charts Professor Hoskins discusses the origins of Devonshire hedge-banks and lanes, the ruined churches in Norfolk and lost villages in Lincolnshire, Somerset's marshland ditches, Cornwall's remote granite farmsteads, and the lonely pastures of upland Northamptonshire. As such, this delightful book, combining scholarship and readability, will appeal to everyone wishing to understand more and thus appreciate fully the varied loveliness of the English landscape.

'An admirable new survey . . . He skilfully mingles the broad generalization with the tiny but relevant detail' – *Sunday Times*

'No one before has ever brought out with quite the same vividness the historical background of the country all round us – *Guardian*

'This is one of those rare books that can produce a permanent and delightful enlargement of consciousness' – *New Statesman*

New Lives, New Landscapes

Nan Fairbrother

'Anyone who cares about the future of Britain will want to read this scholarly, amusing and important book' – *Architects Journal*

New Lives, New Landscapes is concerned with the future of land use in Britain. The countryside is threatened by the economic pressures of mechanized farming, mass car ownership, spreading urbanization and heavy industry – the products of a society which wishes to enjoy both widespread prosperity and the amenities of an unspoiled landscape. Making a realistic assessment of the problems, Nan Fairbrother presents plans to halt haphazard and thoughtless modern development and create a landscape which will preserve rural areas while accommodating the paraphernalia of an industrial democracy.

'She paints the first convincing picture I have ever seen of a new English landscape . . . the great qualities of this book are its sense of hope and its sense of fun. It is wittily and elegantly illustrated' – *Sunday Telegraph*

'Deserves the most serious attention by planners . . . a book remarkable for its breadth and balance of approach' – *Economist*

Geology and Scenery in England and Wales

A. E. Trueman

Why the country looks as it does interests the motorist and the holiday-maker as much as geological structure concerns the student. This Pelican explains very simply why some districts are wooded and others cultivated; why rivers often take the long way round to the sea; why hills may be jagged or rounded; why features like Wenlock Edge or the Chilterns run from north-east to south-west, a thousand other cases where the landscape of England and Wales is decided by the underlying strata, or the superficial deposits.

Geology and its allied sciences have not stood still since Sir Arthur Trueman's study established itself as a little classic many years ago. For this new edition the whole book has been thoroughly revised, to take account of recent findings, by Drs Whittow and Hardy of Reading University. They have completely rewritten several sections, replaced the drawings with appropriate photographs and added many new text figures and maps.

Geology and Scenery in Ireland

J. B. Whittow

In this companion to Sir Arthur Trueman's *Geology and Scenery in England and Wales* Dr Whittow, who helped to revise the earlier volume, now conducts a geological tour from the basalt fantasies of the Giant's Causeway in northern Ulster to the peninsulas, the wooded valleys and the Old Red Sandstone highlands of Munster. Even the featureless stretches of the central lowlands, the eternal bogs and the sluggish procession of the Shannon from lake to lake acquire fresh interest when placed in their geological context.